AF307303

Springer-Verlag Berlin Heidelberg GmbH

Wayne A. Marasco, M.D., Ph.D. (Ed.)

Intrabodies
Basic Research and Clinical Gene Therapy Applications

 Springer

Wayne A. Marasco, M.D., Ph.D.

Division of Human Retrovirology
Dana-Farber Cancer Institute
Boston, Massachusetts, U.S.A.

ISBN 978-3-662-12121-4

Library of Congress Cataloging-in-Publication data

Intrabodies–Basic research and clinical gene therapy applications / Wayne A. Marasco.
    p. cm. — (Biotechnology intelligence unit)
  Includes bibliographical references and index.
  ISBN 978-3-662-12121-4   ISBN 978-3-662-12119-1 (eBook)
  DOI 10.1007/978-3-662-12119-1
  1. Immunoglobulins--Biotechnology. 2. Gene targeting. 3. Gene therapy. 4. Cellular signal
transduction. I. Marasco, Wayne A., 1953- . II. Series.
    [DNLM: 1. Antibodies--physiology. 2. Signal Transduction--physiology. 3. Gene
Expression--physiology. QW 575 I685 1997]
TP248.65.I49I56 1997
616.07'98—dc21
DNLM/DLC   97-36323
for Library of Congress   CIP

Typesetting: R.G. Landes Company Georgetown, TX, U.S.A.

SPIN 10670213   31/3111 - 5 4 3 2 1 0 - Printed on acid-free paper

Antibodies have long been used in biomedical science as in vitro tools for the identification, purification and functional manipulation of target antigens; they have been exploited in vivo for diagnostic and therapeutic applications as well. Recent advances in antibody engineering have allowed the genes encoding antibodies to be manipulated so that the antigen binding domain can be expressed intracellularly. The specific and high-affinity binding properties of antibodies, combined with their ability to be stably expressed in precise intracellular locations inside mammalian cells and to unique epitopes on a target molecule, has provided an extraordinary powerful new family of molecules called intracellular antibodies or "intrabodies."

This book is about intrabodies and contains chapters by leading investigators in the fields of antibody engineering, signals transduction as well as cancer and AIDS research. These investigators are using intrabodies in their laboratories as tools for gene inactivation and/or as therapeutic molecules in clinical gene transfer applications. There are ten chapters in the book.

The first chapter is an introduction to the intrabody technology and the use of intracellular trafficking signals to direct intrabodies to precise subcellular locations. I have tried to briefly review published work on intrabodies in this area as well as to discuss new vector designs that are currently being tested in my laboratory to broaden the subcellular targeting potentials of these molecules. I have also discussed where I believe this technology can go in the future.

The second chapter by Drs. Poul and Marks is an extensive review of antibody phage display technology. Using powerful new immunotechnological tools, the creation of large human immunoglobulin libraries from naive individuals has been achieved and when combined with phage display technology, has allowed investigators to bypass in vivo immunization and produce high affinity human antibodies to human proteins. These powerful new methods have provided a rich source of rearranged human antibody genes and, when combined with the use of classical intracellular trafficking signals, greatly extends and will markedly enable the further development of this new technology.

The third chapter describes studies on the use of intrabodies to obtain phenotypic knockout of the high affinity component of the human interleukin-2 receptor α chain. These studies by Dr. Richardson and myself demonstrate the power of this technology to obtain phenotypic knockout of an integral membrane growth factor receptor on both leukemic T cell lines and PBMCs. The ability to use this technology to systematically examine different components of the IL-2 receptor pathway for their involvement in constitutive signaling and in the growth of HTLV-I transformed cells and adult T cell leukemia cells (ATL) are demonstrated and discussed in detail.

In chapter four, Drs. Beerli, Graus-Porta and Hynes present their elegant biochemical studies on the use of intrabodies as tools to study erbB2 receptor kinases. Epidermal growth factor (EGF) receptor and erbB-2, two receptors whose aberrant expression is frequently involved in human cancer, were chosen as targets. Their studies with ER-directed sFv intrabodies to EGFR could compete with EGF and inhibit EGF receptor function in an autocrine manner while, in contrast, the KDEL containing sFv intrabody inactivated the EGF receptor by ER retention. Additional studies with two KDEL-tagged sFv intrabodies against erbB-2 have demonstrated a marked decrease in the cell-surface expression of erbB2 in NIH3T3 fibroblasts that express an oncogenically activated form of erbB2. Functional inactivation of the receptor is suggested by a reduction in the phosphotyrosine content of the cells and, more importantly, by the reversion of the cells to a non-transformed phenotype. These latter studies also show the importance of receptor transmodulation among the type I RTKs.

Neuroantibodies or recombinant antibody expression in the central nervous system is the topic of chapter five. Here, Drs. Cattaneo, Piccioli and Ruberti present their pioneering work on expressing recombinant antibodies in the central nervous system of transgenic mice. They describe their studies with glial and neuronal cell expression of anti-neurotrophic factor (NGF) and anti-substance P (SP) recombinant antibodies and their neutralizing activity of the corresponding antigens. They also discuss the potential applications of this approach of local production of antibodies within the CNS as a promising therapeutic strategy.

Chapters six and seven contain very valuable contributions from two leading laboratories on the use of intrabodies for the gene therapy of cancer. In chapter six, Dr. Curiel presents a number of studies from his laboratory, complementary to those presented by Drs. Beerli, Graus-Porta and Hynes, on an ER-directed sFv intrabody that has been used to downregulate erbB2 in the human ovarian-carcinoma cell line SKOV3. They demonstrate that constitutive downregulation of erbB2 is incompatible with the long-term survival of these cells which are killed by apoptosis. When ex vivo transduced SKOV3 cells were introduced sub-cutaneously into nude mice, no tumors grew and complete tumor eradi-cation at necropsy was observed. Tumor cell eradication in transduced primary cultures of human ovarian carcinoma cells could also be dem-onstrated. Using a replicative-defective adenovirus to in vivo trans-duce the anti-erbB2 sFv intrabody gene into i.p. transplanted SKOV3 cells, these investigators further demonstrated in vivo tumor cell kill-ing and a significantly prolonged survival of animals compared to con-trol groups. These preclinical studies form the basis of a clinical gene therapy protocol to treat a metastatic ovarian cell carcinoma in patients that have failed conventional chemotherapy.

Chapter seven describes in vitro and in vivo studies by Dr. Cochet and colleagues on an anti-ras sFv intrabody. They demonstrate first in *Xenopus laevis* oocytes that the anti-ras sFv intrabody can inhibit mei-otic maturation in response to insulin which activates the Ras signaling pathway. Germinal vesical breakdown (GVBD), a hallmark of matura-tion and activation of maturation promoting factor or $p34^{cdc2}$ were also inhibited. In mammalian cells, the anti-Ras sFv could also inhibit the activation of the transcriptional AP-1 complex by oncogenic ras. Cell death by apoptosis was observed in human lung carcinoma H460 cells which express the mutant Ki-ras gene. When direct intratumor injec-tion of an adenovirus expressing the anti-ras sFv intrabody was per-formed in nude mice, pronounced tumor regression was observed. Fi-nally, they discuss their proposed studies to determine if selected anti-p53 sFv intrabodies can restore wild-type p53-associated properties such as specific DNA binding and transcriptional activation, which may re-sult in cell cycle arrest and/or apoptosis of certain tumor cells that con-tain point mutations in one of the p53 alleles.

Chapters eight through ten detail studies from both my labora-tory and the laboratories of Drs. Duan and Pomerantz on the use of anti-HIV-1 intrabodies for the gene therapy of HIV-1-infection and AIDS. In chapter eight, Dr. Mhashilkar and I describe studies that evaluate anti-tat and anti-rev sFv intrabodies for "intracellular immunization" against HIV-1 infection. In these studies we present experimental data that demonstrate that the epitope of the target protein is critical for inhibition of HIV-1 replication, both for anti-tat and anti-rev sFv intrabodies.

In chapter nine, Dr. Rondon and I describe studies on intrabodies that are directed against the HIV-1 gag proteins, p17 (matrix (MA)) and p7 (nucleocapsid (NC)). A Fab intrabody against MA was used to block both afferent and efferent arms of the viral life cycle since the matrix protein is both required for nuclear import of the viral preintegration complex and for particle assembly. In these studies we show that a marked inhibition of proviral gene expression occurred in stably transfected T cell lines when single-round HIV-1 CAT virus was used for infections. In challenge experiments using both laboratory strains and syncytium-inducing primary isolates of HIV-1, a substantial reduction in the infectivity of virions released from the cells was also observed. We also present preliminary data with an anti-NC sFv intrabody which suggests that the anti-viral activity that is observed is due to a blockade at both the afferent and efferent arms of the life cycle.

The final chapter prepared by Drs. Duan and Pomerantz details their studies on sFv intrabodies against HIV-1 integrase and reverse transcriptase. These two targets are critical for the afferent arm of the HIV-1 life cycle and it is thought that intracellular immunization which results in preventing the establishment of infection may be more efficient than strategies aimed at inhibiting gene expression after integration of the provirus has occurred. These investigators present data to demonstrate that the anti-IN and anti-RT sFv intrabody transduced cells are protected against HIV-1 challenge. They also discuss their studies on combining sFv intrabodies with ribozymes to increase the anti-viral activity of their retroviral transduced cells.

In summary, this book will broaden the interest and knowledge of scientists already familiar with specific aspects of antibody engineering and gene therapy, for whom the extensive and up-to-date data list will prove particularly useful. It will also be useful to students of cell biology, biochemistry, microbiology, molecular biology and gene therapy. Finally, this book will grab the attention of scientists that are involved in the human genome project and in identifying small molecule drugs for the treatment of human diseases who may wish to utilize intrabodies as a means to achieve their goals. They will surely learn a great deal about intracellular protein trafficking along the way.

CONTENTS

================= EDITOR =================

Wayne A. Marasco, M.D., Ph.D.
Dana-Farber Cancer Institute
Division of Human Retrovirology
Boston, Massachusetts, U.S.A.
Chapters 1, 3, 8, 9

================= CONTRIBUTORS =================

Dr. Antonino Cattaneo
Biophysics Sector
International School
  for Advanced Studies (SISSA)
Trieste, Italy
Chapter 5

Roger R. Beerli
The Scripps Research Institute
La Jolla, California, U.S.A.
Chapter 4

Laurent Bracco
Rhone-Poulenc Rorer
  Central Research
Gene Medicine Department
CRVA
Vitry sur Seine, France
Chapter 7

Oliver Cochet
Laboratoire de Biotechnologie
  des Anticorps
Institut Curie
Paris, France
Chapter 7

Dr. David T. Curiel
Director, Gene Therapy Program
University of Alabama
  at Birmingham
Birmingham, Alabama, U.S.A.
Chapter 6

Isabelle Delumeau
Rhone-Poulenc Rorer
  Central Research
Gene Medicine Department
CRA
Vitry sur Seine, France
Chapter 7

Dr. Lingxun Duan
Division of Infectious Disease
Thomas Jefferson University
Philadelphia, Pennsylvania, U.S.A.
Chapter 10

Diana Graus-Porta
The Scripps Research Institute
La Jolla, California, U.S.A.
Chapter 4

Nadège Gruel
Laboratoire de Biotechnologie
  des Anticorps
Institut Curie
Paris, France
Chapter 7

Dr. Nancy E. Hynes
Department of Molecular Biology
Friedrich Miescher Institute
Basel, Switzerland
Chapter 4

Mireille Kenigsberg
Rhone-Poulenc Rorer
   Central Research
Gene Medicine Department
CRVA
Vitry sur Seine, France
Chapter 7

Dr. James D. Marks
University of California
San Francisco General Hospital
San Francisco, California, U.S.A.
Chapter 2

Dr. Abner M. Mhashilkar
Dana-Farber Cancer Institute
Division of Human Retrovirology
Boston, Massachusetts, U.S.A.
Chapter 8

Dr. Patrizia Piccioll
CNR Institue of Neurobiology
Rome, Itlay
Chapter 5

Dr. Roger J. Pomerantz
Division of Infectious Disease
Thomas Jefferson University
Philadelphia, Pennsylvania, U.S.A.
Chapter 10

Dr. Marie-Alix Poul
Department of Anesthesia
University of California
San Francisco General Hospital
San Francisco, California, U.S.A.
Chapter 2

Dr. Jennifer H. Richardson
Dana-Farber Cancer Institute
Division of Human Retrovirology
Boston, Massachusetts, U.S.A.
Chapter 3

Dr. Isaac J. Rondon
Dana-Farber Cancer Institute
Division of Human Retrovirology
Boston, Massachusetts, U.S.A.
Chapter 9

Francesca Ruberti
International School
   for Advanced Studies (SISSA)
Trieste, Italy
Chapter 5

Fabien Schweighoffer
Rhone-Poulenc Rorer Central
   Research
Gene Medicine Department
CRVA
Vitry sur Seine, France
Chapter 7

Jean Luc Teillaud
Laboratoire de Biotechnologie
   des Anticorps
Institut Curie
Paris, France
Chapter 7

Dr. Bruno Tocqué
Rhône-Poulenc Rorer
Recherche-Development
Centre de Recherche de
   Vitry-Alfortville
Cedex, France
Chapter 7

Designing Intrabodies: Structural Features and the Use of Intracellular Trafficking Signals

Wayne A. Marasco

Introduction

The humoral immune system is incredibly diverse in its ability to produce antibodies to virtually any target molecule whether it be self or foreign, protein or nucleic acid, carbohydrate or lipid. The basic principle behind intrabodies is a marriage between an antibody binding site and the molecular instructions to direct this binding site to a precise intracellular location. By harvesting the power and diversity of the immune system in the form of rearranged antibody genes, the immune system itself accomplishes the binding site design, thereby bypassing some of the technical hurdles involved in other methods of gene inactivation such as targeted gene disruption, dominant negative mutants and RNA based intracellular immunization strategies that use antisense, ribozymes and RNA decoys.[1-3]

At least four mechanisms of inhibition by intrabodies have been described. First, intrabodies can bind to the active site of an intracellular target molecule and block its activity. Second, the intrabody can disrupt intracellular protein-protein interactions. Third, the intrabody can interfere with intracellular target protein transport. Fourth, the binding of the intrabody can accelerate target protein degradation. Examples of these four mechanisms will be described in subsequent chapters.

A key factor contributing to the success of the recent studies has been the use of single-chain antibodies, also known as single-chain variable region fragments or sFv, in which the heavy and light chain variable domains are synthesized as a single polypeptide and are separated by a flexible linker peptide, generally $(Gly_4Ser)_3$. The constant portion of the immunoglobulin molecule, which has no role in antigen binding, is dispensed of entirely. The result is a small ($\approx$ 28 kDa) molecule "sFv intrabody" with high-affinity ligand-binding capability and minimal assembly requirements. Bicistronic expression vectors have also been reported which allow for the Fd heavy and light chain of "Fab intrabodies" (monovalent

Intrabodies: Basic Research and Clinical Gene Therapy Applications, edited by Wayne A. Marasco. © 1998 Springer-Verlag and R.G. Landes Company.

antigen binding fragment) to be expressed in stoichiometrically equivalent amounts.[4] Details of the construction of specific intrabodies will be described by other authors in this book.

Starting Materials For Intrabody Construction

The majority of intrabodies that have been reported to date are derived from murine monoclonal antibodies.[3,5-13] These hybridomas have provided a reliable source of well-characterized mAb reagents for the construction of intrabodies and are particularly useful when their epitope reactivity and affinity have been previously characterized.

Human monoclonal antibody producing cell lines have also been used as starting materials for the construction of intrabodies.[14-16] However, the use of antibody phage display technology[17-19] is rapidly changing the pace at which new intrabodies against different epitopes on a target molecule can be evaluated.[20-23] Indeed, very large naive human sFv libraries have been created and offer a rich source of rearranged antibody genes against an unlimited plethora of target molecules. Smaller immunoglobulin libraries from individuals with autoimmune[24-25] or infectious diseases[26-27] have allowed more disease specific antibodies to be readily isolated. Transgenic mice that contain a human immunoglobulin locus in the absence of the corresponding mouse locus have been produced and stable hybridomas that secrete human antigen-specific antibodies have been reported.[28-29] These transgenic animals should also provide another source of human antibody genes, either through conventional hybridoma technology or in combination with phage display technology. In vitro procedures to manipulate the affinity and fine specificity of the antigen binding site have been reported, including repertoire cloning,[30-32] in vitro affinity maturation,[33-34] semisynthetic libraries,[18,26,35] and guided selection[36] (see chapter 2 for more details). Starting materials for these recombinant DNA based strategies have included RNA from mouse spleens,[30] and human peripheral blood lymphocytes,[24-25,31,37] lymphoid organs and bone marrow from HIV-1-infected donors.[17,38]

Intracellular Trafficking

The selective trafficking and compartmentalization of specific molecules throughout the cell require the recognition of sorting determinants that are most often present in the primary structure of those molecules. The identification of some of these determinants has been one of the major accomplishments of molecular cell biology and this information has been used extensively in the design of intrabodies that can be targeted to precise subcellular locations. In the text that follows, I will review what we have learned to date on the intracellular targeting of intrabodies. In some cases, only the vector designs will be presented since complete biological characterization of these targeted intrabodies has not been completed.

Targeting Intrabodies to the Secretory Pathway

Proteins synthesized in the rough endoplasmic reticulum (ER) of eukaryotic cells use the exocytic pathway for transport to their final destinations (Fig. 1.1). Proteins lacking special sorting signals are vectorially transported along the entire route from the ER through the cis-Golgi network, the cis-, medial- and trans-Golgi, and the trans-Golgi network (TGN) to the plasma membrane. Other pro-

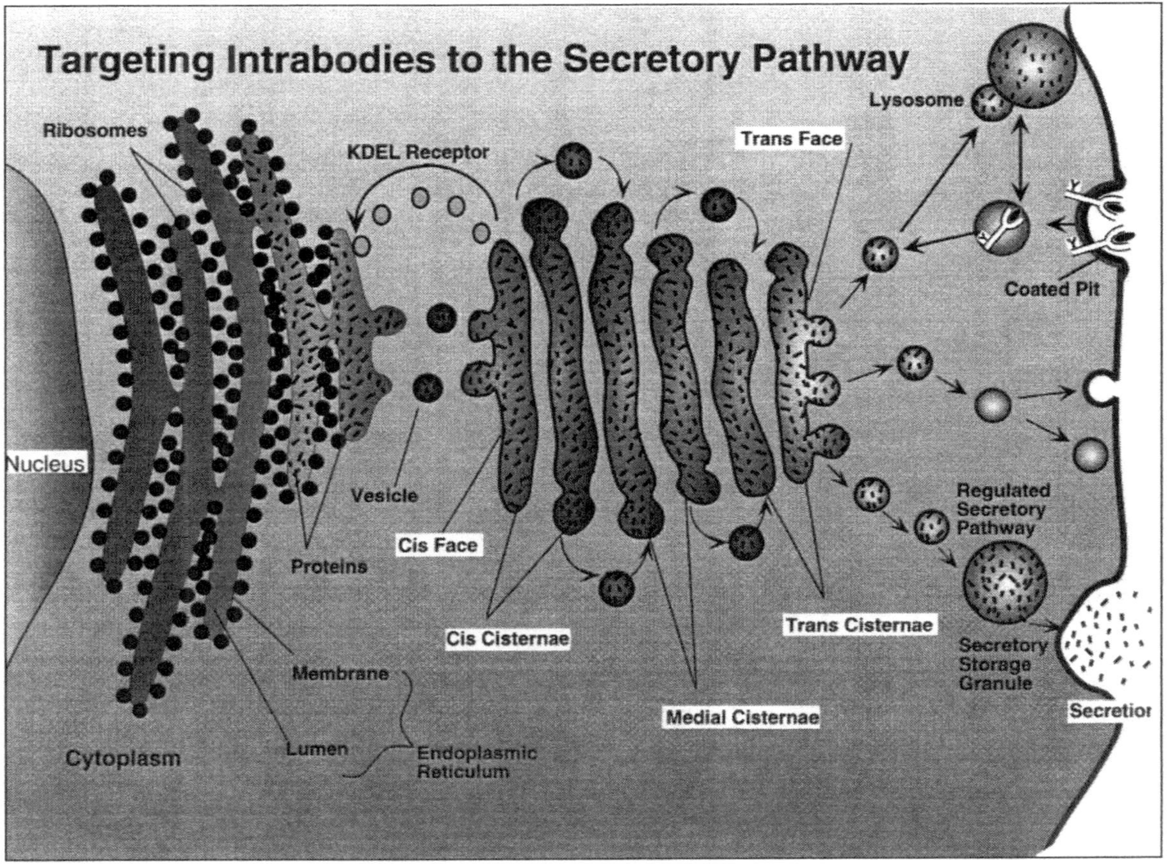

Fig. 1.1. Targeting intrabodies to the secretory pathway. Proteins synthesized in the rough ER of eukaryotic cells use the exocytic pathway for transport to their final destinations. In the TGN, sorting of proteins destined for different plasma membrane domains, lysosomes, and regulated secretory pathway occurs. Following ligand induced internalization of certain cell surface receptors, cytoplasmic motifs on these receptors have been found to be responsible for their intracellular trafficking.

teins have targeting signals for incorporation into specific organelles of the exocytic pathway. Thus, each compartment is equipped with characteristic resident proteins, some of which are enzymes involved in the posttranslational modification of the transit proteins. This understanding forms the basis of targeting intrabodies to precise suborganelle locations along this secretory pathway.

Endoplasmic Reticulum

Leader Sequences

Intrabodies that are targeted to the lumen of the ER provide a simple and effective mechanism for inhibiting the transport of plasma membrane or secreted proteins to the cell surface; even highly abundant cell-surface receptors have been reduced to undetectable levels by FACS analysis using this method.

The endoplasmic reticulum is the location in the cell where native antibodies that are bound for the plasma membrane or secretion are processed. Chaperones such as BiP and GRP94 associate with the native heavy and light chains and assist in their folding and proper assembly into transport competent native immunoglobulins.[39-40] Another ER chaperone, calnexin, has been shown to associate with the membrane-bound multimeric antigen receptor (mIg) on B cells.[41] In addition, the redox state of the endoplasmic reticulum is such that intrachain disulfide bond formation, which is catalyzed by protein disulfide isomerase (PDI) in the presence of glutathione, is optimized.[42-46]

In general, when starting with a hybridoma cell line with the intent of constructing ER-directed sFv intrabodies, our laboratory has taken the approach of using the native heavy chain leader to direct translation of the sFv intrabody (V_H-linker-V_L format) to and across the rough endoplasmic reticulum membrane (RERM). A similar approach of using native heavy and light chain leader sequences was used to construct an ER-directed Fab intrabody against the HIV-1 matrix protein, p17.[4] For this strategy, the investigator can use PCR primers that will anneal to the 5' end of the leader sequence. Degenerate PCR primers that will anneal to human[47-48] and mouse leader sequences have been reported.[49] The Kabat database[50] and Vbase[51] are thorough resources for inspection of a large number of somatically rearranged or germline leader sequences, respectively. Difficulties associated with amplifying by PCR immunoglobulin genes from hybridomas have been reported.[52-55] Eukaryotic expression vectors have also been reported that contain heterologous leader sequences that should allow the RERM targeting of these sFv or Fab intrabodies.[56-57] For a detailed discussion of the use of leader sequences to direct heterologous proteins to the RERM, please see reference 58.

ER Retention Signals

The carboxy-terminal Lys-Asp-Glu-Leu (KDEL), or a closely-related sequence, is important for ER localization of both lumenal as well as type II membrane proteins. This sequence functions as a retrieval signal at postER compartments(s).[59-60] Retrieval occurs through binding to the KDEL receptor, a seven-transmembrane-domain protein that is a temporary resident of the Golgi apparatus: upon binding to KDEL-containing ligands, it moves to the ER where the ligand is released.[61] This KDEL sequence has been introduced into a number of ER-directed intrabodies, and their binding to the target type I membrane proteins such as Tac (CD25, IL-2Ra)[62-64] and gp120[14-16] has resulted in phenotypic knockout of the target surface protein in these cells and in the case of erbB-2[10] has additionally resulted in phenotypic reversion to a nontransformed phenotype.[10] However, the KDEL retention signal is not absolutely required to achieve phenotypic knockout of erbB-2, cell killing or apoptosis (see ref. 11 and chapter 6).

We have reported one case in which instability of the sFv105KDEL that is directed against HIV-1 gp120 was seen. In this report the KDEL containing sFv intrabody could not be detected in stable cell lines until the cells were transfected with an HIV-1 envelope expressor plasmid.[14] We have not seen this phenomenon with any other ER-directed sFv intrabody.

We have also observed ER-retention of a sFv intrabody in the absence of a KDEL retention sequence.[14-15] In these studies, the anti-HIV-1 gp120 sFv intrabody termed sFv105, that did not contain a KDEL retention sequence was also retained intracellularly. The intracellular half-life of sFv105 was > 6 hr when expressed either in the absence or presence of gp160.[14] This sFv intrabody was shown to bind to the envelope glycoprotein within the cells and inhibit processing of the envelope precursor and syncytia formation (Fig. 1.2). The infectivity of the HIV-1 particles produced by the intrabody expressing cells was also substantially reduced. We also demonstrated that sFv105 could be coprecipitated with a rat monoclonal antibody to the ER chaperone protein BiP[14] and concluded that retention of sFv105 in the ER was probably a consequence of association with BiP. However, binding of the sFv intrabody to BiP does not impair its specific binding activity. Other investigators have reported that ER-directed sFv intrabodies without KDEL retention signals can be secreted at different rates from mammalian cells, with the rate-limiting step of secretion being their exit from the ER.[45,65] This variability is likely the result of the primary amino acid sequences of the sFv intrabody V-region genes.[66]

Finally, a comparison of the half-life and fate of the sFvTac or sFvTacKDEL intrabody/human IL-2Rα complexes has been studied in detail. The half-life and fate of the intralumenal complexes once formed appears to be variable. By pulse-chase analysis, sFvTacKDEL had an intracellular half-life of > 30 hr in stimulated Jurkat cells and the intrabody retained IL-2Rα as an immature 40 kDa form of the receptor that was sensitive to endoglycosidase H, consistent with its retention in a pre- or early Golgi compartment.[62] In striking contrast, rapid degradation of sFvTac/IL-2Rα complexes was seen in these same cells. The degradation appeared to occur by a nonlysosomal mechanism.[62] This "ER degradation" of the sFvTac/IL-2Ra complexes may follow a route similar to that of several membrane proteins that misfold in the ER, which involves their transfer to the cytosol followed by attachment of ubiquitin and targeting to the proteasome.[67]

Other ER retrieval/retention motifs have been reported for type I membrane proteins that contain Lys-Lys at positions -3 and -4 from the carboxy-terminal end of the cytoplasmic domain[68] or type II membrane proteins that contain two Arg residues close to the N-terminus.[69] This ER retrieval/retention is most likely the result of binding of these basic motifs to coatomer subunits.[68] The use of these cytoplasmic ER retention motifs may represent an alternative strategy to target proteins in the secretory pathway if chimeric type I or II membrane proteins can be constructed with a cytoplasmic sFv intrabody that contains the appropriately positioned Lys-Lys or double Arg motifs, respectively.

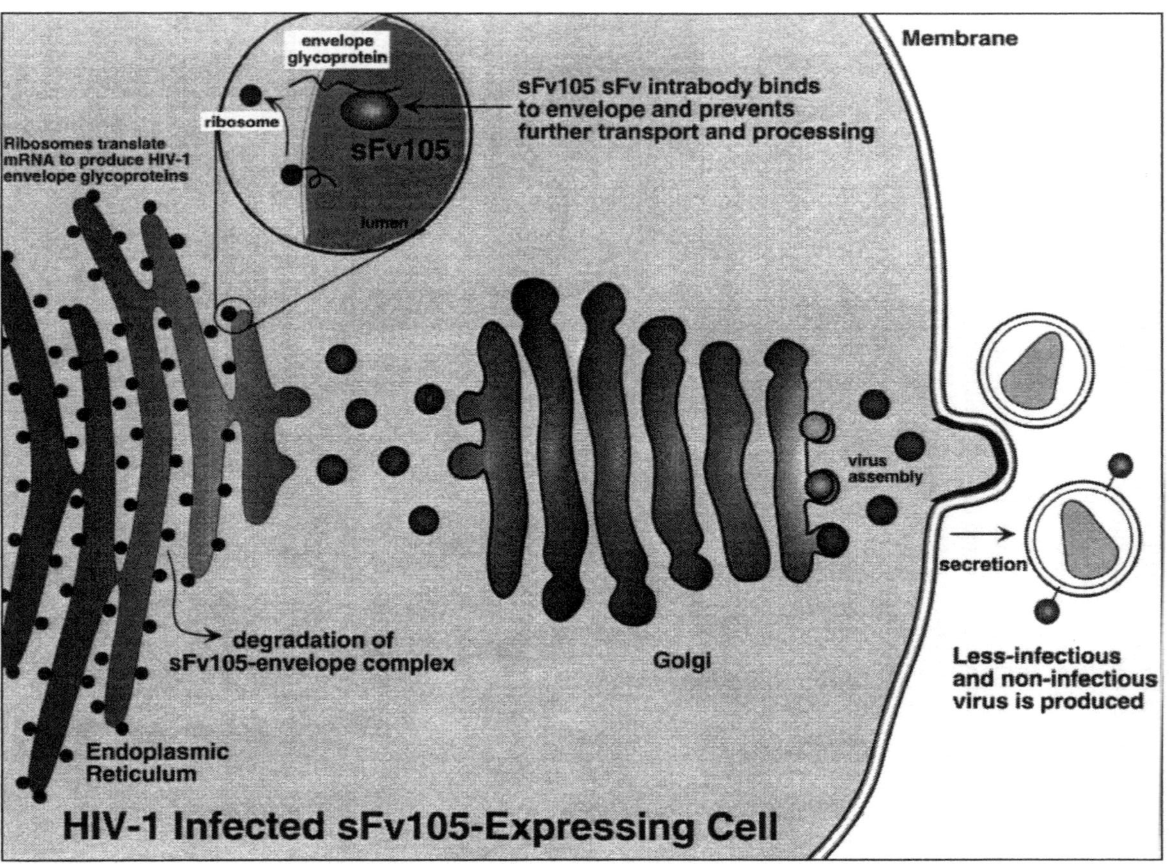

Fig. 1.2. Inhibition of gp160 precursor processing in sFv105 transduced cells leads to inhibition of transport of mature envelope glycoprotein (gp120/gp41) and the release of less-infectious and non-infectious virions from infected cells.

Trans-Golgi Network (TGN) Retention Signals

In some cases, the intrabody target molecule may be a cell surface glycoprotein with a complex oligosaccharide epitope. As a consequence, ER retention of a target molecule may not be successful because the high-mannose containing glycoprotein has not yet been modified, since complex oligosaccharides are added in the medial-Golgi.[70-71] While the KDEL receptor is distributed in a gradient across the Golgi stack, it is most concentrated at the cis face.[59] As a result, it may be preferable or even essential under certain circumstances to retain the intrabody and target protein in the trans-Golgi cisternae.

Sorting of proteins destined for different plasma membrane domains, lysosomes and secretory pathways takes place in the trans-Golgi network (TGN). Targeting of integral membrane proteins to endosomes, lysosomes, the basolateral plasma membrane, and the TGN are largely mediated by sorting signals contained within the transmembrane spanning region and/or within the cytoplasmic domain of the proteins.[72-73] For the latter, many of these sorting signals consist of continuous sequences of four to six amino acids containing a critical tyrosine residue with a reverse-turn conformation or a closely related terminal turn of α helix.[74]

Tyrosine-Based TGN Sorting Signals

Several type I membrane viral[75] and cellular proteins have been identified that are localized to the TGN and contain in the cytoplasmic domains this tyrosine-based motif. Examples include the minimum tyrosine-based internalization signal of the transferrin receptor,[74] the asialo glycoprotein receptor,[76] the mannose-6-phosphate receptor[77] and others (see reviews in refs. 72 and 73).

TGN38 is a resident TGN protein that was isolated from a rat liver cDNA library by an immune-selection technique[78] and the TGN trafficking signals that are responsible for this residency have been studied in detail. The transmembrane spanning region of TGN38 has been shown to be involved in TGN retention[79] while the carboxy-terminal 33 amino acid tail of TGN38, which contains the tyrosine-based YQRL motif, acts as a TGN retrieval signal.[80] Both signals are independently sufficient to allow for TGN targeting to heterologous proteins. The predominant TGN location appears to be due to a steady-state localization between surface TGN38 protein molecules that are internalized and retrieved to the TGN from the cell surface and those molecules that are being routed from the TGN to the cell surface.[79-80] A yeast two-hybrid screen using the TGN38 tyrosine-based motif as bait identified the medium chains (μ_1 and μ_2) of two clathrin-associated protein complexes (AP-1 and AP-2, respectively) as binding specifically to this motif.[81] Thus, it is likely that the medium chains serve as signal-binding components of the clathrin-dependent sorting machinery and perhaps as a component of the TGN retrieval system.

We have constructed a TGN38 based expression vector that should result in targeting to the TGN (Fig. 1.3). This expression vector contains a murine IgG$_3$ hinge region[82] followed by the transmembrane spanning region and complete cytoplasmic tail of TGN38. In an intrabody expression vector of this design, it is likely that the distance from the lumenal membrane to the intrabody binding site may be critical for effective binding to other membrane proteins that traffic through the TGN. Ongoing studies with this and related intrabody expression vectors are underway.

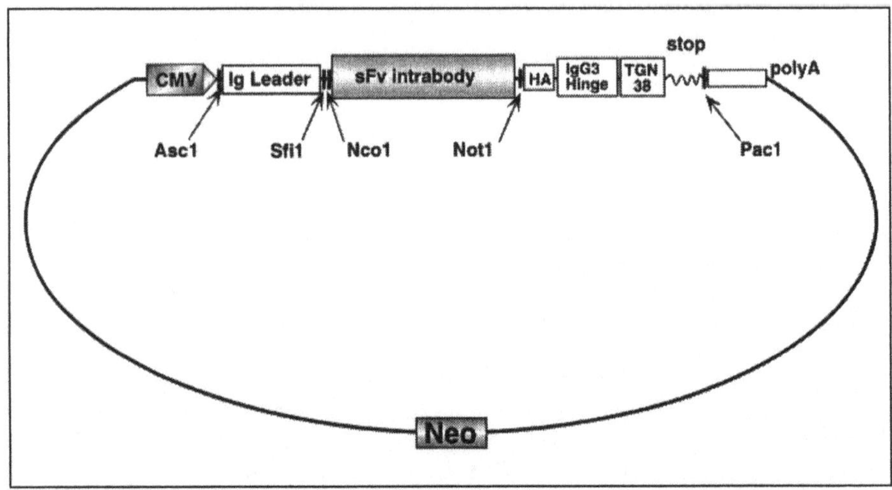

Fig. 1.3. A trans-Golgi network targeting intrabody expression vector contains a HA tag sequence followed by a murine IgG3 hinge region and the transmembrane spanning regions and complete cytoplasmic tail of TGN38.

Dileucine and Acidic Patch Based TGN Sorting Signals

In addition to the tyrosine-based internalization/retrieval motif, two other independent TGN sorting signals have been identified. A dileucine motif in the cytoplasmic domain of the T cell receptor gamma chains and δ chains has been shown to lead to endocytosis and delivery of chimeric transmembrane proteins to lysosomes.[83] Other examples of the dileucine-based motifs have been described as well.[73] A second TGN sorting motif containing acidic amino acids has been described in the cytoplasmic tails of furin[84] and gE of varicella-zoster virus.[75]

Transmembrane Spanning Domain Based TGN Sorting Signals

Several Golgi specific glycosyltransferases with type II orientations have been identified and their transmembrane regions appear to confer Golgi retention. The N-acetyl-glucosaminyltransferase I of the medial Golgi[85] and the β-1,4-galactosyltransferase[86] and α-2,6-sialyl transferases of the trans-Golgi[87] contain retention signals within their hydrophobic transmembrane regions. The first transmembrane spanning domain of the type I Coronavirus E1 protein contains a cis-Golgi retention signal.[88] Whether these or other transmembrane regions of Golgi resident proteins can be used for intrabody targeting remains to be determined.

Lumenal TGN Sorting Signals

I am aware of only one example of a retention signal that has been described for a lumenal Golgi protein, protein kinase Cε.[89] The zinc finger domain of this protein alone localized to the Golgi network. However, both Golgi-specific glycosaminoglycan sulfation and secretion of sulfated glycoaminoglucans into the medium was impaired in cells expressing the PKC zinc finger domain. These inhibitions of Golgi-specific functions raised questions about whether this zinc finger domain will be useful of a intrabody targeting motif for the Golgi network.

Regulated Secretory Pathway

While secretory proteins in general are released from cells via a nonregulated constitutive pathway, neuroendocrine cells of the nervous and endocrine system also have a regulated secretory pathway (RSP) from which hormones, neuropeptides and the granins are secreted in a calcium-dependent pathway. The sorting signal motif for RSP proteins appears to be encoded by a consensus conformation-dependent motif.[90-91] The specific RSP motif has been identified at the N-terminal region of proopiomelanocortin (POMC) and comprises a 13-amino acid amphipathic loop structure that is stabilized by a disulfide bridge.[91] Cross-linking studies utilizing a radiolabeled POMC polypeptide containing the RSP motif led to the recent discovery that carboxypeptidase E is a regulated secretory pathway sorting receptor of the TGN.[92] Whether addition of this sorting signal to the N-terminus of an sFv intrabody can lead to the redirecting of lumenal or membrane bound target proteins to the RSP pathway remains to be determined. An interesting variation of this hypothesis is whether regulated secretion of a sFv intrabody can be achieved. Finally, a 38 amino acid RSP sorting signal has also been identified in the cytoplasmic tail of proprotein convertase PC5.[93]

Transport Competence in the Secretory Compartment

As discussed previously, ER-directed sFv intrabodies without a KDEL retention signal can be retained in the ER, probably by association with BiP or other ER chaperone proteins.[14] Interestingly, when we converted the anti-gp120 sFv105 to a Fab105 intrabody, the Fd heavy and light chains were appropriately assembled, disulfide linked and were secreted from the stably transfected cells.[15] Moreover, the Fab105 intrabodies were effective at binding the HIV-1 envelope glycoproteins throughout the secretory pathway and this led to a potent inhibition of envelope glycoprotein processing and virus infectivity, even of HIV-1 viruses that had escaped extracellular neutralization with the parental F105 mAb (Fig. 1.4). Both the prolonged interaction of the Fab105 intrabodies with their target protein throughout the secretory pathway as well as their high concentrations in the secretory vesicles are probably responsible for this profound inhibitory effect. The role of the CH_1 heavy and V_L domains in allowing the secretion of native, assembled immunoglobulins are well described.[40,94-98]

Targeting Intrabodies to the Cytosol

The reducing environment of the cytosol would be expected not to allow for efficient and proper folding of intrabodies; however, numerous examples now exist to demonstrate that functional intrabodies can be formed. Biocca and Cattaneo[45] demonstrated that the level of expression of an ER-directed, mitochondrial, nuclear and cytosolic sFv intrabody was 100%, 50%, 15% and 5%, respectively. In addition, these investigators demonstrated by alkylation studies that free sulfhydryl groups were available on the cytosolic sFv intrabody, suggesting that intrachain disulfide bonds had not formed or had formed with a low efficiency.

The Peroxisomal Compartment

Peroxisomes are a versatile and ubiquitous subcellular organelle that is involved in numerous catabolic and anabolic pathways: most importantly peroxide metabolism, the β-oxidation of (very long chain) fatty acids, and the biosynthesis of etherphospholipids. In mammalian cells, peroxisomes usually appear as single

Fig. 1.4. HIV-1 envelope processing, transport and associated cytopathic processes of syncytia formation and apoptosis. Syncytia formation (fusion between HIV-1-infected cells expressing surface gp120/gp41 and uninfected cells expressing CD4) is blocked by secreted Fab105 fragments. Apoptosis is proposed to occur by either soluble gp120 or anti-gp120/gp120 complexes binding to CD4 prior to (instead of simultaneous with) engagement of the T cell receptor/antigen/MHC II complex. The combined intra- and extracellular immunization by Fab105 expressing cells is shown.

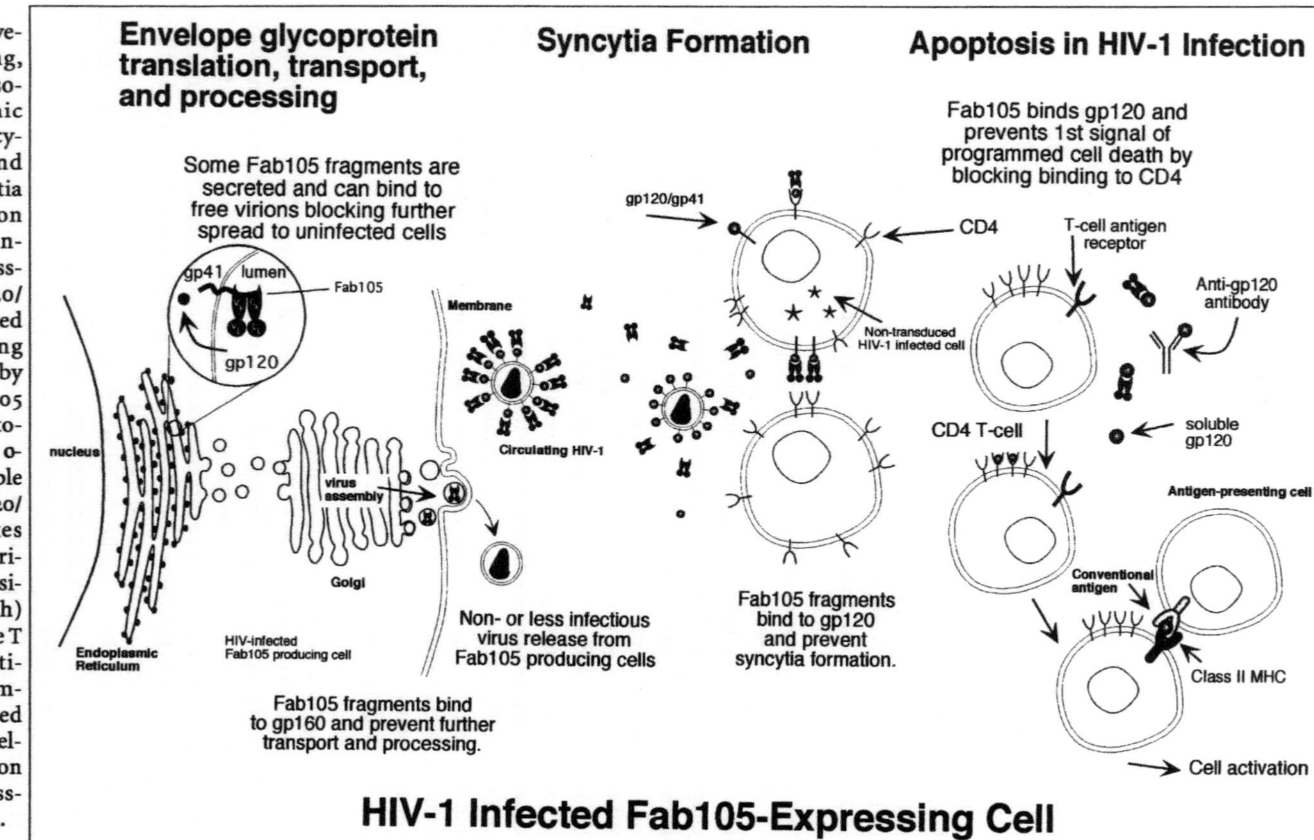

membrane-bound spherical vesicles evenly dispersed throughout the cytoplasm; however, a more complex peroxisomal architecture has been recently described.[99] Unlike mitochondria and chloroplasts, peroxisomes are devoid of DNA and consequently all the peroxisomal membrane and matrix proteins are encoded by nuclear genes. Following translation on free polysomes, both matrix and membrane proteins are targeted posttranslationally to the organelle. Peroxisome proteins contain sequences that route them specifically to the peroxisomal surface, where they recognize and bind to a surface receptor and are subsequently imported. Two pathways for the import of proteins into the peroxisomal matrix have been described, each dependent on the use of a specific peroxisomal targeting signal (PTS) and a cognate receptor.[100] PTS1 has been identified as a carboxy-terminal, tripeptide SKL while a second N-terminal sequence R/K-L/I/V-X_5-H/Q-L/A is used by a smaller subset of proteins.[101-102]

We have constructed the peroxisomal targeting vector shown in Figure 1.5 that directs sFv intrabody expression to the peroxisome (Fig. 1.6, center panel). We anticipate that intrabodies that are directed to this compartment may not only be used for inactivating proteins within this compartment but may also be able to target small molecules (drugs) in this compartment in an effort to detoxify their effects through peroxisomal oxidative reactions.[103]

The Cytosolic Plasma Membrane

Lipidation Signals

Although the classic view of membrane-associated proteins is one in which the protein is inserted into the membrane bilayer so that it spans the membrane, studies over the past decade have highlighted the understanding that additional pathways exist for directing proteins to cell membranes that involve co- or posttranslational modification by specific lipids. This process, termed acylation, is especially evident when one examines the protein machinery involved in transmembrane signaling. Indeed, in order to perform their functions at the correct time and place, many cytoplasmic proteins involved in signal transduction must rely on regulated processes guiding their binding to the cytoplasmic leaflet of the membrane lipid bilayer. In fact, it is now clear that the lipids attached to these signaling molecules play critical roles in their functions. Two main types of protein acylation have been observed in eukaryotic cells: fatty acylation with myristate and palmitate and isoprenoylation with farnesyl or geranylgeranyl moieties.[104-105] The amino acids that are co- or posttranslationally modified have been identified for many cellular and viral proteins.[106-110] These amino acid residues should be easily incorporated into intrabody framework regions to allow these modifications to take place by the cellular enzymes. Several reviews of protein lipidation in cell signaling have been recently published.[104-105]

The CAAX Motif

Many extracellular signals bind to receptors in the plasma membrane and regulate cell proliferation, differentiation, and metabolism by activating different signaling cascades; Ras proteins are critical intermediates between upstream tyrosine kinases and a downstream cascade of protein kinases, including the mitogen-activated protein kinases (MAPKs) and the stress-activated protein kinases or the Jun kinases (SAPKs/JNKs). The C-terminal CAAX motif of Ras proteins undergoes a

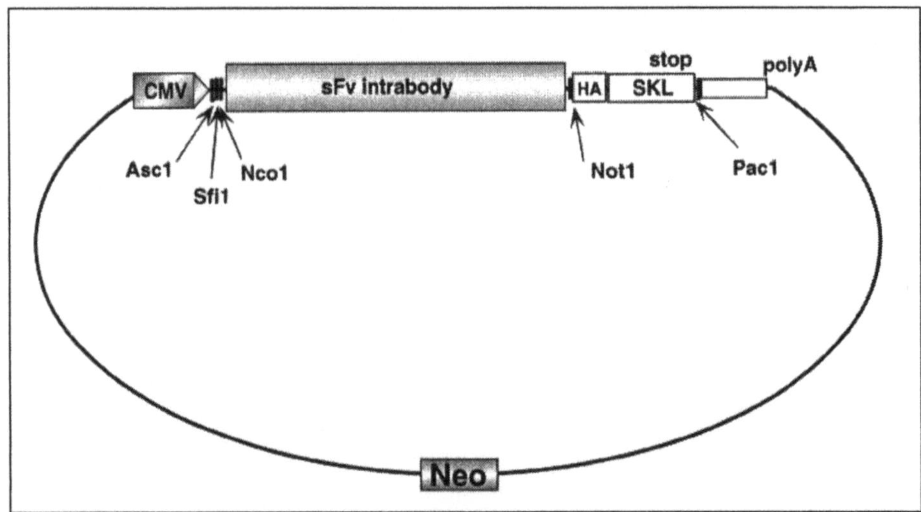

Fig. 1.5. A peroxisomal targeting vector for intrabodies. Following a HA tag sequence is the trip-eptide SKL, which has been identified as a peroxisome targeting signal (PTS1).

triplet of posttranslational modifications that are required for membrane association, and the Ras C-terminal motifs are sufficient to allow plasma membrane targeting of heterologous proteins.[105,111-114]

To target sFv intrabodies to the cytoplasmic side of the plasma membrane, we have introduced the C-terminal CAAX motifs of H-Ras (CMSCKCVLS) or of K-Ras (SKDGKKKKKKSKTKCVIM) into expression vectors to direct the sFv intrabodies to the cytoplasmic side of the plasma membrane (Fig. 1.7). The immunofluorescence data presented in Figure 1.6 (lower panel) demonstrates that this sFv intrabody with the carboxy-terminal K-Ras CAAX motif is localized to both the cytoplasm and cytoplasmic leaflet of the plasma membrane lipid bilayer.

Membrane Anchoring Signals

The syntaxin family of vesicular transport receptors are nervous system-specific proteins implicated in the docking of synaptic vesicles with the pre-synaptic plasma membrane.[115] All members of this family of proteins are oriented toward the cytoplasm and contain a region of highly hydrophobic amino acids at the extreme carboxyl terminus. This domain is of sufficient length and hydrophobicity to serve as a membrane anchor. It is interesting to speculate that these anchoring amino acids could be sufficient to orient a sFv intrabody to the cytoplasmic side of the plasma membrane.

Targeting Intrabodies to the Nucleus

Localization signals for nuclear import and export of proteins have been identified.[116-117] Directing intrabodies to the nucleus has been accomplished in several different manners. We have used a carboxy-terminal nuclear localization signal that is derived from the large T antigen of SV40 virus[118] to direct sFv intrabodies to the nucleus. This TPPKKKRKV sequence has been shown to lead to nuclear import of heterologous proteins.[13,119-120] Biocca and colleagues[121] substituted the

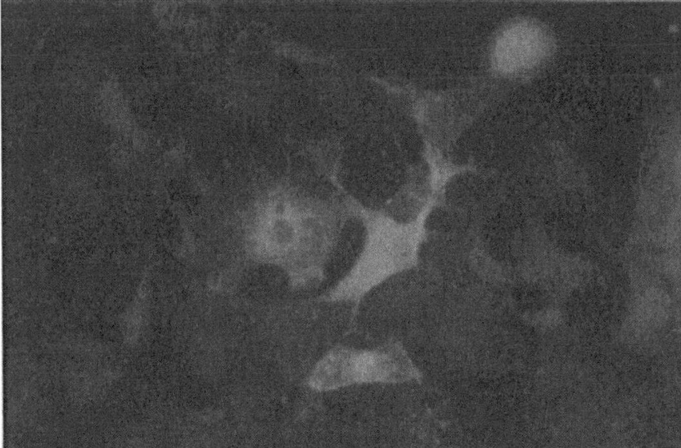

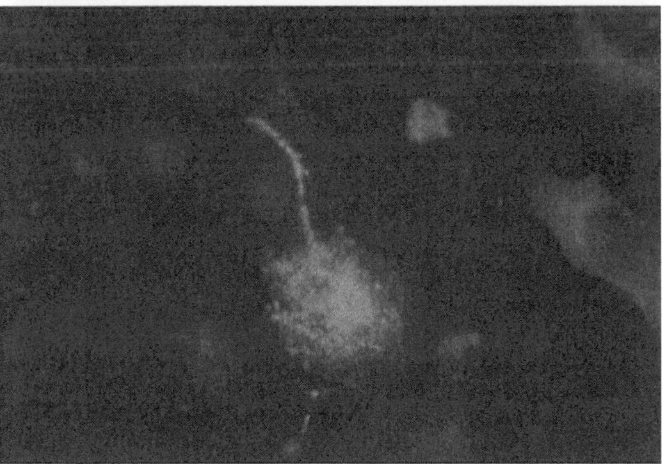

Fig. 1.6. Immunofluorescence staining of an sFv intrabody expressed in COS-7 cells with different intracellular trafficking signals. COS-7 cells were transiently transfected with a vector expressing a cytoplasmic HA tagged sFv intrabody (upper panel) or with a peroxisomal targeting vector as described in Figure 1.5 (center panel) or with a vector in which the intrabody is targeted to the cytoplasmic leaflet of the plasma membrane by the addition of a K-Ras CAAX motif (lower panel—see Figure 1.7 for vector details). Experiments were performed by Dr. Adriaan de Bruïne of the University of Maastricht, The Netherlands.

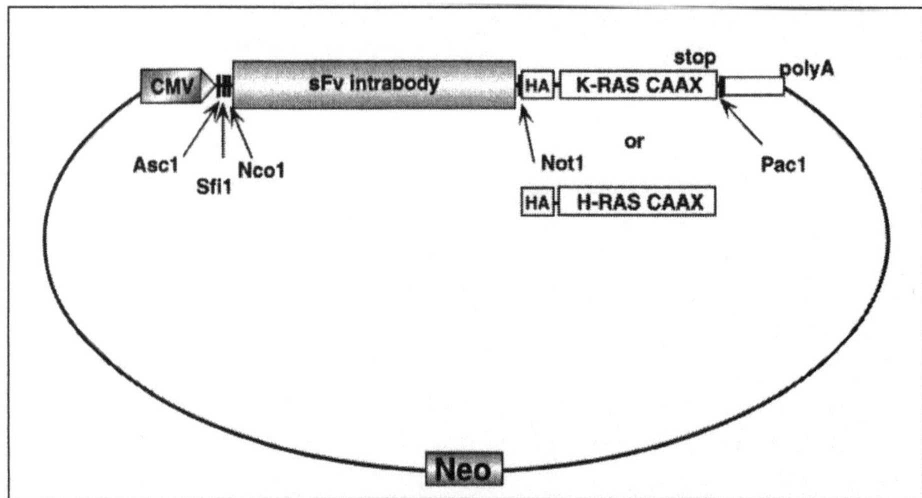

Fig. 1.7. CAAX containing intrabody expression vectors for targeting to the cytoplasmic leaflet of the membrane lipid bilayer. Shown are vectors that contain either the carboxy-terminal H-Ras and K-Ras CAAX motifs.

hydrophobic core of the V-region leader sequence with the same PKKKRKV nuclear localization signal followed by three glycine residues. This mutated leader is not cleaved by the RER associated signal peptidase and also directs the corresponding intrabody (chain) to the nucleus. Levy-Mintz and colleagues,[122] using a V_L-ICL-V_H sFv format have introduced an HIV-1 Tat protein NLS (GRKKRRQRRRAHQN) immediately after the interchain linker region of an anti-HIV-1 integrase sFv intrabody.

Considering that all proteins are made in the cytoplasm, yet many need to be imported into the nucleus, we and others have investigated whether the cytoplasm or the nucleus is the best location to bind to and inhibit a nuclear protein. The best examples are HIV-1 Tat and Rev, which are both critical regulatory proteins of the virus that exert their primary effects by binding to nuclear viral mRNAs (see chapters 8 and 10 for more details). Both cytoplasmic anti-Tat and anti-Rev sFv intrabodies exerted their primary inhibitory effects by blocking nuclear import of the target proteins. With anti-Tat sFv intrabodies, we did not see an additional or improved inhibitory effect by directing the intrabodies to the nucleus with the carboxy-terminal SV40 nuclear localization signal discussed above.

The prototypic nuclear export signals (NES) of HIV-1 Rev and of a polypeptide inhibitor (PKI) of the cAMP-dependent protein kinase (PKA) have been identified as short, hydrophobic polypeptides with a high leucine content.[116-117] Whether these signals could be incorporated into intrabody designs so as to cause nuclear export of target proteins remains to be investigated.

Targeting Intrabodies to the Mitochondria

Mitochondria have four independent compartments, the outer and inner membranes, the intermembrane space, and the matrix. The vast majority of mitochondrial proteins are encoded by nuclear genes, but a limited number are encoded by the mitochondrial chromosome, which is located in the matrix. Some of these proteins are "exported" into or across the inner membrane.

The import of proteins from the cytosol into the mitrochondrial matrix involves the concerted action of two separate import systems: the TOM system in the outer membrane, and the TIM system in the inner membrane.[123] The N-termini of most "imported" mitochondrial proteins include a positively charged signal which routes them to this organelle. Ligation of the N-terminal 25 amino acid pre-sequence of the subunit VIII of human cytochrome c oxidase (COX8.21), plus the first 10 amino acids of the mature human cytochrome c oxidase in frame with a sFv intrabody, has been reported.[45] Targeting studies confirmed that efficient folding and transport had occurred. Unfortunately, the biological activity of this intrabody was not tested. However, these studies do suggest that mitochondrial targeting will be possible.

Summary and Future Prospects

The results obtained to date suggest that intrabodies represent a powerful alternative to other methods of gene inactivation and allow a highly refined approach to the analysis and manipulation of microbial and cellular pathways. Indeed, there are few theoretical constraints to the use of intrabodies, which may represent the only option in circumstances where the desired target has not been cloned or is non-protein in nature (e.g., sugar, DNA, or a soluble metabolite). As a result of these unique properties, intrabodies may find widespread use as high throughput screening tools for functional genomics and target validation, particularly if large human sFv antibody libraries can be used as starting reagents for the screening. While antibody phage display libraries are the most versatile tools for this screening today, high-density screening arrays that rely on novel surface technology may find a use in the future.[124]

Owing to the diversity of the immunoglobulin V gene repertoire, the major hurdle in any given application may be that of identifying an antibody with the desired specificity and affinity; several studies have demonstrated that the epitope on the target molecule is critical to achieve the desired biological effect.[13,122,125] In fact, two reasons for the potent inhibitory activity of the intrabodies thus far reported may reside in the relative ease of directing the intrabodies to relevant subcellular compartments and to precise epitopes on the target protein. Through the development of high throughput screening arrays that will identify sFv intrabodies that are directed to different epitopes on the target molecule and by incorporating different subcellular trafficking signals on these sFv intrabodies, we will undoubtedly achieve the desired biological effect with greater frequency and potency.

For the full potential of intrabodies to be developed, some structural modifications may be required. For example, it is likely that some cytosolic sFv intrabodies may not be active due to the lack of critical intrachain disulfide bond formation. By analogy to studies performed in *E. coli*, antibodies are generally unstable in the absence of disulfide bonds. However, the successful expression and functional activity of cytosolic sFv intrabodies suggest that intact intrachain disulfide bond formation may not be universally required, particularly if the presence of the target molecule can assist in stabilizing intrabody folding.[126] Cytosolic sFv intrabodies could be improved if antibody frameworks more suitable for cytosolic expression could be engineered by design or by selection so that reduced intrabody turnover and increased proper folding efficiency could occur. Indeed, a naturally occurring antibody that folds efficiently even when lacking the highly conserved cysteine residues has been reported.[127] Intrabodies also have the potential, not only to disrupt protein function, but to act in a positive fashion, for example, by enhancing protein activity, by stabilizing protein-protein or protein-nucleic acid interactions, or by performing catalytic functions themselves.

As in the case of other gene-based therapies, certain issues need to be addressed before intrabody technology can be widely applied in the clinical context. These include developing gene-transfer systems that are capable of delivering the intrabody gene into a sufficient number of the correct cell type; sustained and high enough expression of the intrabody gene to achieve the desired clinical benefit; and the intrabody protein should be non-toxic and non-immunogenic unless they are directed to oncoproteins where toxicity and immunogenicity could contribute to their anti-tumor effect. Although the intrabody technology is in its infant stage of development, several clinical gene therapy trials will be performed in the near future, and the results of these studies will surely help to further develop and exploit the use of intrabodies in basic science and gene medicine.

Acknowledgments

I owe a great deal to Dr. Elmer Becker who allowed me the opportunity to become an immunochemist. I would also like to acknowledge the many postdoctoral fellows, past and present, for sharing their enthusiasm for developing the field of intrabodies. Thanks to special colleagues in the field including Drs. Urban Ramstedt and Bruno Tocque and others for sharing information, ideas and reagents over the years. This work was supported by National Institutes of Health grants AI28785, AI33802, and AI31783 and the National Foundation for Cancer Research and the Dana-Farber Cancer Institute Friends 10.

References

1. Marasco WA. Intracellular antibodies (intrabodies) as research reagents and therapeutic molecules for gene therapy. Immunotech 1995; 1:1-19.
2. Marasco WA. Intrabodies: Turning the humoral immune system outside in for intracellular immunization. Gene Therapy 1997; 4:11-15.
3. Richardson JH, Sodroski JG, Waldmann TA et al. Phenotypic knockout of the high-affinity human interleukin 2 receptor by intracellular single-chain antibodies against the α subunit of the receptor. Proc Natl Acad Sci USA 1995; 92:3137-3141.
4. Levin R, Mhashilkar AM, Dorfman T et al. Inhibition of early and late events of the HIV-1 replication cycle by cytoplasmic Fab intrabodies against the matrix protein, p17. Mol Med 1997; 3:96-110.
5. Biocca S, Pierandrei-Amaldi P, Cattaneo A. Intracellular expression of anti-p21ras single chain Fv fragments inhibits meiotic maturation of xenopus oocytes. Biochem and Biophys Res Comm 1993; 197:422-427.
6. Biocca S, Pierandrei-Amaldi P, Campioni N et al. Intracellular immunization with cytosolic recombinant antibodies. Bio/tech 1994; 12:396-399.
7. Duan L, Bagasra O, Laughlin MA et al. Potent inhibition of human immuno-deficiency virus type 1 replication by an intracellular anti-Rev single-chain antibody. Proc Natl Acad Sci USA 1994; 91:5075-5079.

8. Duan L, Zhang H, Oakes JW et al. Molecular and virological effects of intracellular anti-Rev single-chain variable fragments on the expression of various human immunodeficiency virus-1 strains. Hum Gene Therapy 1994; 5:1315-1324.

9. Beerli RR, Wels W, Hynes NE. Autocrine inhibition of the epidermal growth factor receptor by intracellular expression of a single-chain antibody. Biochem and Biophy Res Com 1994; 204:666-672.

10. Beerli RR, Wels W, Hynes NE. Intracellular expression of single chain antibodies reverts ErbB-2 transformation. J Biol Chem 1994; 269:23931-23936.

11. Deshane J, Loechel F, Conry RM et al. Intracellular single-chain antibody directed against erbB2 down-regulates cell surface erbB2 and exhibits a selective anti-proliferative effect in erbB2 overexpressing cancer cell lines. Gene Ther 1994; 1:332-337.

12. Werge TM, Baldari CT, Telford JL. Intracellular single chain Fv antibody inhibits Ras activity in T-cell antigen receptor stimulated Jurkat cells. FEBS Let 1994; 351:393-396.

13. Mhashilkar AM, Bagley J, Chen S-Y et al. Inhibition of HIV-1 Tat-mediated LTR transactivation and HIV-1 infection by anti-Tat single chain intrabodies. EMBO J 1995; 14:1542-1551.

14. Marasco WA, Haseltine WA, Chen S-Y. Design, intracellular expression, and activity of a human anti-human immunodeficiency virus type 1 gp120 single-chain antibody. Proc Natl Acad Sci USA 1993; 90:7889-7893.

15. Chen SY, Khouri Y, Bagley J et al. Combined intra- and extracellular immunization against human immunodeficiency virus type 1 infection with a human anti-gp120 antibody. Proc Natl Acad Sci USA 1994; 91:5932-5936.

16. Chen SY, Khouri Y, Bagley J et al. Combined intra- and extracellular immunization against human immunodeficiency virus type 1 infection with a human anti-1gp120 antibody. Proc Natl Acad Sci USA 1994; 91:5932-5936.

17. Burton DR, Barbas CF III, Persson MA et al. A large array of human monoclonal antibodies to type 1 human immunodeficiency virus from combinatorial libraries of asymptomatic seropositive individuals. Proc Natl Acad Sci USA 1991; 88:10134-10137.

18. Hoogenboom HR, Marks JD, Griffiths AD et al. Building antibodies from their genes. Immunol Rev 1992; 130:41-68.

19. Winter G, Griffiths AD, Hawkins RE et al. Making antibodies by phage display technology. Annu Rev Immunol 1994; 12:433-455.

20. Marks JD, Hoogenboom HR, Griffiths AD et al. Molecular evolution of proteins on filamentous phage: mimicking the strategy of the immune system. J Biol Chem 1992; 267:16007-16010.

21. Nissim A, Hoggenboom HR, Tomlinson IM et al. Antibody fragments from a 'single pot' phage display library as immunochemical reagents. EMBO J 1994; 13:692-698.

22. Vaughan TJ, Williams AJ, Pritchard K et al. Human antibodies with sub-nanomolar affinities isolated from a large non immunized phage display library. Nature Bio 1996; 14:309-314.

23. Marks C, Marks JD. Phage libraries—a new route to clinically useful antibodies. New Eng J Med 1996; 335:730-733.

24. Portolano S, McLachlan SM, Rapoport B. High affinity, thyroid-specific human auto antibodies displayed on the surface of filamentous phage use V genes similar to other auto antibodies. J Immunol 1993; 151:2839-2851.

25. Barbas SM, Ditzel HJ, Salonen EM et al. Human auto antibody recognition of DNA. Proc Natl Acad Sci USA 1995; 92:2529-2533.

26. Barbas CF 3d, Bjorling E, Chiodi F et al. Recombinant human Fab fragments neutralize human type 1 immunodeficiency virus in vitro. Proc Natl Acad Sci USA 1992; 89:9339-9343.

27. Zebedee SL, Barbas CF 3d, Hom YL et al. Human combinatorial antibody libraries to hepatitis B surface antigen. Proc Natl Acad Sci USA 1992; 89:3175-3179.

28. Lonberg N, Taylor LD, Harding FA et al. Antigen-specific human antibodies from mice comprising four distinct genetic modifications. Nature 1994; 368:856-859.

29. Green LL, Hardy MC, Maynard-Currie CE et al. Antigen-specific human monoclonal antibodies from mice engineered with human Ig heavy and light chain YACs. Nat Genet 1994; 7:13-21

30. Clackson T, Hoogenboom HR, Griffiths AD et al. Making antibody fragments using phage display libraries. Nature 1991; 352:624-628.

31. Marks JD, Hoogenboom HR, Bonnert TP et al. By-passing immunization. Human antibodies from V-gene libraries displayed on phage. J Mol Biol 1991; 222:581-597.

32. Griffiths AD, Malmqvist M, Marks JD et al. Human anti-self antibodies with high specificity from phage display libraries. EMBO J 1993; 12:725-734.

33. Marks JD, Griffiths AD, Malmqvist M et al. By-passing immunization: building high affinity human antibodies by chain shuffling. Biotech 1992; 10:779-783.

34. Gram H, Marconi L-A, Barbas CF 3d et al. In vitro selection and affinity maturation of antibodies from a naive combinatorial immunoglobulin library. Proc Natl Acad Sci USA 1992; 89:3576-3580.

35. Akamatsu Y, Cole MS, Tso JY et al. Construction of a human Ig combinatorial library from genomic V segments and synthetic CDR3 fragments. J Immunol 1993; 151:4631-4659.

36. Jespers LS, Roberts A, Mahler SM et al. Guiding the selection of human antibodies from phage display repertoires to a single epitope of an antigen. Bio Tech 1994; 12:899-903.

37. Barbas CF 3d, Kang AS, Lerner RA et al. Assembly of combinatorial antibody libraries on phage surfaces: the gene III site. Proc Natl Acad Sci USA 1991; 88: 7978-7982.

38. Barbas CF 3d, Björling E, Chiodi F et al. Recombinant human Fab fragments neutralize human type 1 immunodeficiency virus in vitro. Proc Natl Acad Sci USA 1992b; 89:9339-9343.

39. Melnick J, Aviel S, Argon Y. The endoplasmic reticulum stress protein, GRP94, in addition to BiP, associates with unassembled immunoglobulin chains. J Biol Chem 1992; 2657:21303-21306.

40. Melnick J, Argon Y. Molecular chaperones and the biosynthesis of antigen receptors. Immunol Today 1995; 16:243-250.

41. Hochstenbach F, David V, Watkins S et al. Endoplasmic reticulum resident protein of 90 kilodaltons associates with the T- and B-cell antigen receptors and major histocompatibility complex antigens during their assembly. Proc Natl Acad Sci USA 1992; 89:4734-4738.

42. Freedman RB, Hirst TR, Tuite MF. Protein disulfide isomerase: building bridges in protein folding. Trends Biochem 1994; 19:331-336.

43. Freedman RB. The formation of protein disulfide bonds. Curr Op Struct Biol 1995; 5:85-91.

44. Hwang C, Sinskey AJ, Lodish HF. Oxidized redox state of glutathione in the endoplasmic reticulum. Science 1992; 257:1496-1502.

45. Biocca S, Ruberti F, Tafani M et al. Redox state of single chain Fv fragments targeted to the endoplasmic reticulum, cytosol and mitochondria. BioTech 1995; 13:1110-1115.

46. Ryabova LA, Desplancq D, Spirin AS et al. Functional antibody production using cell-free translation: effects of protein disulfide isomerase and chaperones. Nature Biotech 1997; 15:79-84.

47. Larrick JW, Danielsson L, Brenner CA et al. Rapid cloning of rearranged immu-

noglobulin genes from human hybridoma cells using mixed primers and the polymerase chain reaction. Biochem and Biophy Res Comm 1989; 160:1250-1256.

48. Larrick JW, Danielsson L, Brenner CA et al. Polymerase chain reaction using mixed primers: cloning of human monoclonal antibody variable region genes from single hybridoma cells. Bio/Tech 1989; 7:934-938.

49. Jones ST, Bendig MM. The design and use of redundant PCR-primers to rapidly clone mouse immunoglobulin variable domains. Poster presentation for the First Conference on Antibody Engineering, San Diego CA, 1990; 10-11 December.

50. Kabat EA, Wu TT, Perry HM, et al. Tabulation and analysis of amino acid and nucleic acid sequences of precursors, V-regions, C-regions, J-chain, T-cell receptors for antigen, T-cell surface antigens, Thy-1, complement, c-reactive protein, thymopoietin, integrins, postgamma globulin, α-macroglobulins, and other related proteins. In: Sequences of Proteins of Immunological Interest. U.S. Department of Health and Human Services, 1991. NIH Publication No. 91-3242.

51. Tomlinson IM, Cox JPL, Gherardi E et al. The structural repertoire of the human V domain. EMBO J 1995; 14:4628-4638.

52. Carroll WL, Mendel E, Levy S. Hybridoma fusion cell lines contain an aberrant kappa transcript. Molec Immunol 1988; 25:991-995.

53. Duan L, Pomerantz RJ. Elimination of endogenous aberrant kappa chain transcripts from sp2/30-derived hybridoma cells by specific ribozyme cleavage: utility in genetic therapy of HIV-1 infections. Nucl Acids 1994; 22:5432-5438.

54. Kipriyanov SM, Kupriyanova OA, Little M et al. Rapid detection of recombinant antibody fragments directed against cell-surface antigens by flow cytometry. J Immunol Meth 1996; 196:51-62.

55. Krebber A, Bornhauser S, Burmester J et al. Reliable cloning of functional antibody variable domains from hybridomas and spleen cell repertoires employing a re-engineered phage display system. J Immunol Meth 1997; 201:35-55.

56. Chu T-H T, Martinez I, Olson P et al. Highly efficient eukaryotic gene expression vectors for peptide secretion. BioTech 1995; 18:890-899.

57. Persic L, Righi M, Roberts A et al. Targeting vectors for intracellular immunization. Gene 1997; 187(1):1-8.

58. Pugsley AP. Protein Targeting. Academic Press 1989 San Diego CA.

59. Griffiths G, Ericsson M, Krijnse-Locker J et al. Localization of the Lys, Asp, Glu, Leu tetra peptide receptor to the Golgi complex and the intermediate compartment in mammalian cells. J Cell Biol 1994; 127:1557-1574.

60. Miesenböck G, Rothman JE. The capacity to retrieve escaped ER proteins extends to the trans-most Cisterna of the Golgi stack. J Cell Biol 1995; 129:309-319.

61. Townsley FM, Wilson DW, Pelham HRB. Mutational analysis of the human KDEL receptor distinct structural requirements for Golgi retention, ligand binding and retrograde transport. EMBO J 1993; 12:2821-2829.

62. Richardson JH, Marasco WA. Intracellular antibodies: development and therapeutic potential. Trends Biotech 1995; 13:306-310.

63. Richardson JH, Waldmann TA, Sodroski JG et al. Inducible knockout of the interleukin-2 receptor α chain: Expression of the high-affinity IL-2 receptor is not required for the in vitro growth of HTLV-I transformed T cell lines. Virology 1997; submitted.

64. Richardson JH, Hofmann W, Sodroski JG et al. Intrabody-mediated knockout of the high-affinity IL-2 receptor in primary human T cells using a bicistronic HIV-1-based vector. Nature Biotechnology 1997, submitted.

65. Jost CR, Kurucz I, Jacobus CM. Mammalian expression and secretion of functional single-chain Fv molecules. J Biol Chem 1994; 269:26267-26273.

66. Flynn GC, Pohl J, Flocco MT et al. Peptide-binding specificity of the molecular chaperone BiP. Nature 1991; 353:726-730.

67. Wiertz HJ, Tortorella D, Bogyo M et al. Sec61-mediated transfer of a membrane protein from the endoplasmic reticulum to the proteasome for destruction. Nature 1996; 384:432-438.

68. Cosson P, Letourneur F. Coatomer interaction with Di-lysine endoplasmic reticulum retention motifs. Science 1994; 263:1629-1631.

69. Schutze M-P, Peterson PA, Jackson MR. An N-terminal double-arginine motif maintains type II membrane proteins in the endoplasmic reticulum. EMBO J 1994; 13:1696-1705.

70. Rothman JE. The compartmental organization of the Golgi apparatus. Sci Am 1985: 253(3):74-89.

71. Dunphy WG, Rothman JE. Compartmental organization of the Golgi stack. Cell 1985; 42:13-21.

72. Trowbridge IS, Collawn JF. Signal-dependent membrane protein trafficking in the endocytic pathway. Annu Rev Cell Biol 1993; 9:129-161.

73. Sandoval IV, Bakke O. Targeting of membrane proteins to endosomes and lysosomes. Trends in Cell Biol 1994; 4:292-297.

74. Collawn JF, Stangel M, Kuhn LA et al. Transferrin receptor internalization sequence YXRF implicates a tight turn as the structural recognition motif for endocytosis. Cell 1990; 63:1061-1072.

75. Zhu Z, H Y, Gershon MD et al. Targeting of glycoprotein I (gE) of Varcicella-Zoster virus to the trans-Golgi network by an AYRV sequence and an acidic amino acid-rich patch in the cytosolic domain of the molecule. J Virol 1996; 70:6563-6575.

76. Drickamer K, Mamon JF, Binns G et al. Primary structure of the rat liver asialoglycoprotein receptor. J Biol Chem 1984; 259:770-778.

77. Canfield WM, Johnson KF, Ye RD et al. Localization of the signal for rapid internalization of the bovine cation-independent mannose 6-phosphate/insulin-like growth factor-II receptor to amino acids 24-29 of the cytoplasmic tail. J Biol Chem 1991; 266:5682-5688.

78. Luzio JP, Brake B, Banting G et al. Identification, sequencing and expression of an integral membrane protein of the trans-Golgi network (TGN38). Biochem J 1990; 270:97-102.

79. Ponnambalam S, Rabouille C, Luzio JP et al. The TGN38 glycoprotein contains two non overlapping signals that mediate localization to the trans-Golgi network. J Cell Biol 1994 125:253-268.

80. Bos K, Wraight C, Stanley KK. TGN38 is maintained in the trans-Golgi network by a tyrosine-containing motif in the cytoplasmic domain. EMBO J 1993; 12:2219-2228.

81. Ohno H, Stewart J, Fournier M-C et al. Interaction of tyrosine-based sorting signals with clathrin-associated proteins. Science 1995; 269:1872-1875.

82. Pack P, Plückthun A. Mini antibodies: Use of amphipathic helices to produce functional, flexibly linked dimeric F_v fragments with high avidity in *Escherichia coli*. Biochem 1992; 31:1579-1584.

83. Letourneur F, Klausner RD. A novel Di-Leucine motif and a tyrosine-based motif independently mediate lysosomal targeting and endocytosis of CD3 chains. Cell 1992; 69:1143-1157.

84. Schäfer W, Stroh A, Berghöfer S et al. Two independent targeting signals in the cytoplasmic domain determine trans-Golgi network localization and endosomal trafficking of the proprotein convertase furin. EMBO J 1995; 14:2424-2435.

85. Tang BL, Wong SH, Low SH et al. The transmembrane domain of *N*-glucosaminyltransferase I contains a Golgi retention signal. J Biol Chem 1992; 267:10122-10126.

86. Nilsson T, Lucocq JM, Mackay D et al. The membrane spanning domain of β-1,4-galactosyltransferase specifies trans Golgi localization. EMBO J 1991; 10:3567-3575.

87. Munro S. Sequences within and adjacent to the transmembrane segment of α-2,6-sialytransferase specify Golgi retention. EMBO J 1991; 10:3577-3588.

88. Swift AM, Machamer CE. A Golgi retention signal in a membrane-spanning domain of coronavirus E1 protein. J Cell Biol 1991; 115:19-30.

89. Lehel C, Olah Z, Jakab G et al. Protein kinase C is localized to the Golgi via its zinc-finger domain and modulates Golgi function. Proc Natl Acad Sci USA 1995; 92:1406-1410.

90. Tam WWH, Andreasson KI, Loh YP. The amino-terminal sequence of pro-opiomelanocortin directs intracellular targeting to the regulated secretory pathway. Eur J Cell Biol 1993; 62:294-306.

91. Cool DR, Fenger M, Snell CR et al. Identification of the sorting signal motif within pro-opiomelanocortin for the regulated secretory pathway. J Biol Chem 1995; 270:8723-8729.

92. Cool DR, Normant E, Shen F-S et al. Carboxypeptidase E is a regulated secretory pathway sorting receptor: genetic obliteration leads to endocrine disorders in *Cpe^fat* mice. Cell 1997; 88:73-83.

93. De Bie I, Marcinkiewicz M, Malide D et al. The isoforms of proprotein convertase PC5 are sorted to different subcellular compartments. J Cell Biol 1996; 135:1261-1275.

94. Wu GE, Hozumi N, Murialdo H. Secretion of a λ immunoglobulin chain is prevented by a single amino acid substitution in its variable region. Cell 1983; 33:77-83.

95. Hendershot L, Bole D, Köhler G et al. Assembly and secretion of heavy chains that do not associate posttranslationaly with immunoglobulin heavy chain-binding protein. J Cell Biol 1987; 104:761-767.

96. Dul JL, Argon Y. A single amino acid substitution in the variable region of the light chain specifically blocks immunoglobulin secretion. Proc Natl Acad Sci USA 1990; 87:8135-8139.

97. Knittler MR, Haas IG. Interaction of BiP with newly synthesized immunoglobulin light chain molecules: cycles of sequential binding and release. EMBO J 1992; 11:1573-1581.

98. Kaloff CR, Haas IG. Coordination of immunoglobulin chain folding and immunoglobulin chain assembly is essential for the formation of functional IgG. Immunity 1995; 2:629-637.

99. Wiemer EAC, Wenzel T, Deerinck TJ et al. Visualization of the peroxisomal compartment in living mammalian cells: Dynamic behavior and association with micro tubules. J Cell Biol 1997; 136:71-80.

100. Subramini S. Protein import into peroxisomes and biogenesis of the organelle. Annu Rev Cell Biol 1993; 9:445-478.

101. Osumi T, Tsukamoto T, Hata S et al. Amino-terminal pre sequence of the precursor of peroxisomal 3-ketoacyl-CoA thiolase is a cleavable signal peptide for peroxisomal targeting. Biochem and Biophy Res Comm 1991; 181:947-954.

102. Swinkels BW, Gould SJ, Bodnar AG et al. A novel, cleavable peroxisomal targeting signal at the amino-terminus of the rat 3-ketoacyl-CoA thiolase. EMBO J 1991; 10:3255-3262.

103. van den Bosch H, Schutgens RBH, Wanders RJA et al. Biochemistry of peroxisomes. Annu Rev Biochem 1992; 61:157-197.

104. Casey PJ. Protein lipidation in cell signaling. Science 1995; 268:221-225.

105. Gelb MH. Protein prenylation, et cetera: Signal transduction in two dimensions. Science 1997; 275:1750-1751.

106. Saermark T, Bex F. Acylation of HIV proteins. Biochem Soc Trans 1989; 17:869-871.
107. Aitken A. Structure determination of acylated proteins. Biochem Soc Trans 1989; 17:871-875.
108. Towler DA, Gordon JI, Adams SP et al. The biology and enzymology of eukaryotic protein acylation. Ann Rev Biochem 1988; 57:69-99.
109. Timson Gauen LK, Linder ME, Shaw AS. Multiple features of the p59^{fyn} src homology 4 domain define a motif for immune-receptor tyrosine-based activation motif (ITAM) binding and for plasma membrane localization. J Cell Biol 1996; 133:1007-1015.
110. van't Hof W, Resh MD. Rapid plasma membrane anchoring of newly synthesized p59^{fyn}: Selective requirement for NH$_2$-terminal myristoylation and palmitoylation at cysteine-3. J Cell Biol 1997; 136:1023-1035.
111. Hancock JF, Paterson H, Marshall CJ. A polybasic domain or palmitoylation is required in addition to the CAAX motif to localize p21^{ras} to the plasma membrane. Cell 1990; 63:133-139.
112. Hancock JF, Cadwallader K, Paterson H et al. A CAAX or a CAAL motif and a second signal are sufficient for plasma membrane targeting of ras proteins. EMBO J 1991; 10:4033-4039.
113. Huang DCS, Marshall CJ, Hancock JF. Plasma membrane-targeted Ras GTPase-activating protein is a potent suppressor of p21^{ras} function. Molec Cell Biol 1993; 13:2420-2431.
114. Stokoe D, Macdonald SG, Cadwallader K et al. Activation of Raf as a result of recruitment to the plasma membrane. Science 1994; 264:1463-1467.
115. Bennett MK, Garcia-Arrarás JE, Elferink LA et al. The syntaxin family of vesicular transport receptors. Cell 1993; 74:863-873.
116. Görlich D, Mattaj IW. Nucleocytoplasmictransport. Science 1996; 271:1513-1518.
117. Nigg EA. Nucleocytoplasmic transport: signals, mechanisms and regulation. Nature 1997; 386:779-787.
118. Kalderon D, Roberts BL, Richardson WD et al. A short amino acid sequence able to specify nuclear location. Cell 1984; 39:499-509.
119. Lanford RE, Kanda P, Kennedy RC. Induction of nuclear transport with a synthetic peptide homologous to the SV40 T antigen transport signal. Cell 1986; 46:575-582.
120. Goldfarb DS, Gariépy J, Schoolnik G et al. Synthetic peptides as nuclear localization signals. Nature 1986; 322:641-644.
121. Biocca S, Neuberger MS, Cattaneo A. Expression and targeting of intracellular antibodies in mammalian cells. EMBO J 1990; 9:101-108.
122. Levy-Mintz P, Duan L, Zhang H et al. Intracellular expression of single-chain variable fragments to inhibit early stages of the viral life cycle by targeting human immunodeficiency virus type 1 integrase. J Virol 1996; 70:8821-8832.
123. Pfanner N, Meijer M. Mitochondrial biogenesis: The Tom and Tim machine. Curr Biol 1997; 7:R100-R103.
124. Lockhart DJ, Dong H, Byrne MC et al. Expression monitoring by hybridization to high-density oligonucleotide arrays. Nature Biotech 1996; 14:1675-1680.
125. Wu Y, Duan L, Zhu M et al. Binding of intracellular anti-Rev single chain variable fragments to different epitopes of human immunodeficiency virus type 1 rev: Variations in viral inhibition. J Virol 1996; 70:3290-3297.
126. Glöckshuber R, Malia M, Pfitzinger I et al. A comparison of strategies to stabilize immunoglobulin F$_v$-Fragments. Biochem 1990; 29:1362-1367.
127. Rudikoff S, Pumphrey JG. Functional antibody lacking a variable-region disulfide bridge. Proc Natl Acad Sci USA 1986; 83:7875-7878.

CHAPTER 2

Phage Libraries for Generation of Single Chain Fv Antibodies for Intracellular Immunization

Marie Alix Poul and James D. Marks

Introduction

Antibodies are heterodimers that specifically bind to a target molecule (antigen). Their ability to interfere with antigen function or synthesis make them useful tools for the in situ analysis of intracellular molecules. Antibodies can be directly introduced into the cell or they can be expressed from their genes in a specific cellular compartment where they will bind their target. The introduction of antibodies intracellularly was first successfully achieved by micro injection (in the cytoplasm or in the nucleus)[1] or more recently, by electroporation.[2] The specific (but transient) effects that were obtained by these methods opened the way to a large number of applications. Expression was also obtained by injection of poly A+ mRNA purified from a hybridoma secreting an antibody of interest in xenopus oocytes.[3] For example, an anti-golgi protein antibody mRNA was able to inhibit the intracellular trafficking of a viral protein.[4] These experiments showed clearly that a specific antibody/antigen interaction was possible in eukaryotic cells.

Using the injection techniques described above, the intracellular use of antibodies was limited to a small number of cells. With the progress in recombinant antibody technology, intracellular "immunization" became a more feasible strategy. Stable expression of IgG antibody in non lymphoid cells was first obtained in 1987 by Cattaneo and Neuberger[5] using vectors containing immunoglobulin heavy and light chain genes. Biocca et al, further showed that the assembly of functional heterodimeric antibody could occur not only in the secretory pathway, but also in the cytoplasm.[6-7] Moreover, the addition of a nuclear targeting sequence was able to drive the newly synthesized antibody to the nucleus. Constitutive or inducible expression of antibodies can now be obtained theoretically in any cell type and provides a powerful tool to study the function of a protein in the intracellular context. While initial experiments used complete IgG molecules, more recent experiments have utilized antibody fragments that can be encoded in a single gene, greatly facilitating the ease of transfection. To understand this approach, antibody structure and function will be briefly reviewed in the next section.

Intrabodies: Basic Research and Clinical Gene Therapy Applications, edited by Wayne A. Marasco. © 1998 Springer-Verlag and R.G. Landes Company.

Antibody Structure and Function

An antibody is composed of two heavy (H) chains and two light (L) chains (Fig. 2.1A and B). Each L chain is composed of an N-terminal variable (V) domain (V_L) and a constant (c) domain (C_L). Each H chain is composed of an N-terminal V domain, 3 or 4 C domains, and a hinge region. The Fv (variable fragment) consists of the paired V_H and V_L domains, and comprises the antigen binding portion of the molecule. The V_H and V_L domains consist of four regions of relatively conserved sequence called framework regions (FR1, FR2, FR3, and FR4) which form a scaffold for three regions of hypervariable sequence (complementarity determining regions, CDRs)[8] (Fig. 2.1C, D). The CDRs contain most of the antigen binding residues and antibodies are able to recognize virtually any shape antigen because of the tremendous sequence diversity in the CDRs.

This diversity is generated at the genetic level by gene recombination (reviewed in ref. 9). The rearranged V_H gene consists of a V-gene segment, which encodes the first 2 CDRs, a D or diversity segment which encodes most of the third CDR, and a J or joining segment which encodes the remainder of CDR3 and all of framework 4 (Fig. 2.2A). The rearranged V_L gene consists of a V-gene segment, encoding all of CDR1 and 2 and most of CDR3, and a J-gene segment encoding the remainder of the CDR3 and all of framework 4 (Fig. 2.2B). Recombination permits a relatively small number of gene segments to generate tremendous diversity. For example, in humans, there are approximately 51 functional V_H gene segments,[10] 33 D gene segments,[11] and 6 J_H gene segments.[11] Since the D gene segments can be read in one of three reading frames, the combinatorial diversity (V x D x J) is 3.0 x 10^4. Variable H chain diversity is increased further by random joining anywhere in the D-segment (junctional diversity) and by random nucleotide addition (N-segment addition) both V-D and D-J.[12] In humans, N-segment addition averages approximately 2 amino acids V-D and 2 amino acids D-J.[13] This results in a potential V_H diversity of $> 10^9$. Variable L chain diversity is considerably less, due to the absence of D segments and less extensive N segment addition. In humans there are 40 functional V_κ,[14] 31 functional V_λ,[15] 5 J_κ,[11] and 5 J_λ,[11] yielding a combinatorial L chain diversity of approximately 10^3 -10^4, depending on the amount of N-segment addition. Total diversity is the product of V_H and V_L diversity. It should be noted that as a result of the large number of D-segments, junctional diversity, and N segment addition, diversity is concentrated in V_H CDR3 which is located in the center of the antibody combining site.

Single Chain Fv Antibody Fragments

Antibody fragments which retain the ability to bind antigen can be constructed from the genes of complete antibodies. The smallest antigen binding fragment is the Fv (variable fragment), which consists of non covalently associated V_H and V_L domains. The single chain Fv (scFv) (Fig. 2.1D) consists of the V_H and V_L domains joined by a short and flexible linker peptide[16-17] which overcomes the tendency of many Fvs to dissociate at physiologic concentrations. The V_H and V_L domains fold to form a functional antigen binding site which typically binds antigen with similar affinity to the IgG from which the V-regions were derived.[18] The peptide linker can connect either the C-terminus of the V_H to the N-terminus of the V_L or the C-terminus of the V_L to the N-terminus of the V_H. As long as the linker is of adequate length, a wide range of different linker sequences can be utilized.[19]

Single chain Fvs have general features that make them attractive compared to complete antibodies for intracellular use. First, they are encoded by a single gene, thus bypassing the problem of H and L chain expression ratio and correct assembly in nonlymphoid cells. Unlike IgG,[20,21] scFv can also be rapidly expressed in bacteria as functional native protein[22] (see below). This facilitates in vitro characterization, such as affinity measurement and epitope mapping.

Traditional Approach for Production of Recombinant ScFv Antibody Fragments

V-Gene Isolation and Cloning

Single chain Fvs have traditionally been constructed from the V_H and V_L genes of immunoglobulins produced from hybridomas. Until the advent of PCR, the cloning of antibody genes was a laborious and time-consuming process involving the creation and screening of genomic or cDNA libraries. The polymerase chain reaction (PCR) has greatly simplified the task of obtaining and cloning immunoglobulin DNA. After amplification, the DNA can be easily cloned, particularly if restriction sites have been incorporated into the primers.

For PCR amplification, V_H and V_L first strand cDNA is easily obtained from hybridoma cells by reverse transcription of total RNA or mRNA with oligonucleotides that prime in the C_H region ($C_\gamma 1, C_\gamma 2, C_\gamma 3, C_\gamma 4, C_\mu, C_\epsilon, C_\alpha$ or C_δ) for V_H amplification or in the C_L (C_κ or C_λ) region for V_L amplification. This is straightforward since the sequences of the constant domain exons are known.[11] Design of PCR primers for the 3' end of rearranged murine[23] or human[24] V_H and V_L gene is also straightforward since primers can be based on the J gene segments, which have been sequenced.

Design of primers for the 5' end of the V gene was thought to be less straightforward due to the sequence variability of different V-genes. In the earliest attempt to use PCR to amplify V-genes, N terminal protein sequencing was done on purified antibody from a hybridoma and the sequence used to assign the V_H and V_L (V_λ or V_κ) gene families.[11] The V_H and V_L gene assignments were used to design degenerate primers based in FR1.[25] A generally applicable approach was taken by Orlandi et al.[26] The nucleotide sequences of murine V_H and V_L genes were extracted from the Kabat database,[11] aligned, and the frequency of the commonest nucleotide plotted for each position. Conserved regions were identified at the 5' region of the V_H and V_L genes and the sequences used to design degenerate oligonucleotide primers containing restriction sites for directional cloning. Additional groups have designed sets of universal V-gene primers containing internal or appended restriction sites suitable for amplification of murine[27-31] human,[13,24,32,33] chicken,[34] and rabbit[35] V-genes.

Single chain Fvs are constructed either by sequential cloning of V_H and V_L genes[16,17] or by PCR splicing by overlap extension (PCR assembly) of V_H and V_L genes into an scFv gene[23,24] (Fig. 2.3). With PCR assembly, only two restriction sites are required to clone the scFv. These can be appended to the 5' and 3' end of the scFv gene cassette and expression systems have been described where octanucleotide cutters can be utilized. This markedly decreases the likelihood that a restriction enzyme will cut internally in the V-gene. Sequential cloning requires four restriction sites and increases the chances of restriction enzymes cutting internally.[36,37]

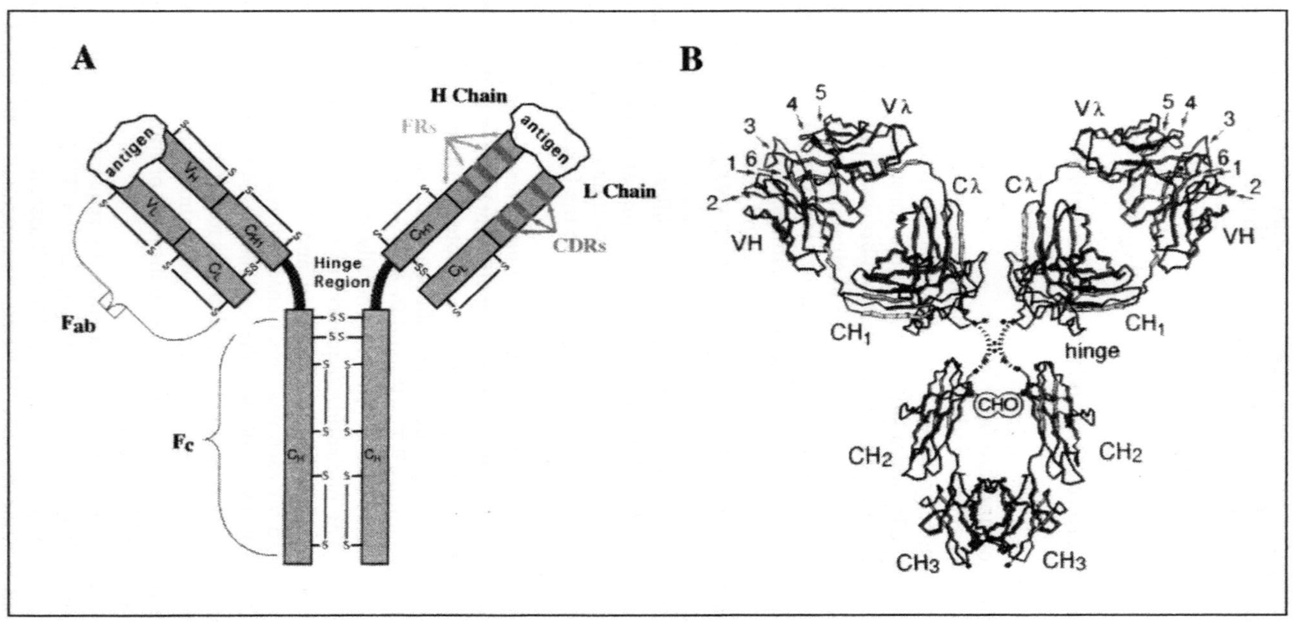

A
antigen
VH
VL
CH1
CL
Fab
Hinge Region
H Chain
FRs
antigen
CH1
L Chain
CL
CDRs
SS
SS
Fc
CH
CH
B
5
4
3
1 6
2
Vλ
Cλ
VH
CH₁
Cλ
Vλ
5 4
3
6 1
2
VH
CH₁
hinge
CHO
CH2
CH2
CH3
CH3

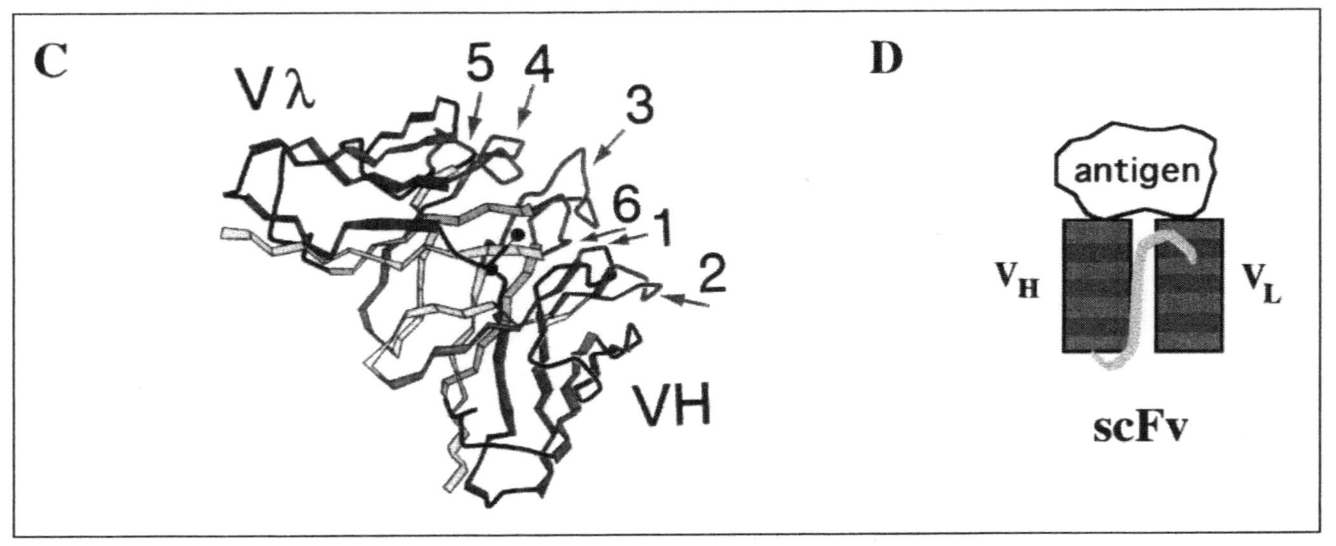

Fig. 2.1. Structure of immunoglobulins and derived fragments. (A, opposite) The basic antibody (IgG1) is composed of 2 heavy (H) chains and 2 light (L) chains. Each L chain is composed of a N-terminal variable domain (V_L) and a constant domain (C_L). Each H chain is composed of a N-terminal variable domain (V_H), 3 C domains (C_{H1}, C_{H2}, C_{H3}) and a hinge region. The V_H and V_L domains consist of relatively conserved framework regions (FR1, FR2, FR3 and FR4) separated by three areas of hypervariable sequence or complementary determining regions (CDR1, CDR2 and CDR3, which contain most of the antigen binding residues. (B, opposite) The α-carbon backbone of a human IgG1 demonstrates the domain architecture of antibodies. Each domain has a similar structural motif: the immunoglobulin fold. (C, above) Enlargement of the α-carbon backbone of the variable fragment (Fv). Each domain consists of a β-sheet framework which serves as the scaffold for three antigen binding loops (1, 2 and 3 for the V_H and 4, 5 and 6 for the V_λ) located within the CDRs. (D, above) The single chain variable fragment (scFv) consists of V_H and V_L domains linked by a short and flexible linker peptide. B and C are reprinted with permission from: Winter G and Milstein C, Nature 349:293-299. Copyright 1991 Macmillan Magazines Ltd.

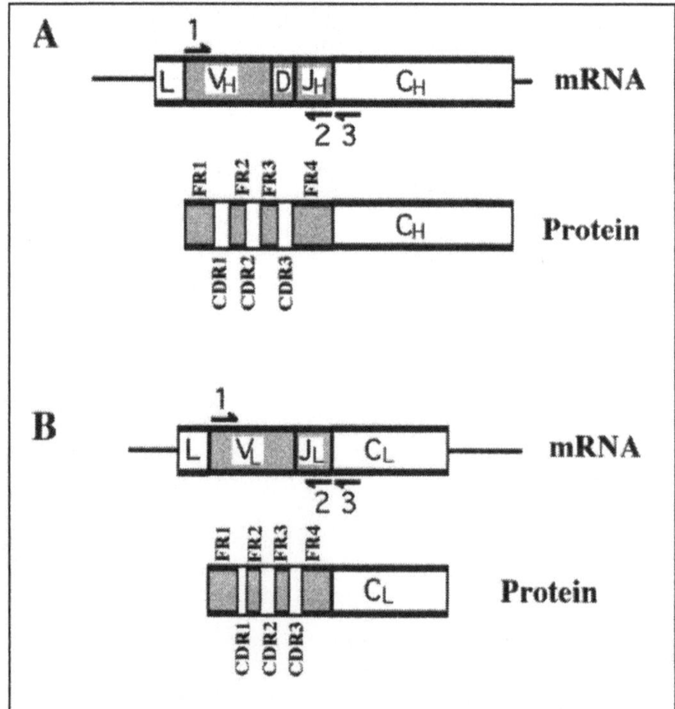

Fig. 2.2. Organization of the rearranged V_H (A) and V_L (B) domains at the RNA and protein levels. Constant region primers (3) can be used to prime RNA for first strand cDNA synthesis. Primer 1 (annealing to the 5' end of the V gene encoding FR1) and 2 (annealing to the J gene segment encoding FR4) are used for amplification of full length rearranged V-genes. The dark arrows represent the primers, L; leader exon, V_H; heavy variable germline gene segment, V_L; light variable germline gene segment; D; diversity segment, J; joining segment, C_H; heavy gene constant region, C_L; light gene constant region.

Expression of ScFv in E. Coli

Expression of scFv in bacteria is ideal for creating and analyzing large numbers of different antibodies due to the ease of genetic manipulation, high transformation efficiencies, rapid growth, simple fermentation and favorable economics. Initial attempts to express IgG in *E. coli* produced very low yields of functional antibody when the H and L chains were expressed intracellularly.[20-21] The proteins were contained in 'inclusion bodies' and generation of functional native antibody required in vitro denaturation and refolding, and the yields were very low. A significant breakthrough was achieved in 1988 when two groups simultaneously described the expression of native, correctly folded antibody fragments in *E. coli* in high yields. One group[38] attached the *E. coli* signal sequences ompA (outer membrane protein A) and phoA (alkaline phosphatase) to the V_H and V_κ domains of an antibody. The signal sequences directed the expressed domains into the periplasmic

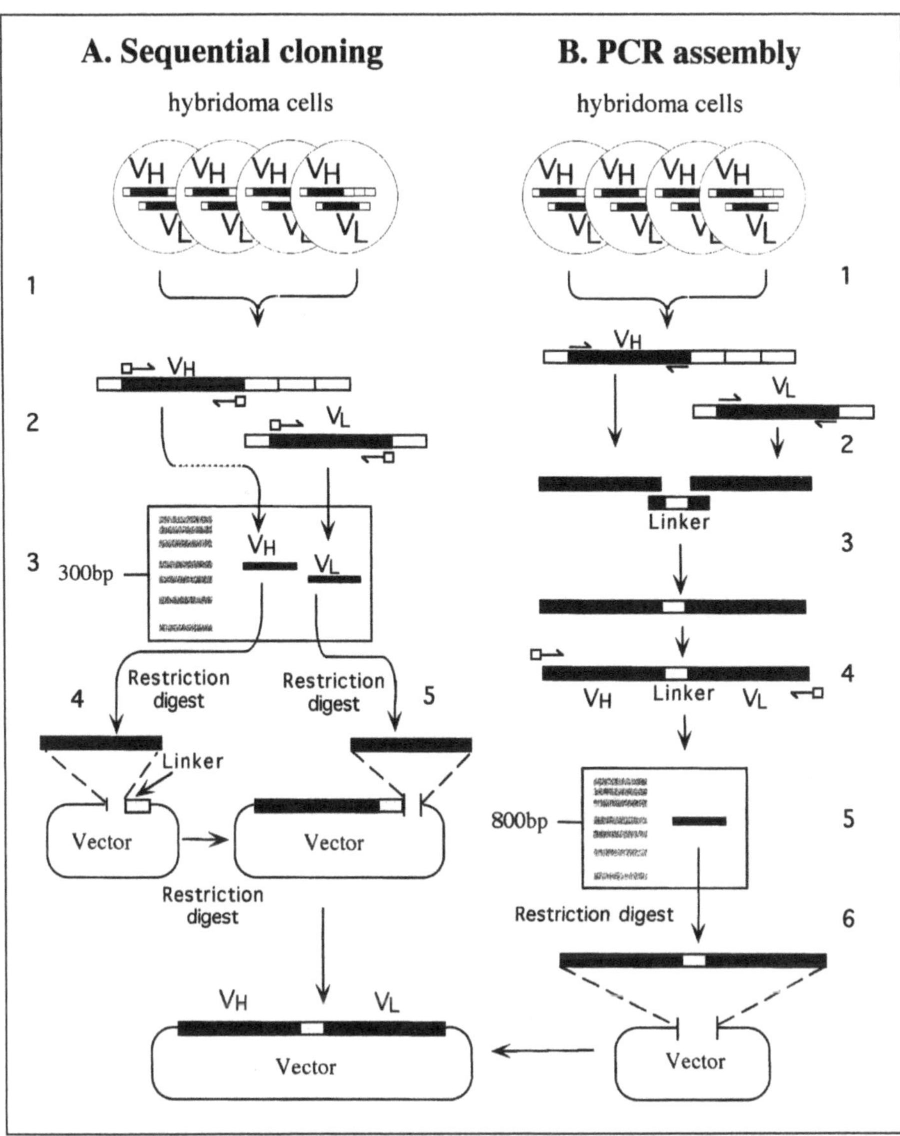

Fig. 2.3. Comparison of sequential vs PCR assembly strategies for cloning V-genes. A. Sequential cloning: After RNA preparation from hybridoma cells (1), rearranged V_H and V_L genes are PCR amplified (2), gel purified (3), digested with two restriction enzymes each and cloned sequentially (4 and 5) into a vector containing the sequence for the linker. B. PCR assembly: After RNA preparation from hybridoma cells (1), rearranged V_H and V_L genes are PCR amplified (2), and linked together with linker DNA using PCR (3). The assembled genes are reamplified with primers containing restriction sites (4). After gel purification (5), the assembled genes are digested with two enzymes and cloned (6). The arrows represent the PCR primers, boxes at the ends of arrows denote restriction sites.

space where they folded correctly into functional heterodimeric Fv fragments. Another group used two copies of a different signal sequence, pelB (pectate lyase), to express a functional Fab antibody fragment.[39]

Similarly, native functional scFv can be secreted in *E. coli* by attachment of a signal sequence (Fig. 2.4).[22,24,40] Because scFv expression is toxic to *E. coli*, the vectors are usually engineered for expression using the Lac, Trp, alkaline phosphatase, or other inducible promoters. The newly synthesized scFv is directed to the bacterial periplasm where the leader sequence is cleaved. This makes it possible to harvest native properly folded scFv directly from the bacterial periplasm.[41] Growth of cultures for more than 4 to 5 hours leads to cell death from the toxic scFv. The scFv will then leak out into the culture media, which can then be harvested and used for assays such as ELISA. Detection of scFv expression and binding to antigen by ELISA are facilitated by fusion of a C-terminal or N-terminal epitope tag, such as the c-myc epitope[24,42] or E-tag.[43] While the epitope tags can also be used for affinity purification,[24,40] this requires use of expensive antibody columns and stringent elution conditions which may damage the scFv. The technique currently used is to fuse a C-terminal hexahistidine tag and purify the scFv from the bacterial periplasm using immobilized metal affinity chromatography.[44,45]

Limitations of Generating ScFv From Hybridomas

While in theory it is simple to produce an scFv from a hybridoma, there are many factors that limit the probability of success. First, a hybridoma producing a monoclonal antibody of the desired specificity must exist. Murine monoclonal antibodies are classically obtained from mice immunized with a purified antigen. Besides being time-consuming, only a relatively small number of antibodies are

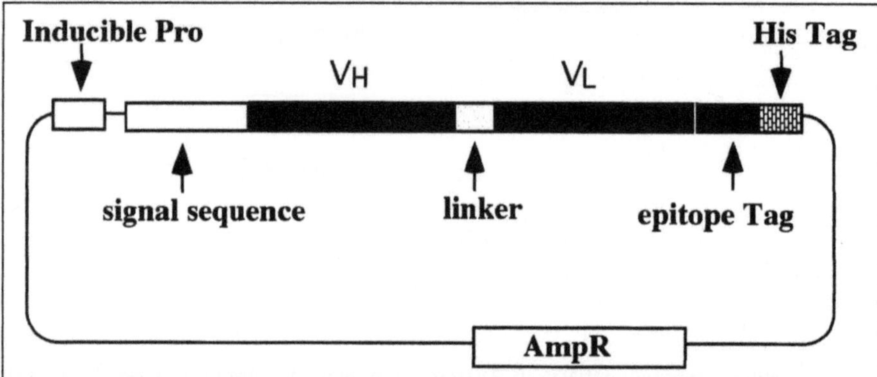

Fig. 2.4. An example of a generic vector for the expression of scFv in *E. coli*. Native functional scFv can be expressed in the bacterial periplasm by attachment of a leader sequence (e.g., PelB) at the N-terminus. The vectors, containing a selectable antibiotic resistant gene (e.g., AmpR), are usually engineered for expression using an inducible promoter (e.g., Lac, Trp, alkaline phosphatase) and allow for the production of C-terminus tagged-scFv. The epitope tag (e.g., myc-Tag, E-Tag) consists of a short peptide recognized by a monoclonal antibody which permits detection of the scFv in various assays including Western blotting and ELISA. The addition of a hexahistidine Tag (His-Tag) at the C-terminus permits purification by immobilized metal affinity chromatography.

typically produced against a few dominant immunogenic epitopes. This may result in failure to isolate the precise specificity desired for a particular aim. Furthermore, production of antibodies against proteins conserved between species is difficult or impossible. For each hybridoma, the V_H and V_L genes corresponding to the monoclonal antibody have to be successfully isolated and cloned as a scFv DNA construct, a relatively time-consuming process. Mut-ations introduced by the somatic hypermutation machinery into the regions where the primers anneal may make PCR amplification difficult or impossible, necessitating another amplification approach such as RACE or oligoligation PCR.[46-48] Cloning the correct V_H and V_L can also be complicated by the presence of several immunoglobulin transcripts, some of them arising from the fusion partner.[49] PCR may also introduce mutations coding for stop codons or destabilizing amino acids, necessitating the sequencing of multiple clones. Furthermore, the V-genes may be cut internally by restriction enzymes during cloning, especially when cloned sequentially using hexanucleotide recognition sites.[37,50]

Once the genes have been successfully cloned, the scFv should be expressed and characterized biophysically and biochemically prior to use intracellularly. This is necessary for the reasons described above, and also because some recombinant scFv are not able to recognize the antigen previously bound by the parental antibody. The most rapid method for expression is in *E. coli*, however, expression levels in bacteria vary considerably[51] due to scFv toxicity and poor folding kinetics.[52] These differences are sequence dependent, differ dramatically between antibodies, and in many instances result in failure to produce adequate quantities of scFv for further in vitro characterization.[52] Thus, despite the large number of well characterized hybridoma cell lines, very few of these antibodies can be successfully reconstructed as scFv which express at high levels in *E. coli*. Thus secretion in a eukaryotic cell line is necessary to prove the scFv is functional and for biochemical and biophysical characterization.[53,54]

Finally, scFv differ considerably in their stability with respect to solubility, immunoreactivity, and extent of dimerization and aggregation.[55-57] Single chain Fv dimerization and aggregation occur from the V_H domain of one scFv molecule pairing with the V_L domain of a second scFv molecule.[56,58] The tendency of scFv to dimerize is sequence dependent, with some scFv existing as stable monomer[45,58-60] and others as mixtures of monomeric and oligomeric scFv.[56,59-62]

All of the above limitations can be partially overcome by taking advantage of recent advances in biotechnology to produce scFv directly in bacteria (reviewed in refs. 63-66). Antigen specific scFv are directly selected from scFv gene repertoires expressed on the surface of bacteriophage, viruses which infect bacteria (phage display)[67,68] (Fig. 2.5). Single chain Fv produced in this manner almost invariably express at high level in bacteria.[24,69] Higher affinity scFv can be selected from phage antibody libraries created by mutating the scFv genes of initial isolates.[70,71] Moreover, the approach can be used to produce human antibodies, which are difficult to produce using conventional hybridoma technology, and to produce antibodies without immunization.

Antibody Phage Display

To display antibody fragments on the surface of phage (phage display), an antibody fragment gene is inserted into the gene encoding a phage surface protein, resulting in expression of the antibody fragment on the phage surface (Fig. 2.6). While fusions have been made with the major phage coat protein pVIII,[72,73] the

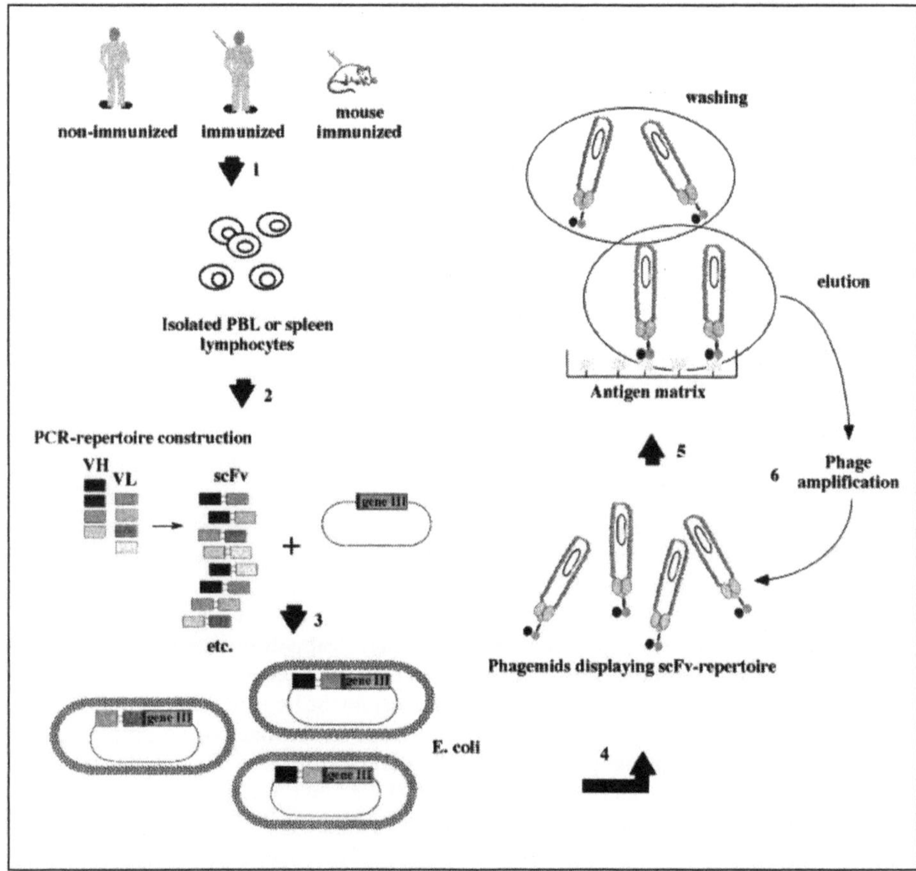

Fig. 2.5. Overview of the antibody phage display technique. Lymphocytes are isolated from mice or humans (Step 1) and the V_H and V_L genes amplified using PCR (Step 2). scFv gene repertoires are created by PCR assembly of V_H and V_L genes (see Fig. 2.3B) (Step 2) or by sequential cloning of V_H and V_L gene repertoires (not shown, see Fig. 2.3A for example). The scFv repertoire is cloned into a phage display vector and the ligated DNA used to transform *E. coli* (Step 3). Phage displaying scFv are prepared by growing the transformed *E. coli* (Step 4) and incubated with immobilized antigen (Step 5). Nonspecifically bound phage are removed by washing (see section 5.1) and antigen specific phage eluted by acid or alkali (Step 5). Eluted phage are used to infect *E. coli*, to produce more phage for the next round of selection (Step 6). Repetition of the selection process results in the isolation of antigen specific phage antibodies present at frequencies as low as $1/10^{10}$ in the original library.

most widely used vectors result in fusion with the minor coat protein pIII.[67,68,74-77] There are three to five copies of pIII per phage. In phage vectors,[67] each pIII exists as an antibody fusion (Fig. 2.6A), while in phagemid vectors[68,74] there is typically only one fusion per phage due to competition with wild type pIII provided by helper phage (Fig. 2.6B). Advantages of phagemid systems, the most widely used vectors, include higher transformation efficiencies and better affinity discrimination due to monovalent display and the absence of an avidity effect.

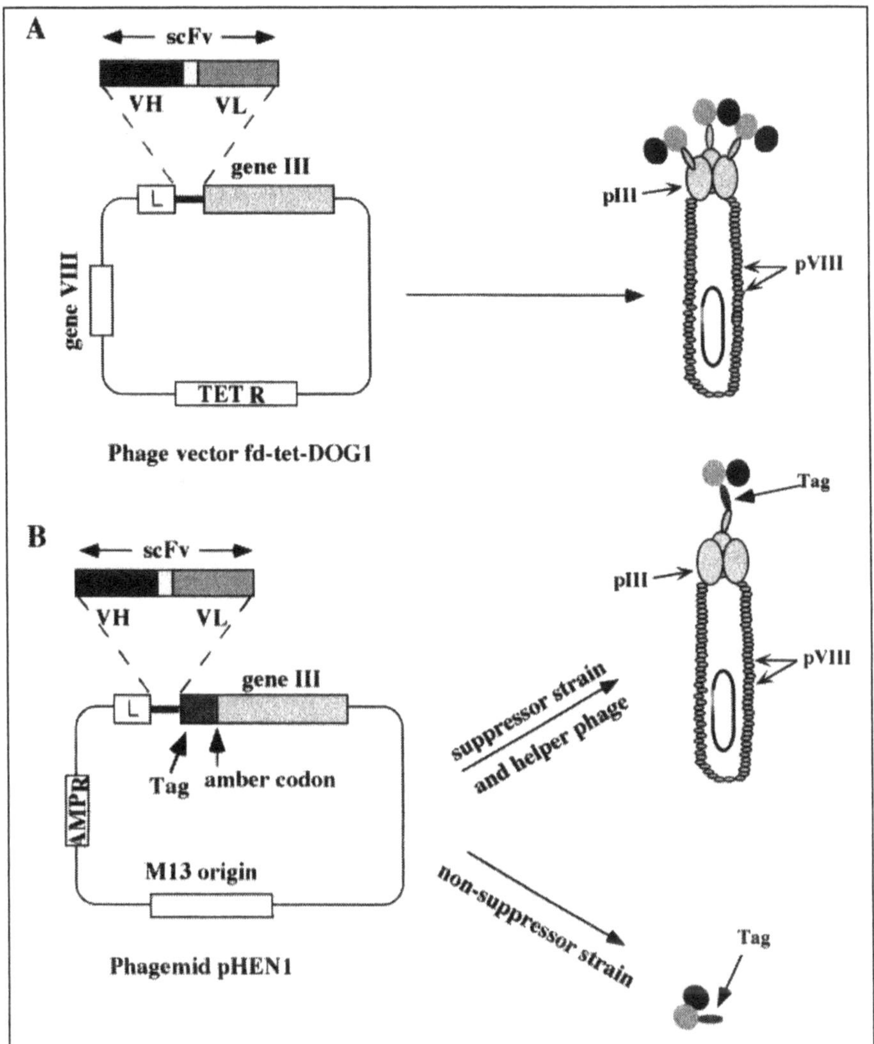

Fig. 2.6. Display of scFv on the surface of filamentous bacteriophage. Insertion of an scFv gene in frame with bacteriophage gene III results in expression of a scFv-pIII fusion protein which is incorporated into the phage surface. The gene encoding the scFv is contained within the phage. Thus the phage serves as a vehicle which physically links genotype with phenotype. A. In phage vectors, three to five copies of the scFv fragment are displayed (one on each of the pIII molecules). B. In phagemid vector, less than three copies are displayed because pIII from the helper phage competes with the fusion protein for incorporation into the phage surface. Phagemid vectors can contain an amber codon between the antibody fragment gene and gene III. This makes it possible to easily switch between displayed and soluble antibody fragment simply by changing the host bacterial strain (see details in text). L; leader exon, V_H; heavy variable region, V_L; light variable region; gene III; encoding the minor coat protein pIII; gene VIII; encoding the major coat protein pVIII, AMP_R; ampicillin resistance gene TET_R; tetracycline resistance gene.

Since the antibody fragments on the surface of the phage are functional,[67] phage bearing antigen binding antibody fragments can be separated from nonbinding or lower affinity phage by antigen affinity chromatography.[67] Mixtures of phage are allowed to bind to the affinity matrix, nonbinding or lower affinity phage are removed by washing, and bound phage are eluted by treatment with acid or alkali (Fig. 2.5). Depending on the affinity of the anti-body fragment, enrichment factors of 20-fold-1,000,000-fold are obtained for a single round of affinity selection. By infecting bacteria with the eluted phage, however, more phage can be grown and subjected to another round of selection (Fig. 2.5). In this way, an enrichment of 1000-fold in one round can become 1,000,000-fold in two rounds of selection.[67] Thus even when enrichments are low,[24] multiple rounds of affinity selection can lead to the isolation of rare phage and the genetic material contained within which encodes the sequence of the binding antibody. The physical link between genotype and phenotype provided by phage display makes it possible to test every member of an antibody fragment library for binding to antigen, even with libraries as large as 10^{10} clones.[69,78] For example, after multiple rounds of selection on antigen, a binding scFv that occurred with a frequency of only 1/30,000,000 clones could be recovered.[24]

Analysis for phage binding to antigen can be performed in an ELISA format using anti-phage antibodies.[40] Analysis for binding is simplified by including an amber codon between the antibody fragment gene and gene III[68] (Fig. 2.6B). This makes it possible to easily switch between displayed and soluble antibody fragment simply by changing the host bacterial strain. When phage is grown in a supE suppressor strain of *E. coli*, the amber stop codon between the antibody gene and gene III is read as glutamine and the antibody fragment is displayed on the surface of the phage. When eluted phage is used to infect a nonsuppressor strain, the amber codon is read as a stop codon and soluble antibody is secreted from the bacteria into the periplasm and culture media.[68] Binding of soluble scFv to antigen can be detected by ELISA using epitope tags, such as c-myc[24] or E-tag,[43] incorporated into the vectors 5' to the amber codon. Inclusion of a hexahistidine tag makes it possible to easily purify native scFv without the need for subcloning.[69]

Selection Strategies

The first phage antibody libraries were selected in columns using antigen covalently linked to a matrix such as cyanogen bromide activated Sepharose.[40,67] The easiest technique is merely to adsorb the antigen of interest onto a polystyrene surface, such as an ELISA plate well[33] or immunotube.[24] Some proteins will partially or completely denature when adsorbed, resulting in the selection of scFv which bind antigen on the plate, but not native antigen, for example as presented on the surface of a cell.[45] This difficulty can be overcome by selecting on soluble antigen in solution. The antigen can be chemically tagged, for example by biotinylation, and captured along with bound phage using avidin or streptavidin magnetic beads.[71,79] Alternatively, the antigen can be genetically tagged with a hexahistidine tag with capture on Ni-NTA agarose.[80] Not only will this approach increase the likelihood of obtaining antibodies that recognize native antigen, it is also extremely useful for limiting the antigen concentration for selection of antibodies on the basis of affinity (see below). Furthermore, selecting in solution reduces the avidity effect of dimeric scFv, thus biasing for the selection of stable monomeric scFv.[79] Selections can also be driven to specific epitopes on an antigen either by competitive elution with a known ligand, for example an antibody[81] or by blocking irrelevant epitopes with an antibody.[82,83]

An elegant selection technique involves the generation of phage antibodies that lack functional pIII.[84-86] This either requires the use of a gene III deleted helper phage, for phagemid systems, or use of a phage vector.[86] Antigen binding phage are rendered infectious by binding to an antigen-pIII fusion, constructed either chemically[84,85] or genetically.[86] For genetic fusions, placing the phage antibody gene and antigen-pIII gene on the same replicon permits the selection of antibody-antigen pairs which bind to each other.[86]

Selections can also be performed on more complex mixtures of antigen, including intact cells. Cell surface antigen specific antibodies have been isolated by selecting phage antibody libraries on erythrocytes,[87] lymphocyte,[88] and melanoma cells.[89,90] For densely expressed antigens, such as blood group B, specific scFv were isolated by direct selection on erythrocytes.[87] For antigens expressed at lower density, it was necessary to deplete the library of binders to other antigens using a closely related cell type.[87] Antigens can also be displayed as fusion proteins on the surface of bacteria, resulting in the creation of 'living columns'.[91] The ultimate selection technique has been recently described, where a peptide phage library was selected in vivo, yielding tissue specific phage peptides.[92] The widespread applicability of this approach is unknown.

Use of Phage Display to Bypass Hybridoma Technology

Phage display can be used to produce monoclonal murine and human scFv from immunized mice and humans. The primer systems described above can be used to produce diverse murine or human V_H and V_L gene repertoires from spleen, bone marrow, or peripheral blood lymphocytes.[13,24,40,80] The V_H and V_L genes can be randomly spliced together using PCR,[24,40] or cloned sequentially to create scFv gene repertoires displayed on the surface of phage (Fig. 2.5). In the first example, a scFv phage antibody library was created from the V-genes of a mouse immunized with the hapten phenyloxazolone.[40] After selection, more than 20 scFv were isolated which had affinities comparable to the affinities of IgG from hybridomas constructed from mice immunized with the same hapten. Similar panels of scFv have been obtained using phage display and mice immunized with EGF receptor[93] and Botulinum neurotoxin type A.[80] This approach has also been used to produce monoclonal chicken[34] and rabbit[35] antibody fragments using species specific primers.

A greater number of examples exist of phage libraries constructed from the V-genes of immunized humans. In most of these examples, Fab libraries were constructed, however, similar results would be expected if the scFv format had been used. Variable region genes are obtained either from peripheral blood lymphocytes, bone marrow, lymph nodes, and rarely, spleen. In the area of infectious disease, human monoclonal scFv or Fab antibody fragments have been isolated from immunized volunteers or infected patients against tetanus toxin,[74] HIV-1 gp120,[94] HIV-1 gp41,[95] hepatitis B surface antigen,[96] hepatitis C,[97] respiratory syncytial virus,[98,99] and hemophilus influenza.[100] Autoimmune antibodies have been isolated from patients with systemic lupus erythematosus (SLE) (anti-DNA),[101] Hashimoto's disease (anti-thyroid),[102] myasthenia gravis (anti-acetylcholine receptor),[103] as well as other diseases.

Antibodies from these libraries share a number of characteristics. Typically, a panel of antibodies are obtained. Some of these are clonally related, being derived from the same V-D-J rearrangement with differences due to somatic hypermutation. Others are clearly derived from different V-D-J rearrangements. The antibodies are highly specific for the antigen used for selection, and bind antigen with affinities typical of IgG produced from hybridomas derived from immunized mice or humans. Moreover, the ability to select rather than screen for binding properties (see above) permits the isolation of antibodies to rare epitopes,[83,104,105] cell surface markers[89] or unstable antigens[106] which have proven difficult to produce using hybridoma technology. The antibody fragments can be used for the same purposes as IgG derived from hybridomas including Western blotting,[61] epitope mapping,[95] cell agglutination assays,[60,87] cell staining, FACS,[88] and immunohistochemistry.[90]

The use of immune phage antibody libraries allows the construction of a large number of scFv antibody fragments of high specificity and affinity. This approach, however, is subject to the limitations associated with immunization. First, with many antigens the immune response is directed to only a few immunodominant epitopes.[80,95] This may result in failure to isolate the precise specificity desired for a particular aim. Furthermore, production of antibodies against proteins conserved between species may prove difficult or impossible. Production of nonimmunogenic human antibodies is also limited by the inability to immunize with many antigens, especially those of human origin. Finally, for each desired specificity, a new library must be constructed, a relatively time-consuming process.

Use of Phage Display to Bypass Immunization

As an alternative to immune phage antibody libraries, monoclonal antibody fragments can be produced without prior immunization by displaying very large and diverse scFv gene repertoires on phage.[24] This results in the ability to isolate antibodies to any desired specificity from a single phage antibody library. In the first example, rearranged human V_H, V_κ, and V_λ gene repertoires were amplified from the mRNA of peripheral blood lymphocytes obtained from two nonvaccinated volunteers.[24] Diverse scFv gene repertoires were obtained by randomly joining the V_H and V_L gene repertoires with DNA encoding the 15 amino acid linker sequence $(G_4S)_3$. The human scFv gene repertoire was cloned into the phagemid pHEN-1 to create a nonimmune phage antibody library of 3.0×10^7 members. From this single nonimmune phage antibody library, scFv were isolated against more than 20 different antigens, including a hapten, 3 different polysaccharides and 16 different proteins.[24,45,59,87,107] The random splicing of V_H and V_L gene repertoires creates new V_H-V_L pairings that did not exist in vivo. Thus it is possible to isolate scFv that bind self proteins. For example, from the same library human scFv have been produced against human proteins, including thyroglobulin, immunoglobulin, tumor necrosis factor, CEA, c-erbB-2 and human blood group antigens.[24,45,59,87,107] Selections can also be performed on complex antigen mixtures, including on cells, as described above. For example, scFv were produced against the blood group B antigen (a polysaccharide antigen with 500,000 sites/cell), Rh(D) and Rh(E) antigens (20,000-30,000 sites/cell) and the Kpb antigen (5,000 sites/cell) by direct selection of the phage antibody library on erythrocytes.[60,87] From a library of this size, at least 1 scFv can be isolated against a purified protein antigen 80% of the time. While scFv can be isolated against multiple epitopes on the same antigen, on average only two to three scFv are isolated per antigen. The scFv are highly specific for the antigen used for selection and have affinities typical of the primary immune response (1 μM to 100 nM range).[24,45,59]

Larger or more diverse phage antibody libraries should theoretically provide higher affinity antibodies against a greater number of epitopes on all antigens used for selection.[108,109] One approach to increase diversity is to rearrange V-genes in vitro using cloned V-gene segments and random oligonucleotides encoding V_H CDR3. With this approach it is possible to ensure that the potential diversity is large enough so that a library will only contain a single member of each sequence. In contrast, there may be considerable duplication of sequences using V-gene segments rearranged in vivo. In the first example of such a 'semisynthetic' phage antibody library, a V_H gene repertoire was constructed from 51 cloned germline human V_H gene segments[10] and oligonucleotides designed to yield a CDR3 of 5 or 8 residues in which 5 residues were of random sequence. The repertoire was cloned into a vector containing a single rearranged V_l light chain gene to create an scFv phage library of 2.2 x 10^7 members.[64] This library yielded large panels of monoclonal scFv to hapten antigens, but not to protein antigens.[64] Increasing the length of the synthetic V_H CDR3 to between 4 and 12 residues resulted in a library from which antibodies could be isolated to protein antigens as well. The affinities of the antibodies, however, were not reported.

Increasing library size several orders of magnitude results in libraries capable of yielding panels of high affinity antibody fragments against any antigen. Two published examples of such libraries exist, a 6.5 x 10^{10} member Fab library constructed using semisynthetic V-genes[78] and a 1.3 x 10^{10} member scFv library constructed from human V-genes rearranged in vivo (obtained from healthy human volunteers).[69] In the case of the Fab library, combinatorial infection was used to overcome limitations imposed on library size by transformation efficiency.[110] From these two libraries, panels of antigen specific antibody fragments were isolated against 43 different antigens, including haptens and proteins. Affinities of the antibody fragments were typical of the secondary immune response (K_d ranging between 4.1 x 10^{-8} M and 3.0 x 10^{-10} M for antibody fragments directed against 11 different antigens). Similarly, we have constructed a 6.7 x 10^9 member scFv phage antibody library from the V_H and V_L genes of healthy humans (M.D. Sheets and JDM, unpublished data). An average of 9.2 scFv with K_d as high as 3.7 x 10^{-10} M were isolated to 10 different protein antigens. These libraries represent the current state of the art; the ability to generate high affinity human monoclonal scFv within weeks and without immunization.

In Vitro Affinity Maturation Using Phage Display

The relationship between antibody fragment affinity and efficiency of intracellular phenotypic knockout is unknown. In theory, higher affinity antibodies could lead to more efficient antigen neutralization. Phage display technology permits the facile engineering of antibody affinity[70] to values rarely achieved by immunization.[111] The antibody sequence is diversified, a phage antibody library constructed, and higher affinity binders selected on antigen.[70,112-114] When designing a mutant phage antibody library, decisions must be made as to how and where to introduce mutations. Mutations can be randomly introduced using error prone PCR,[71] or mutator strains,[115] thus apparently mimicking the process of somatic hypermutation. Alternatively, random mutations can be introduced using chain shuffling,[40,116] where the V_H or V_L gene of a binding scFv is replaced with a V-gene repertoire. These approaches have yielded large increases in affinity for hapten antigens (>100-fold),[70,115] but results with protein binding antibody fragments have been more modest (<10-fold).[71,79] Moreover, the relatively random distribution of mutations in higher affinity clones provides little useful information as to where to direct additional mutations. Alternatively, knowledge of the general structure of the Fv fragment and its complexes with antigen can be used to direct mutagenesis to the CDRs that form the contact interface between antibody and antigen.

Targeting mutations to the CDRs has been shown to be a very effective approach for increasing antibody affinity. Yang et al increased the affinity of an anti-HIV gp120 Fab 420-fold (to a $K_d = 1.5 \times 10^{-11}$ M) by mutating four CDRs in five libraries and combining independently selected mutations.[113] Similarly, Schier et al increased the affinity of an anti-ErbB-2 scFv more than 1200-fold (to a $K_d = 1.3 \times 10^{-11}$ M) by mutating V_H and V_L CDR3.[114] Analysis of a number of reports suggests that mutation of the CDRs located in the center of the antigen combining site (V_H and V_L CDR3) yields a greater increase in affinity than mutating other CDRs.[113,114,117]

Efficient selection of higher affinity phage antibodies is less than straightforward due to sequence dependent differences in phage antibody expression and in toxicity to *E. coli*. These differences can lead to selection for increased expression levels, or decreased toxicity, rather than for higher affinity. In the case of scFv phage antibodies, selection is also complicated by the tendency of some scFv to dimerize. Dimeric scFv can form on the phage surface by noncovalent association of the V-domains of the scFv-pIII fusion with the V-domains of native scFv in the periplasm. Native scFv appears in the periplasm both from incomplete suppression of the amber codon between the scFv gene and gene III, as well as by proteolysis. Dimeric scFv exhibit increased apparent affinity due to avidity and are preferentially enriched over monomeric scFv when selections are performed on antigen immobilized on a solid phase.[62,79] Thus selections must be carefully designed to ensure enrichment based on affinity, rather than expression level, toxicity to *E. coli*, or avidity.

Selection for monomeric higher affinity scFv is optimal when selections are performed in solution on biotinylated antigen with subsequent capture on streptavidin-coated magnetic beads.[71,79] For the initial round of selection, an antigen concentration greater than the K_d of the wild type scFv is used in order to capture rare, or poorly expressed, phage antibodies. In subsequent rounds, the antigen concentration is reduced to significantly less than the desired K_d.[79] Failure to adequately reduce the antigen concentration results in failure to sort on the basis of affinity, while a too large reduction results in loss of binding phage.[79,118] Moreover, the optimal antigen concentration cannot be predicted a priori, due to variability in phage antibody expression levels and uncertainty regarding the highest affinities present in the mutant phage antibody library. An elegant technique for monitoring the stringency of selections is by measuring the concentration and percentage of binding phage present in polyclonal phage prepared after each round of selection using surface plasmon resonance in a BIAcore.[118] The results can then be used to determine the antigen concentration used for the next round of selection. Alternatively, selection can be monitored using the titer of eluted phage.[118]

It is also necessary to ensure that all specifically bound phage are eluted for selection of the highest affinity antibodies. Solutions used for elution include competition with soluble antigen,[40,71] 100 mM triethylamine,[119] glycine, pH 2.2,[74]

100 mM NaOAc, pH 2.8 containing 500 mM NaCl,[62] or 76 mM citric acid, pH 2.8.[87] Alternatively, antigen-bound phage can be directly used infect to *E. coli*.[120] However, it has recently been shown that eluting with soluble antigen or by adding *E. coli* directly to antigen bound phage leads to selection of scFv phage antibodies with the fastest dissociation rate constant (k_{off}) (lower affinity).[118] This is because the phage must dissociate from antigen before infection can occur. As affinity increases, more stringent eluents are required to ensure that the highest affinity phage are eluted.[118] Choice of eluent can be determined by measuring their ability to dissociate polyclonal phage antibodies from antigen using surface plasmon resonance in a BIAcore.[118]

Future Uses

Phage display technology is already replacing hybridoma technology for the production of therapeutic antibodies. Completely human antibody fragments can be isolated from immune or nonimmune libraries and should prove less immunogenic than humanized rodent monoclonal antibodies. For some applications, for example tumor targeting, the small size, excellent penetration, and rapid clearance of scFv or Fab antibody fragments may be desirable. For applications where a longer serum half life is desired, complete IgG can be constructed from the V-genes.[121] Furthermore, antibody affinity can be increased to values not obtainable by immunization. This should open up new therapeutic and diagnostic applications, allowing the detection of antigen concentrations not previously possible, for example.

As detailed above, phage libraries are an ideal source of scFv genes for intracellular phenotypic knockout. Combining these two technologies will vastly expand the application of intracellular immunization for studying the cellular function of proteins. The single gene scFv may also see wide application for in vivo targeted gene therapy, for example by fusing them to viral envelope proteins[122,123] or DNA binding proteins.[124]

Nonimmune phage antibody libraries will also prove tremendously powerful as drug discovery technologies. They will allow the rapid production of antibodies to any antigen for use in the research laboratory. It may also prove possible to develop approaches to pan phage antibody libraries against cDNA expression libraries and simultaneously isolate a cDNA and antibodies which bind the gene product. The antibodies can then be used for determination of the location of cellular and tissue gene expression, as well as for study of function, for example using intracellular immunization. This will vastly facilitate functional genomics projects. Ultimately, the use of phage antibody libraries will only be limited by the investigator's imagination.

References

1. Rungger-Brandle E, Chaponnier C, Garbiani G. Intracellular injection of anti-actin antibodies into Xenopus oocytes blocks chromosome condensation. Nature 1979; 282(5736):320-321.
2. Lukas J, Bartek J, Strauss M. Efficient transfer of antibodies into mammalian cells by electroporation. J Immunol Meth 1994; 170(2):255-259.
3. Valle G, Jones E, Colman A. Anti-ovalbumin monoclonal antibodies interact with their antigen in internal mem-branes of Xenopus oocytes. Nature 1982; 176(5887): 221-233.

4. Burke B, Warren G. Microinjection of a mRNA coding for an anti-Golgi antibody inhibits intracellular transport of a viral membrane protein. Cell 1984; 36(4):847-856.

5. Cattaneo A, Neuberger MS. Polymeric immunoglobulin M is secreted by transfectants of nonlymphoid cells in the absence of immunoglobulin J chain. EMBO J 1987; 6(9):2753-2758.

6. Biocca S, Neuberger MS, Cattaneo A. Expression and targeting of intracellular antibodies in mammalian cells. EMBO J 1990; 9(1):101-108.

7. Biocca S, Pierandrei-Amaldi P, Campioni N et al. Intracellular immunization with cytosolic recombinant antibodies. Biotechnology 1994; 12(14):396-399.

8. Kabat EA, Wu TT. Attempts to locate complementarity-determining residues in the variable positions of light and heavy chains. Ann NY Acad Sci 1971; 190(382): 382-393.

9. Alt FW, Blackwell TK, Yancopoulos GD. Development of the primary antibody repertoire. Science 1987; 238(4830):1079-1087.

10. Tomlinson IM, Walter G, Marks JD et al. The repertoire of human germline VH sequences reveals about fifty groups of VH segments with different hypervariable loops. J Mol Biol 1992; 227(3):776-798.

11. Kabat EA, Wu TT, Perry HM et al. Sequences of proteins of immunological interest. U.S. Department of Health and Human Services, U.S. Government Printing Office, 1991.

12. Alt FW, Baltimore D. Joining of immunoglobulin heavy chain gene segments: implications from a chromosome with evidence of three D-JH fusions. Proc Natl Acad Sci USA 1982; 79(13):4118-4122.

13. Marks JD, Tristrem M, Karpas A et al. Oligonucleotide primers for polymerase chain reaction amplification of human immunoglobulin variable genes and design of family-specific oligonucleotide probes. Eur J Immunol 1991; 21(4):985-991.

14. Tomlinson IM, Cox JP, Gherardi E et al. The structural repertoire of the human V_κ domain. EMBO J 1995; 14(18):4628-4638.

15. Williams SC, Frippiat J-P, Tomlinson IM et al. Sequence and evolution of the human germline V_λ repertoire. J Mol Biol 1996; 264(2):220-232.

16. Bird RE, Hardman KD, Jacobson JW et al. Single-chain antigen-binding proteins. Science 1988; 242(4877):423-426.

17. Huston JS, Levinson D, Mudgett HM et al. Protein engineering of antibody binding sites: recovery of specific activity in an anti-digoxin single-chain Fv analogue produced in *Escherichia coli*. Proc Natl Acad Sci USA 1988; 85(16):5879-83.

18. Bird RE, Walker BW. Single chain antibody variable regions. Trends Biotech 1991; 9(4):132-138.

19. Huston JS, Mudgett-Hunter M, Tai M et al. Protein engineering of single-chain Fv analogs and fusion proteins. Meth Enzymol 1991; 203:46-88.

20. Cabilly S, Riggs AD, Pande H et al. Generation of antibody activity from immunoglobulin polypeptide chains produced in *Escherichia coli*. Proc Natl Acad Sci USA 1984; 81(11):3273-7.

21. Boss MA, Kenten JH, Wood CR et al. Assembly of functional antibodies from immunoglobulin heavy and light chains synthesized in *E. coli*. Nucl Acids Res 1984; 12(9):3791-806.

22. Glockshuber R, Malia M, Pfitzinger I et al. A comparison of strategies to stabilize immunoglobulin Fv-fragments. Biochemistry 1990; 29(6):1362-7.

23. Clackson T, Hoogenboom HR, Griffiths AD et al. Making antibody fragments using phage display libraries. Nature 1991; 352(6336):624-628.

24. Marks JD, Hoogenboom HR, Bonnert TP et al. By-passing immunization: Human antibodies from V-gene libraries displayed on phage. J Mol Biol 1991; 222(3):581-597.

25. Larrick JW, Chiang YL, Sheng-Dong R et al. Generation of specific human monoclonal antibodies by in vitro expansion of human B cells: a novel recombinant DNA approach. In: Borrebaeck CAK, ed. In Vitro Immunization in Hybridoma Technology. Amsterdam: Elsevier Science Publishers B.V., 1988: 231-246.

26. Orlandi R, Gussow DH, Jones PT et al. Cloning immunoglobulin variable domains for expression by the polymerase chain reaction. Proc Natl Acad Sci USA 1989; 86(10):3833-7.

27. LeBoeuf RD, Galin FS, Hollinger SK et al. Cloning and sequencing of immunoglobulin variable-region genes using degenerate oligodeoxyribonucleotides and polymerase chain reaction. Gene 1989; 82(2):371-377.

28. Iverson SA, Sastry L, Huse WD et al. A combinatorial system for cloning and expressing the catalytic antibody repertoire in Escherichia coli. Cold Spring Harbor Symposium Quant Biol 1989; 1(273):273-81.

29. Sastry L, Alting MM, Huse WD et al. Cloning of the immunological repertoire in *Escherichia coli* for generation of monoclonal catalytic antibodies: construction of a heavy chain variable region-specific cDNA library. Proc Natl Acad Sci USA 1989; 86(15):5728-5732.

30. Kettleborough CA, Saldanha J, Ansell KH et al. Optimization of primers for cloning libraries of mouse immunoglobulin genes using the polymerase chain reaction. Eur J Immunol 1993; 23(1):206-211.

31. Orum H, Andersen PS, Oster A et al. Efficient method for constructing comprehensive murine Fab antibody libraries displayed on phage. Nucl Acids Res 1993; 21(19):4491-4498.

32. Persson MA, Caothien RH, Burton DR. Generation of diverse high-affinity human monoclonal antibodies by repertoire cloning. Proc Natl Acad Sci USA 1991; 88(6): 2432-2436.

33. Burton DR, Barbas CF, Persson MAA et al. A large array of human monoclonal antibodies to type 1 human immunodeficiency virus from combinatorial libraries of asymptomatic seropositive individuals. Proc Natl Acad Sci USA 1991; 88(22): 10134-10137.

34. Davies E, Smith J, Birkett C et al. Selection of specific phage-display antibodies using libraries derived from chicken immunoglobulin genes. J Immunol Meth 1995; 186(1): 125-135.

35. Lang I, Barbas Cr, Schleef R. Recombinant rabbit Fab with binding activity to type-1 plasminogen activator inhibitor derived from a phage-display library against human alpha-granules. Gene 1996; 172(2):295-298.

36. Chaudhary VK, Batra JK, Gallo MG et al. A rapid method of cloning functional variable-region antibody genes in *Escherichia coli* as single-chain immunotoxins. Proc Natl Acad Sci USA 1990; 87(3):1066-1070.

37. Persic L, Roberts A, Wilton J et al. An integrated vector system for the eukaryotic expression of antibodies or their fragments after selection from phage display libraries. Gene 1997; 187:9-18.

38. Skerra A, Pluckthun A. Assembly of a functional immunoglobulin Fv fragment in *Escherichia coli*. Science 1988; 240(4855):1038-1041.

39. Better M, Chang CP, Robinson RR et al. *Escherichia coli* secretion of an active chimeric antibody fragment. Science 1988; 240(4855):1041-1043.

40. Clackson T, Hoogenboom HR, Griffiths AD et al. Making antibody fragments using phage display libraries. Nature 1991; 352(6336):624-628.

41. Breitling F, Dubel S, Seehaus T et al. A surface expression vector for antibody screening. Gene 1991; 104(2):147-153.

42. Munro S, Pelham HR. An Hsp-like protein in the ER: Identity with the 78kd glucose regulated protein and immunoglobulin heavy chain binding protein. Cell 1986; 46(2): 291-300.

43. Schier R, Balint RF, McCall A et al. Identification of functional and structural amino acid residues by parsimonious mutagenesis. Gene 1996; 169(2):147-155.

44. Hochuli E, Bannwarth W, Dobeli H et al. Genetic approach to facilitate purification of recombinant proteins with a novel metal chelate adsorbent. Bio/Tech 1988; 6: 1321-1325.

45. Schier R, Marks JD, Wolf EJ et al. In vitro and in vivo characterization of a human anti-c-erbB-2 single-chain Fv isolated from a filamentous phage antibody library. Immunotech 1995; 1(1):73-81.

46. Edwards JB, Delort J, Mallet J. Oligodeoxyribonucleotide ligation to single stranded cDNAs: a new tool for cloning 5' ends of mRNAs and for constructing cDNA libraries by in vitro amplification. Nucl Acids Res 1991; 19(19):5227-5232.

47. Heinrichs A, Milstein C, Gherardi E. Universal cloning and direct sequencing of rearranged antibody V genes using C region primers, biotin captured cDNA and one sided PCR. J Immunol Meth 1995; 178(2):241-251.

48. Ruberti F, Cattaneo A, Bradbury A. The use of the RACE method to clone hybridoma cDNA when V region primers fail. J Immunol Meth 1994; 173(1):33-39.

49. Kipriyanov SM, Kupriyanova OA, Moldenhauer G. Rapid detection of recombinant antibody fragments directed against cell-surface antigens by flow cytometry. J Immunol Meth 1996; 13(1):51-62.

50. Ge L, Knappik A, Pack P et al. Expressing antibodies in *Escherichia coli.* In: Borrebaeck CAK, ed. Antibody Engineering. Oxford: Oxford University Press, 1995; 2:229-266.

51. Knappik A, Krebber C, Pluckthun A. The effects of folding catalysts on the in vivo folding process of different antibody fragments expressed in *Escherichia coli.* Bio/ Tech 1993; 11(1):77-83.

52. Knappik A, Pluckthun A. Engineered turns of a recombinant antibody improve its in vivo folding. Protein Eng 1995; 8(1):81-89.

53. Chesnut JD, Baytan AR, Russell M et al. Selective isolation of transiently transfected cells from a mammalian cell population with vectors expressing a membrane anchored single-chain antibody. J Immunol Meth 1996; 193(1):17-27.

54. Kretzschmar T, Aoustin L, Zingel O et al. High-level expression in insect cells and purification of secreted monomeric single-chain Fv. J Immunol Meth 1996; 195(1-2):93-101.

55. Weidner KM, Denzin LK, Voss EW. Molecular stabilization effects of interactions between anti-metatype antibodies and liganded antibodies. J Biol Chem 1992; 267(15):10281-10288.

56. Whitlow M, Filpula D, Rollence ML et al. Multivalent Fvs: characterization of single-chain Fv oligomers and preparation of a bispecific Fv. Protein Eng 1994; 7(8):1017-1026.

57. Reiter Y, Brinkmann U, Jung S et al. Disulfide stabilization of antibody Fv: computer predictions and experimental evaluation. Protein Eng 1995; 8(12):1323-1331.

58. Holliger P, Prospero T, Winter G. 'Diabodies': small bivalent and bispecific antibody fragments. Proc Natl Acad Sci 1993; 90:6444-6448.

59. Griffiths AD, Malmqvist M, Marks JD et al. Human anti-self antibodies with high specificity from phage display libraries. EMBO J 1993; 12(2):725-734.

60. Hughes-Jones N, Bye JM, Marks JD et al. Blood group antibodies. Brit J Hemaetol 1994; 88(1):180-186.

61. Nissim A, Hoogenboom HR, Tomlinson IM et al. Antibody fragments from a 'single pot' phage display library as immunochemical reagents. EMBO J 1994; 13(3):692-698.

62. Deng S-J, MacKenzie CR, Hirama T et al. Basis for selection of improved carbohydrate-binding single-chain antibodies from synthetic gene libraries. Proc Natl Acad Sci USA 1995; 92(3):4992-4996.

63. Marks JD, Hoogenboom HR, Griffiths AD et al. Molecular evolution of proteins on filamentous phage: mimicking the strategy of the immune system. J Biol Chem 1992; 267(23):16007-16010.

64. Hoogenboom HR, Marks JD, Griffiths AD et al. Building antibodies from their genes. Immunol Rev 1992; 130:41-68.

65. Winter G, Griffiths A, Hawkins R et al. Making antibodies by phage display technology. Ann Rev Immunol 1994; 12:433-455.

66. Marks C, Marks JD. Phage libraries: a new route to clinically useful antibodies. N Eng J Med 1996; 335(10):730-733.

67. McCafferty J, Griffiths AD, Winter G et al. Phage antibodies: filamentous phage displaying antibody variable domains. Nature 1990; 348(6301):552-4.

68. Hoogenboom HR, Griffiths AD, Johnson KS et al. Multi-subunit proteins on the surface of filamentous phage: methodologies for displaying antibody (Fab) heavy and light chains. Nucl Acids Res 1991; 19(15):4133-4137.

69. Vaughan TJ, Williams AJ, Pritchard K et al. Human antibodies with sub-nanomolar affinities isolated from a large nonimmunized phage display library. Nature Biotech' 1996; 14(3):309-314.

70. Marks JD, Griffiths AD, Malmqvist M et al. Bypassing immunization: high affinity human antibodies by chain shuffling. Bio/Tech 1992; 10(7):779-783.

71. Hawkins RE, Russell SJ, Winter G. Selection of phage antibodies by binding affinity: mimicking affinity maturation. J Mol Biol 1992; 226(3):889-896.

72. Kang AS, Barbas CF, Janda KD et al. Linkage of recognition and replication functions by assembling combinatorial antibody Fab libraries along phage surfaces. Proc Natl Acad Sci USA 1991; 88(10):4363-4366.

73. Huse W, Stinchcombe T, Glaser S et al. Application of a filamentous phage pVIII fusion protein system suitable for efficient production, screening, and mutagenesis of F(ab) antibody fragments. J Immunol 1992; 149(12):3914-3920.

74. Barbas CF, Kang AS, Lerner RA et al. Assembly of combinatorial antibody libraries on phage surfaces: The gene III site. Proc Natl Acad Sci USA 1991; 88(18):7978-7982.

75. Engelhardt O, Grabher R, Himmler G et al. Two-step cloning of antibody variable domains in a phage display vector. Bio/Techniques 1994; 17(1):44-46.

76. Dubel S, Breitling F, Fuchs P et al. A family of vectors for surface display and production of antibodies. Gene 1993; 128(1):97-101.

77. Lah M, Goldstraw A, White J et al. Phage surface presentation and secretion of antibody fragments using an adaptable phagemid vector. Human Antibodies and Hybridomas 1994; 5(1-2):48-56.

78. Griffiths AD, Williams SC, Hartley O et al. Isolation of high affinity human antibodies directly from large synthetic repertoires. EMBO J 1994; 13(14):3245-3260.

79. Schier R, Bye JM, Apell G et al. Isolation of high affinity monomeric human anti-c-erbB-2 single chain Fv using affinity driven selection. J Mol Biol 1996; 255(1):28-43.

80. Wong C, Chen S, Amersdorfer P et al. Molecular characterization of the murine humoral immune response to Botulinum neurotoxin type A binding domain as assessed using phage antibody libraries. Infect Immun 1997; 65:3743-3752.

81. Meulemans EV, Slobbe R, P. W et al. Selection of phage-displayed antibodies specific for a cytoskeletal antigen by competitive elution with a monoclonal antibody. J Mol Biol 1994; 224(4):353-360.

82. Sanna P, Williamson R, De Logu A et al. Directed selection of recombinant human monoclonal antibodies to herpes simplex virus glycoproteins from phage display libraries. Proc Natl Acad Sci USA 1995; 92(14):6439-6443.

83. Ping T, Tornetta M, Ames R et al. Isolation of a neutralizing human RSV antibody from a dominant, nonneutralizing immune repertoire by epitope-blocked panning. J Immunol 1996; 157(2):772-780.

84. Duenas M, Ca B. Clonal selection and amplification of phage displayed antibodies by linking antigen recognition and phage replication. Biotech 1994; 12(10):999-1002.

85. Duenas M, Malmborg A, Casalvilla R et al. Selection of phage displayed antibodies based on kinetic constants. Mol Immunol 1996; 33279-85(3):279-285.

86. Krebber C, Spada S, Desplancq D et al. Coselection of cognate antibody-antigen pairs by selectively-infective phages. FEBS Lett 1995; 377(2):227-231.

87. Marks JD, Ouwehand WH, Bye JM et al. Human antibody fragments specific for blood group antigens from a phage display library. Bio/Tech 1993; 11(10):1145-1149.

88. de Kruif J, Terstappen L, Boel E et al. Rapid selection of cell subpopulation-specific human monoclonal antibodies from a synthetic phage antibody library. Proc Natl Acad Sci USA 1995; 92(6):3938-3942.

89. Cai X, Garen A. Anti-melanoma antibodies from melanoma patients immunized with genetically modified autologous tumor cells: selection of specific antibodies from single-chain Fv fusion phage libraries. Proc Natl Acad Sci USA 1995; 92(14):6537-6541.

90. Cai X, Garen A. A melanoma-specific VH antibody cloned from a fusion phage library of a vaccinated melanoma patient. Proc Natl Acad Sci USA 1996; 93(13):6280-6285.

91. Bradbury A, Persic L, Werge T et al. From gene to antibody: the use of living columns to select phage antibodies. Bio/Tech 1993; 11(13):1565-1569.

92. Pasqualini R, Ruoslahti E. Organ targeting in vivo using phage display peptide libraries. Nature 1996; 380(6572):364-366.

93. Kettleborough C, Ansell K, Allen R et al. Isolation of tumor cell-specific single-chain Fv from immunized mice using phage-antibody libraries and the re-construction of whole antibodies from these antibody fragments. Eur J Immunol 1994; 24(4):952-958.

94. Barbas CF, Collet TA, Amberg W et al. Molecular profile of an antibody response to HIV-1 as probed by combinatorial libraries. J Mol Biol 1993; 230(3):812-823.

95. Binley JM, Ditzel HJ, Barbas III CF et al. Human antibody responses to HIV type 1 glycoprotein 41 cloned in phage display libraries suggest three major epitopes are recognized and give evidence for conserved antibody motifs in antigen binding. AIDS Res Hum Retroviruses 1996; 12(10):911-924.

96. Zebedee SL, Barbas CF, Hom Y-L et al. Human combinatorial antibody libraries to hepatitis B surface antigens. Proc Natl Acad Sci USA 1992; 89(8):3175-3179.

97. Chan S, Bye J, Jackson P et al. Human recombinant antibodies specific for hepatitis C virus core and envelope E2 peptides from an immune phage display library. J Gen Virol 1996; 10:2531-2539.

98. Barbas C, Crowe J, Cababa D et al. Human monoclonal Fab fragments derived from a combinatorial library bind to respiratory syncytial virus F glycoprotein and neutralize infectivity. Proc Natl Acad Sci USA 1992; 89(21):10164-10168.

99. Crowe J, Murphy B, Chanock R et al. Recombinant human respiratory syncytial virus (RSV) monoclonal antibody Fab is effective therapeutically when introduced directly into the lungs of RSV-infected mice. Proc Natl Acad Sci USA 1994; 91(4):1386-1390.

100. Reason D, Wagner T, Lucas A. Human Fab fragments specific for the Haemophilis influenzae b polysaccharide isolated from a bacteriophage combinatorial

library use variable region gene combinations and express an idiotype that mirrors in vivo expression. Infection and Immunity 1997; 65(1):261-266.

101. Barbas S, Ditzel H, Salonen E et al. Human auto antibody recognition of DNA. Proc Natl Acad Sci USA 1995; 92(7):2529-2533.

102. Portolano S, McLachlan S, Rapoport B. High affinity, thyroid-specific human auto-antibodies displayed on the surface of filamentous phage use V genes similar to other auto-antibodies. J Immunol 1993; 151(5):2839-2851.

103. Graus Y, de Baets M, Parren P et al. Human anti-nicotinic acetylcholine receptor recombinant Fab fragments isolated from thymus-derived phage display libraries from myasthenia gravis patients reflect predominant specificities in serum and block the action of pathogenic serum antibodies. J Immunol 1997; 158(4):1919-1929.

104. Griffin H, Ouwehand W. A human monoclonal antibody specific for the leucine-33 (P1A1, HPA-1a) form of platelet glycoprotein IIIa from a V gene phage display library. Blood 1995; 86(12):4430-4436.

105. Williamson R, Peretz D, Smorodinsky N et al. Circumventing tolerance to generate autologous monoclonal antibodies to the prion protein. Proc Natl Acad Sci USA 1996; 93(14):7279-7282.

106. Bruggeman Y, Boogert A, van Hock A et al. Phage antibodies against an unstable hapten: oxygen sensitive reduced flavin. FEBS Lett 1996; 388(2-3):242-244.

107. Finnern R, Bye JM, Dolman KM et al. Molecular characteristics of anti-self antibody fragments against neutrophil cytoplasmic antigens from human V gene phage display libraries. Clin Exp Immunol 1995; 102(3):566-574.

108. Perelson AS, Oster GF. Theoretical studies of clonal selection: Minimal antibody repertoire size and reliability of self nonself discrimination. J Theor Biol 1979; 81:645-670.

109. Perelson AS. Immune network theory. Immunol Rev 1989; 110(5):5-36.

110. Waterhouse P, Griffiths A, Johnson K et al. Combinatorial infection and in vivo recombination: a strategy for making large phage antibody repertoires. Nucl Acid Res 1993; 21(9):2265-2266.

111. Foote J, Eisen HN. Kinetic and affinity limits on antibodies produced during immune responses. Proc Natl Acad Sci USA 1995; 92(2):1254-1256.

112. Barbas CF, Hu D, Dunlop N et al. In vitro evolution of a neutralizing human antibody to human immunodeficiency virus type 1 to enhance affinity and broaden strain cross-reactivity. Proc Natl Acad Sci USA 1994; 91(9):3809-3813.

113. Yang W-P, Green K, Pinz-Sweeney S et al. CDR walking mutagenesis for the affinity maturation of a potent human anti-HIV-1 antibody into the picomolar range. J Mol Biol 1995; 254(3):392-403.

114. Schier R, McCall A, Adams GP et al. Isolation of high affinity anti-c-erbB2 single-chain Fv by molecular evolution of the complementarity determining regions in the center of the antibody combining site. J Mol Biol 1996; 263(4):551-567.

115. Low N, Holliger P, Winter G. Mimicking somatic hypermutation: affinity maturation of antibodies displayed on bacteriophage using a bacterial mutator strain. J Mol Biol 1996; 260(3):359-368.

116. Kang AS, Jones TM, Burton DR. Antibody redesign by chain shuffling from random combinatorial immunoglobulin libraries. Proc Natl Acad Sci USA 1991; 88(24):11120-11123.

117. Yelton DE, Rosok MJ, Cruz G et al. Affinity maturation of the BR96 anti-carcinoma antibody by codon-based mutagenesis. J Immunol 1995; 155(4):1994-2004.

118. Schier RS, Marks JD. Efficient in vitro selection of phage antibodies using BIAcore guided selections. Human Antibodies and Hybridomas 1996; 7(3):97-105.

119. Marks JD, Hoogenboom HR, Bonnert TP et al. By-passing immunization: Human antibodies from V-gene libraries displayed on phage. J Mol Biol 1991; 222(3):581-597.
120. Figini M, Marks JD, Winter G et al. In vitro assembly of repertoires of antibody chains on the surface of phage by renaturation. J Mol Biol 1994; 239(1):68-78.
121. Bender E, Woof J, Atkin J et al. Recombinant human antibodies: linkage of an Fab fragment from a combinatorial library to an Fc fragment for expression in mammalian cell culture. Human Antibodies and Hybridomas 1993.
122. Russell SJ, Hawkins RE, Winter G. Retroviral vectors displaying functional antibody fragments. Nucl Acids Res 1993; 21(5):1081-1085.
123. Somia NV, Zoppe M, Verma IM. Generation of targeted retroviral vectors by using single-chain variable fragment: an approach to in vivo gene therapy. Proc Natl Acad Sci USA 1995; 92(16):7570-7574.
124. Chen SY, Zani C, Khouri Y et al. Design of a genetic immunotoxin to eliminate toxin immunogenicity. Gene Ther 1995; 2(2):116-123.

CHAPTER 3

Phenotypic Knockout of the Human Interleukin-2 Receptor α Chain on Primary and HTLV-I Transformed T Cells

Jennifer H. Richardson and Wayne A. Marasco

Introduction

Interleukin-2 (IL-2) is the major T cell growth factor that functions in both an autocrine and paracrine fashion to promote the expansion of activated T cells.[1,2] Two functional forms of IL-2 receptor have been identified and are distinguished by their subunit compositions and affinities for IL-2 (Table 3.1).[3,4] The high-affinity receptor (K_d 10^{-11} M) present on activated T cells is composed of α, β and γ chains. Quiescent T lymphocytes and natural killer cells lack the α chain but express an intermediate-affinity receptor (K_d 10^{-9} M) composed of β and γ chains. A third, low-affinity receptor (K_d 10^{-8} M) detected on activated T cells consists of the α chain alone. The functional significance of the low-affinity receptor, which has no known signaling capability, is presently unclear. The 55 kDa α chain (IL-2Rα, CD25, Tac antigen, p55) is unique to IL-2R whereas the 75 kDa β chain (IL-2Rβ, CD122, p75) is also a component of IL-15R and the 64 kDa IL-2Rγ or 'common' chain ($γ_c$), is shared by the receptors for IL-4, 7, 9 and 15.[5,6] Large granular lympho-cytes respond to IL-2 through the intermediate-affinity (βγ) receptor, however, the proliferation of T lymphocytes in response to IL-2 requires the high-affinity (αβγ) receptor and hence IL-2Rα expression.[7-9] IL-2Rα is undetectable on resting T cells but is rapidly induced following antigenic stimulation of the cell. Transient ap-pearance of the high-affinity receptor then allows a brief proliferative burst of the antigen-activated cells.[10-12]

In addition to functioning as a growth factor for T cells, IL-2 stimulates the growth and/or functional maturation of B cells, natural killer cells, lymphokine-activated killer (LAK) cells, monocytes and endothelial cells. Both in vitro and in vivo studies have recently indicated an additional role for IL-2 in terminating the response of activated T cells. In vitro, IL-2 renders T cells sensitive to activation-induced cell death (AID), a phenomenon whereby previously activated cells undergo

Table 3.1. IL-2 receptor forms

Receptor	Affinity	Kd (M)	IL-2 Signal
αβγ	High	10^{-11}	+
βγ	Intermediate	10^{-9}	+
α	Low	10^{-8}	−

"suicide" upon secondary exposure to antigen.[13] The in vivo significance of this pathway is evidenced by the spontaneous activation and massive expansion of lymphocytes that occurs in IL-2Rα-deficient mice.[14] IL-2 therefore has an essential homeostatic function in addition to its role as a growth factor, and—depending upon the context—can induce either cell prolifera-tion or cell death.

Biochemistry of IL-2r Signaling

Signaling events that occur following ligand binding have recently been reviewed[15,16] and are summarized in Figure 3.1. The cytosolic domains of IL-2Rβ and γ_c are absolutely required for IL-2 signal transduction and the hererodimerization of these molecules by extracellular ligand is both required and sufficient for signal transduction.[17,18] One of the earliest events to occur following IL-2 binding is the activation of cytosolic protein tyrosine kinases (PTKs) which are physically associated with the IL-2 receptor. These include the *src*-family kinase p56^{lck} which binds to IL-2Rβ and the Janus-family PTKs Jak1 and Jak3 which associate with IL-2Rβ and γ_c, respectively. Once activated, the receptor associated PTKs phosphorylate substrates critical for signal transduction, including the receptor chains themselves. Several *src*-homology 2 (SH2) domain proteins including *shc*, phosphatidyl-inositol 3-kinase and STAT5, are recruited to the phosphorylated receptor where they become activated, and on to mediate downstream events which include the activation of p21^{ras}, p70^{s6k} and nuclear gene expression.

The Role of IL-2rα in Adult T-Cell Leukemia

Certain T cell, B cell, monocytic and granulocytic leukemias are characterized by the constitutive upregulation of IL-2Rα and the high-affinity IL-2R. The most widely studied example of these is adult T-cell leukemia (ATL), a malignancy that is highly associated with infection by the exogenous retrovirus, human T-cell leukemia virus type I (HTLV-I).[19,20] Consistent with its leukemogenic properties, HTLV-I can immortalize human T cells in vitro.[21-23] In the process of T cell immortalization by HTLV-I, there is an initial phase of IL-2-dependent cell growth, which in vitro can progress to an IL-2-independent phase.[21,24,25] A similar progression from IL-2 dependence to independence is thought to occur during the evolution of ATL, as the leukemic cells from a majority of patients with acute ATL are unresponsive to IL-2 in vitro.[26-28] Since the HTLV-I genome does not contain a dominant oncogene or integrate at a specific chromosomal locus in the leukemic cells, the mechanism of HTLV-I-oncogenesis is not fully understood. The Tax protein of HTLV-I is thought to play a role in the early stages of leukemogenesis, as it activates a variety of genes involved in cellular growth including IL-2Rα and IL-2 itself.[29-32] However, the ex-

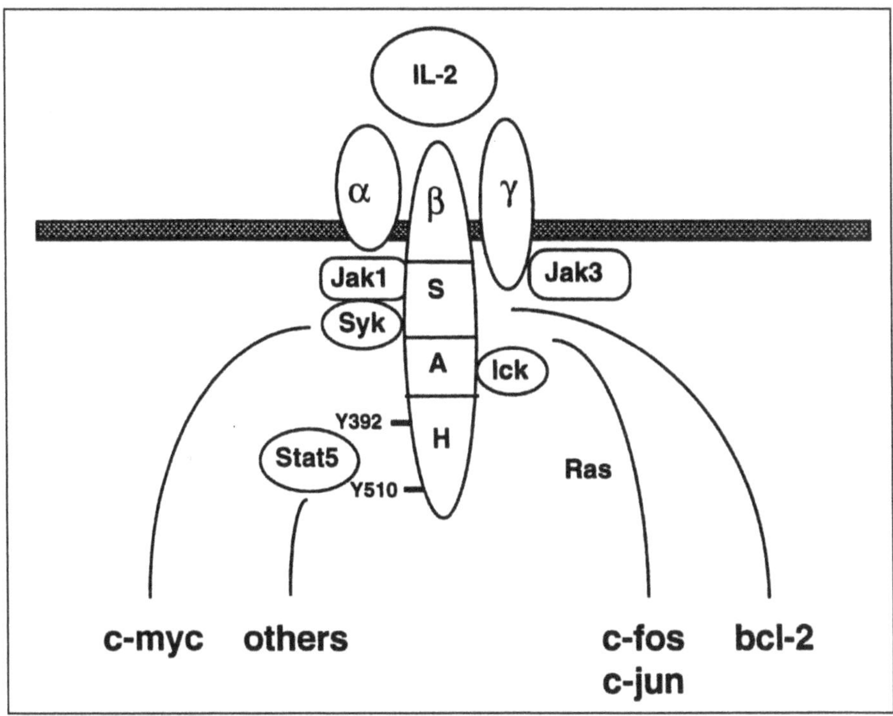

Fig. 3.1. Schematic illustration of the IL–2R-associated signaling pathways. Phosphorylated tyrosines 392 and 510 of IL–2Rβ serve as docking sites for STAT5. Non-illustrated proteins that bind to the phosphorylated IL–2Rβ chain are shc, Raf–1 and PI–3K.

pression of *tax* is generally undetectable in freshly isolated leukemic cells,[33,34] suggesting that growth of the end-stage leukemic cell is driven by an alternative, *tax*-independent, mechanism.

One of the phenotypic hallmarks of leukemic (ATL) cells and T cells immortalized by HTLV-I in vitro is the constitutive upregulation of IL-2Rα.[21,35,36] It has been hypothesized that constitutive signaling by the high-affinity IL-2R may play a role in the growth of HTLV-I transformed T cells. A simple autocrine IL-2 mechanism can be ruled out in most cases, as a majority of fresh ATL cells and many HTLV-I immortalized lines neither produce nor respond to IL-2.[26-28,34,37,38] An alternative hypothesis of ligand (IL-2)-independent signaling by an aberrant or constitutively activated IL-2R has therefore been proposed.[39] Indirect support for this model has come from reports documenting the constitutive phosphorylation of IL-2Rβ in the HTLV-I immortalized cell line HUT102,[40] and constitutive activation of the IL-2Rγ-associated kinase Jak3 in several IL-2-independent HTLV-I transformed cell lines.[41,42] To test the hypothesis that IL-2Rα upregulation plays a direct role in the growth of HTLV-I transformed cells, we constructed a single-chain antibody which, when expressed in the endoplasmic reticulum, causes intracellular retention of the IL-2Rα chain. Our hypothesis was that the loss of cell surface IL-2Rα would prevent aberrant signaling by the high-affinity (αβγ) receptor, possibly achieving a reversal of the transformed state.

Functional Studies

Generation of an ER-Targeted Single-Chain Antibody Against Human IL-2rα

We constructed a single-chain variable region fragment (sFv) of the well-characterized monoclonal antibody anti-Tac, which binds the extracellular domain of human IL-2Rα with high affinity.[43] Heavy and light chain variable regions were obtained from the hybridoma by PCR amplification and the single-chain antibody, sFvTac, was assembled by three-piece overlap assembly.[44] The interchain linker is $(Gly_4Ser)_3$ and the format of the sFv is V_H-ICL-V_L. The native heavy chain leader sequence was included to direct the sFv into the lumen of the endoplasmic reticulum (ER). Two versions of the sFvTac gene, differing only in the presence or absence of a C-terminal ER retention signal (amino acids KDEL) were stably expressed in the human T cell line Jurkat. Pulse-chase analysis revealed the sFv molecules to be stable, exhibiting intracellular half-lives of 4-6 hr for sFvTac and >30 hr for sFvTacKDEL.[44] Only trace amounts of sFv were detectable in the culture supernatants, indicating that neither sFvTac nor sFvTacKDEL was secreted to a significant extent. The basis for intracellular retention of the sFvTac intrabody is not clear, but the same phenomenon has been observed with other sFv molecules that lack the KDEL retention signal.[45-47]

SFvTacKDEL Inhibits Cell Surface Expression of IL-2rα

The ability of sFvTac intrabodies to cause intracellular retention of IL-2Rα was initially assessed in Jurkat cells, which do not express IL-2Rα under normal circumstances but can be induced to do so by treatment with PMA and PHA.[48] Jurkat clones stably expressing sFvTacKDEL demonstrated a complete absence of surface IL-2Rα following PHA/PMA stimulation, consistent with a block to the cell surface transport of IL-2Rα in these cells (Fig. 3.2). A proportion of clones expressing the non ER-targeted sFvTac gene also showed downregulation of IL-2Rα, but this was incomplete, as evidenced by some breakthrough expression of IL-2Rα at the cell surface (Fig. 3.2). The IL-2Rα downregulation observed in all sFvTacKDEL and some sFvTac expressing clones was specific for IL-2Rα as the cell surface expression of other plasma membrane proteins, including MHC class I CD2, CD3 and CD4, was unaffected.[44]

IL-2rα Is Retained Intracellularly as an Endoglycosidase H-Sensitive Precursor Complexed with SFvFacKDEL

Two forms of IL-2Rα are seen in PHA/PMA-treated Jurkat cells: a 55-kDa form (p55), representing the mature receptor, and a less abundant 40-kDa form (p40), which has previously been identified as an immature form of IL-2Rα.[44,49] In Jurkat cells expressing sFvTacKDEL, PHA/PMA treatment led to intracellular accumulation of the p40 precursor, suggesting that the maturation of p40 into p55 was blocked (Fig. 3.3). sFvTacKDEL coprecipitated with p40 in these cells, consistent with a physical association between the two proteins. The accumulated p40 was sensitive to endoglycosidase H digestion, and is therefore a high-mannose glycoprotein typical of glycoproteins found in the ER (Fig. 3.3). Together, these data strongly suggest that the absence of p55 is due to retention of p40 in the ER as a complex with sFvTacKDEL.

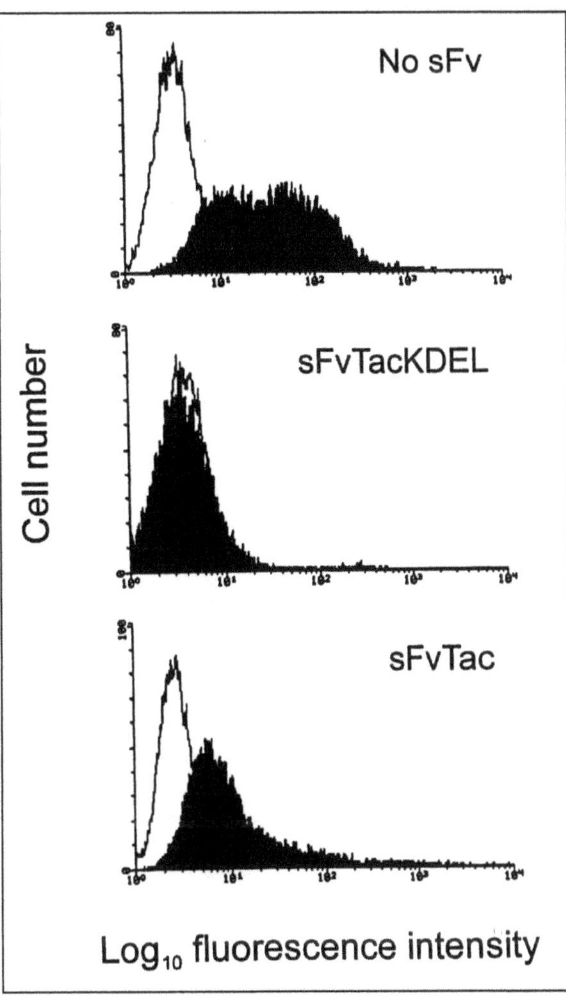

Fig. 3.2. Flow cytometric analysis of IL–2Rα expression in Jurkat clones. Jurkat cells expressing no sFv, sFvTacKDEL, or sFvTac, were stained with anti-Tac mAb after 18 hr of stimulation with PHA and PMA (solid area). Open area represents unstimulated cells.

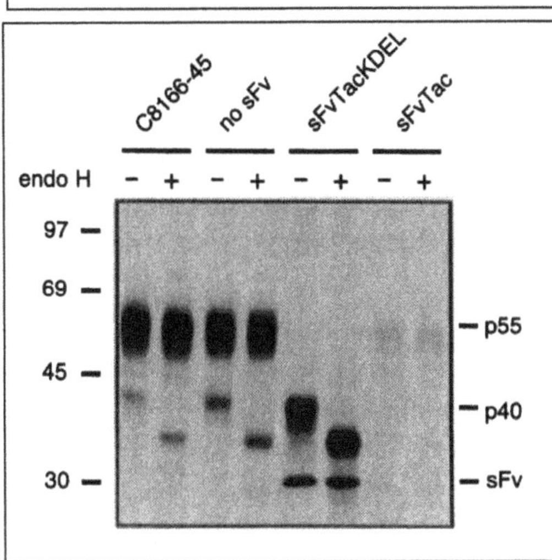

Fig. 3.3. Endoglycosidase (endo) H sensitivity of IL–2Rα immunoprecipitated with mAb 7G7/B6 from C8166-45 cells or Jurkat cells expressing no sFv, sFvTacKDEL, and sFvTac. SFvTacKDEL and the p55 and p40 forms of IL–2Rα are indicated.

Rapid Degradation of IL-2Rα in Cells Expressing sFvTac

In contrast to the stable accumulation of p40 in Jurkat cells expressing sFvTacKDEL, very little IL-2Rα (either p55 or p40) could be detected by immunoprecipitation in sFvTac-expressing clones that showed downregulation of IL-2Rα. This suggested rapid degradation of the sFvTac/IL-2Rα complexes (Fig. 3.3). The degradation appeared to occur by a nonlysosomal mechanism, as the presence of the lysosomal inhibitors methionine methyl ester and ammonium chloride did not prevent the rapid disappearance of IL-2Rα in these cells.[44] In addition, IL-2Rα degradation was not prevented by treatment of the sFv-Tac-expressing cells with brefeldin A, which causes disassembly of the Golgi apparatus and prevents proteins from exiting the ER.[44,50] These data strengthen the suggestion that degradation of the sFvTac/IL-2Rα complexes occurs either within the ER or in the cytosol by proteasomes following retrograde transport of the sFvTac/IL-2Rα complexes across the ER membrane, as has been recently demonstrated for nascent MHC I heavy chains.[51]

Although both sFvTac and sFvTacKDEL were retained intracellularly, the presence of a specific ER-retention signal had a significant impact upon sFv stability, efficacy and the fate of sFv/IL-2Rα complexes. A functional comparison of the two sFvs is provided in Table 3.2.

Phenotypic Knockout of IL-2Rα in HTLV-I Transformed Cell Lines

To examine the consequences of IL-2Rα downregulation in HTLV-I transformed cells, the sFvTac and sFvTacKDEL genes were stably introduced into the cell lines C8166-45 and HUT102 using a bicistronic lentiviral vector (Fig. 3.4).[52] Flow cytometric analysis of the vector-transduced cell indicated that complete downregulation of IL-2Rα by sFvTacKDEL could be achieved in both C8166-45 and HUT102 cells (Fig. 3.4). The sFvTac intrabody, in contrast, was unable to downregulate IL-2Rα in these cells (JHR and WAM, unpublished data). To eliminate the possibility that IL-2Rα was present at the cell surface of sFvTacKDEL-expressing cells but masked by the sFv itself (which might be cotransported to the cell surface or released by dead cells), the cells were stained with the monoclonal antibody 7G7/B6, which recognizes a different IL-2Rα epitope.[53] The lack of staining with 7G7/B6 confirmed that IL-2Rα was indeed not present (Fig. 3.4 and data not shown). Immunoprecipitation studies further confirmed the absence of mature p55 and the accumulation of p40 in the sFvTacKDEL-expressing cells (not shown). The growth rate of the IL-2Rα-negative cells was in each case comparable

---

Table 3.2. *Functional characteristics of ER-targeted and nontargeted sFvTac*

Property	sFvTacs	FvTacKDEL
Half life	4 - 6 h	> 24 h
Secreted	No	No
ER retention of IL-2Rα (Jurkat)	Leaky	Yes
ER retention of IL-2Rα (C8166-45)	No	Yes
sFv/IL-2Rα complex	Rapidly Degraded	Stable

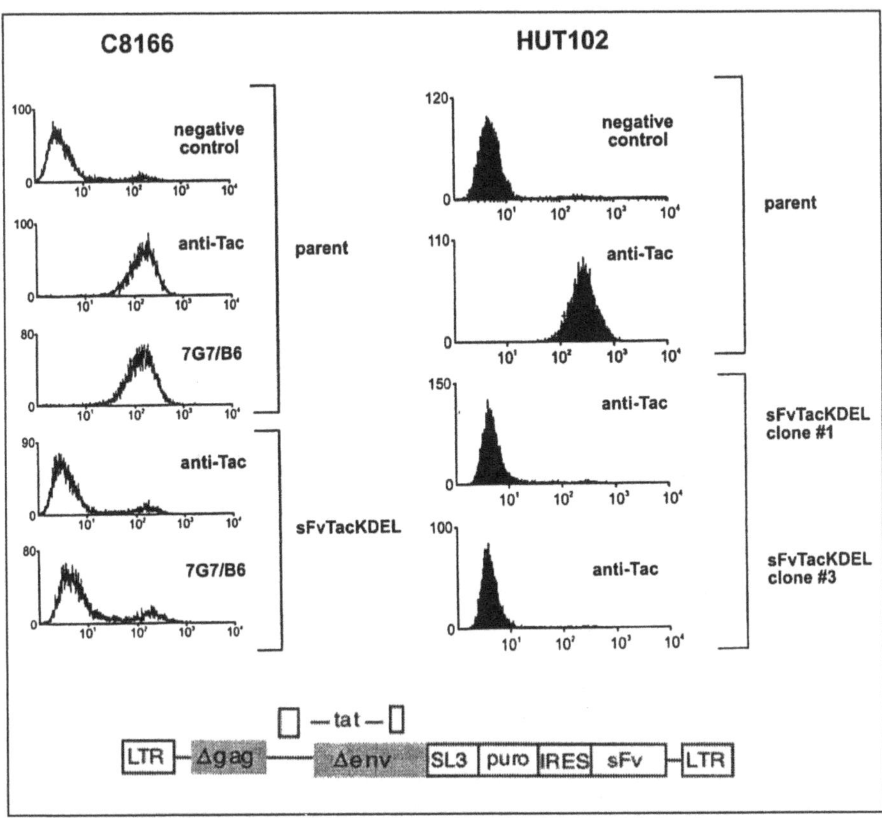

Fig. 3.4. Flow cytometric analysis of C8166-45 and HUT102 cells stably transduced with an HIV–1–based retrovirus vector encoding the sFvTacKDEL gene. Parental or transduced cells were stained with anti-Tac, 7G7/B6 or a control antibody, as indicated. Relative fluorescence intensity is shown on the X axis. Shown below is the HIV–1 based vector used to express sFvTacKDEL.

with that of the parental cell line, and we conclude that cell surface expression of IL-2Rα is dispensable for the in vitro growth of these HTLV-I transformed cells. These studies do not rule out the possibility that p40 molecules retained in the ER may somehow function to promote cell growth. This seems unlikely, however, in view of the fact that IL-2Rα is required for efficient IL-2 binding but does not appear to have any intracellular signaling function.[15]

Inducible Knockout of the IL-2rα: A Model System to Analyze the IL-2 Receptor Signal Transduction Pathway

When using intracellular sFvs to examine signal transduction pathways or genes that are involved in activation-induced (AID) or programmed cell death (apoptosis), tight regulation of sFv expression would be desirable, if not essential. Using the IL-2 receptor as a model system, we have demonstrated inducible downregulation of IL-2Rα by placing the sFvTacKDEL gene under the control of a tetracycline-repressible promoter.[54] In this system, a hybrid transactivator (the tTA protein) composed of the DNA-binding domain of the *E. coli* tet repressor fused to the activation domain of VP16, is used to activate transcription from a minimal promoter [phCMV*-1] which contains multiple binding sites for the tet repressor (Fig. 3.5). Tetracycline prevents tTA from interacting with the promoter and thereby inhibits transcriptional activation of the sFvTacKDEL gene.

Tetracycline-repressible "knockout" of IL-2Rα was demonstrated using the cell line Kit225, an IL-2Rα positive, IL-2 dependent T-cell line which can also be maintained using IL-7.[55] Kit225 cells doubly transfected with the tTA and sFvTacKDEL genes and maintained in the presence of IL-7 demonstrated complete downregulation of IL-2Rα when tetracycline was removed (Fig. 3.5). Immunoprecipitation studies showed full induction of sFvTacKDEL expression within 24 hours of tetracycline withdrawal[56] and no new IL-2Rα was expressed at the cell surface from this time, as judged by the absence of the mature p55 in cell lysates.[56] Instead, the cells accumulated the p40 precursor of IL-2Rα.

Despite the rapid shutoff in IL-2Rα maturation and transport, significant levels of IL-2Rα persisted on the Kit225 cell surface for up to 72 hours following tetracycline withdrawal (Fig. 3.5). This probably reflects the long half life of IL-2Rα molecules already present at the time of sFvTacKDEL induction, as previous studies have shown IL-2Rα to be recycled to the cell surface following endocytosis, resulting in a half life of greater than 24 hours.[57]

In summary, the tetracycline system allows tight regulation of sFvTacKDEL expression and should prove valuable when expressing an sFv that interferes with cell growth, viability or function. One potential use of this system is in studies of IL-2R signaling in HTLV-I transformed cells, where intracellular antibodies that inhibit β or γ chain function may induce growth arrest or apoptosis.

Constitutive Phenotypic and Functional Knockout of IL-2rα in Human PBMCs

To establish the feasibility of using intracellular antibodies to downregulate growth factor receptors, costimulatory molecules, etc. on primary human lymphocytes, the sFvTacKDEL gene was introduced into PHA-activated human peripheral blood mononuclear cells (PBMC) using the lentiviral vector depicted in Figure 3.4. Vector-transduced cells generated by puromycin selection and maintained in human rIL-7 showed a marked reduction in IL-2Rα expression when compared to cells transduced with a control (empty) vector (Fig. 3.6). Five of ten single-cell clones that were subsequently generated from the sFvTacKDEL-transduced population by subcloning in the presence of IL-7 did not express IL-2Rα at a detectable level (Fig. 3.6). These data indicate that complete phenotypic knockout of a cell surface receptor can be achieved in primary human T cells using an ER-targeted sFv.

A tritiated thymidine incorporation assay was used to compare the IL-2 responsiveness of an IL-2Rα negative and an IL-2Rα positive clone (clones 2 and 5, in Fig. 3.6). The IL-2Rα negative clone did not respond to low doses of IL-2 (Table 3.3). Some proliferation was seen at IL-2 concentrations of 10 and 100 units/ml but when compared to the IL-2Rα-positive clone, ~10 times more IL-2 was required to achieve an equivalent proliferative response. These data demonstrate functional as well as phenotypic evidence for the absence of high-affinity IL-2 receptors in the sFvTacKDEL-transduced cells.

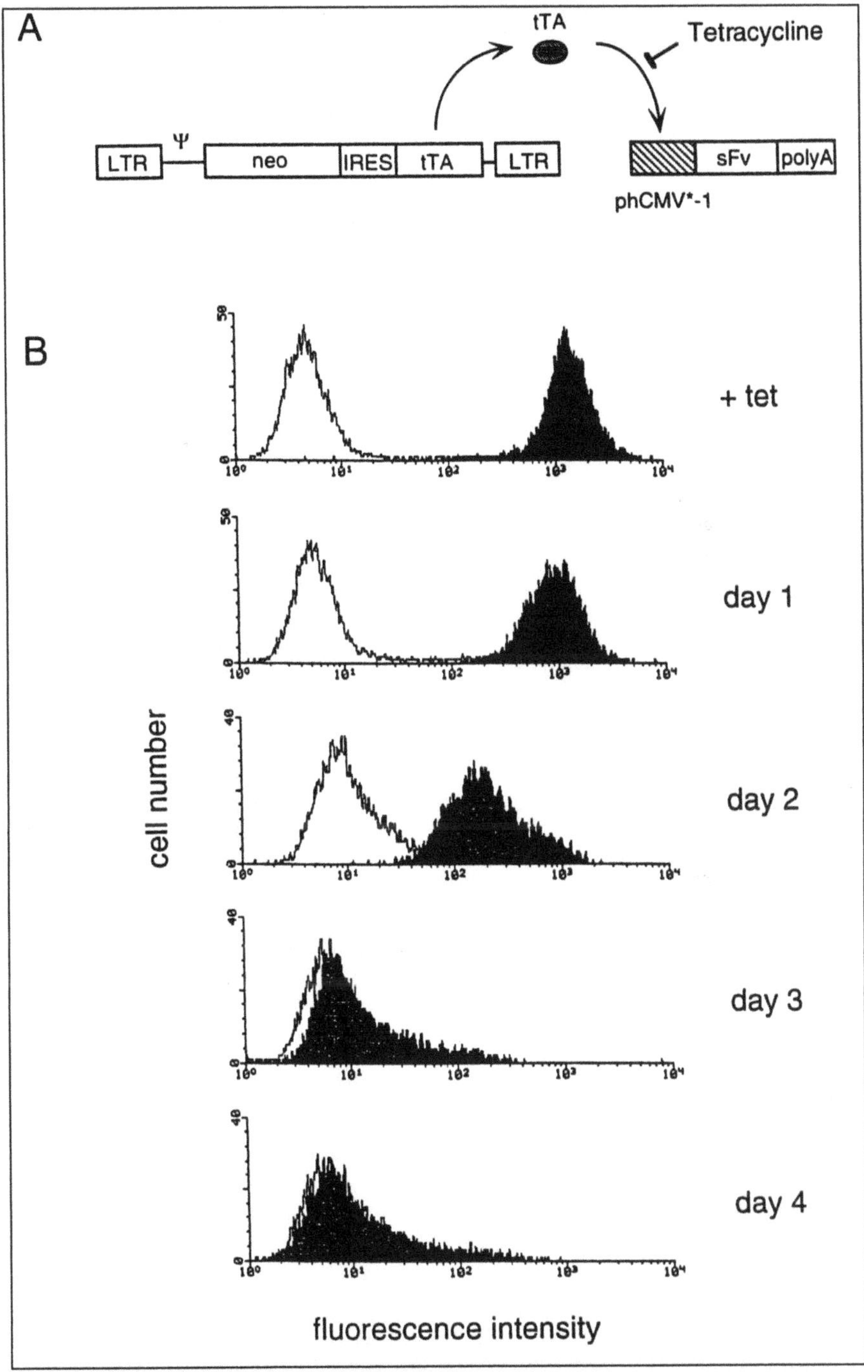

Fig. 3.5. Tetracycline-repressible expression of sFvTac in Kit225 cells. (A) Diagram illustrating the tetracycline repressible expression system. (B) Flow cytometric analysis of Kit225 cells stained with anti-Tac. Cells were analyzed 1-4 days after tetracycline withdrawal, as indicated. Top panel shows cells maintained in the presence of tetracycline.

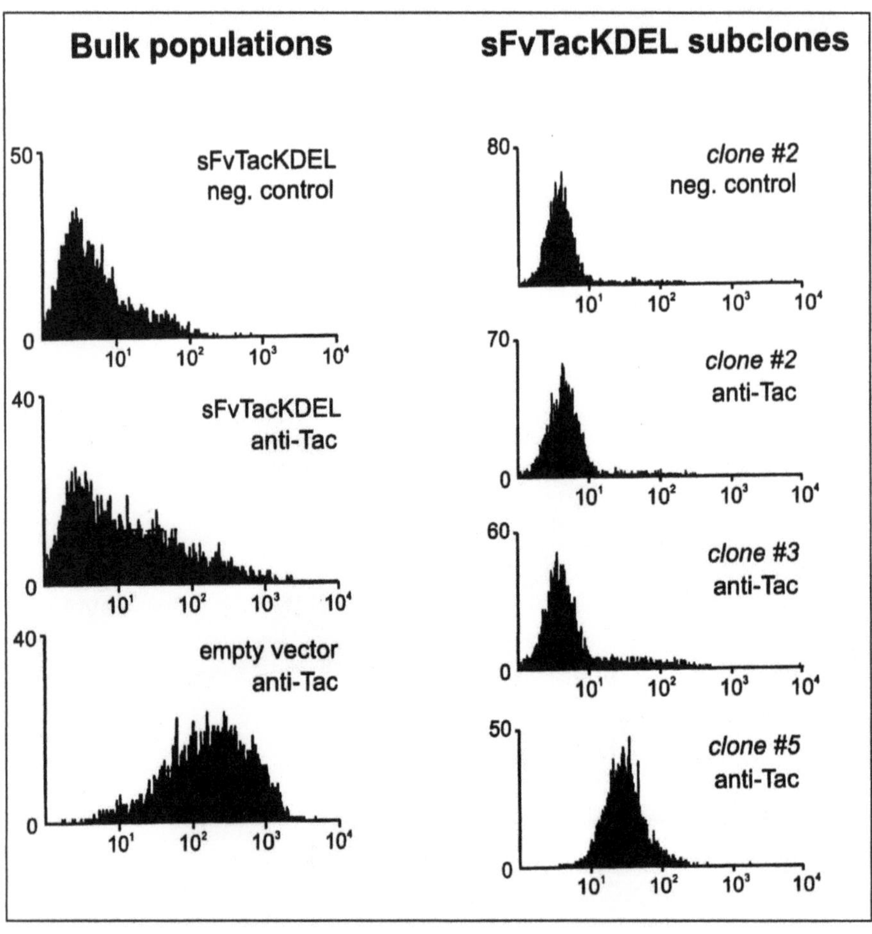

Fig. 3.6. IL–2Rα expression on PHA-activated peripheral blood mononuclear cells transduced with a control (empty) vector or a vector expressing sFvTacKDEL. The cells were stained with anti-Tac mAb followed by FITC-conjugated goat anti-mouse IgG. Negative controls were stained with the secondary antibody alone.

Conclusions

An ER-targeted single-chain antibody, sFvTacKDEL, has been used to block the cell surface transport of IL-2Rα in both established human T cell lines and primary human T cells. Intracellular retention of IL-2Rα by sFvTacKDEL is associated with functional and phenotypic loss of the high-affinity IL-2R on these cells. The presence of the ER retention signal (KDEL) significantly improved sFv stability and efficacy and was critical for the downregulation of IL-2Rα on cell lines which express this receptor at high levels. Inducible knockout of IL-2Rα has been achieved by using a tetracycline-regulated system to express the sFvTacKDEL gene. Using sFvTacKDEL to knockout of IL-2Rα on HTLV-I transformed cell lines, we demonstrate that expression of the high-affinity ($\alpha\beta\gamma$) IL-2R is dispensable for the in vitro growth of these cell lines. We speculate that the high affinity receptor may

Table 3.3. IL-2 induced proliferation of IL-2Rα-positive and negative T-cell clones

	^{3}H-thymidine Incorporation (cpm)	
IL-2 (u/ml)	Clone 2 (IL-2Rα⁻)	Clone 5 (IL-2Rα⁺)
0	235 (59)	419 (183)
0.1	168 (83)	378 (64)
1.0	342 (62)	4724 (328)
10	2739 (78)	20354 (1147)
100	20778 (947)	55843 (3098)

10^5 cells were incubated in triplicate for 72 hours with the indicated concentration of IL-2, then pulsed for 8 hours with 1 µCi of tritiated-thymidine. Standard error is shown in parentheses.

have been required for IL-2-mediated cell proliferation at an earlier stage in the immortalization process. Although the IL-2Rα subunit is not required for the growth of HTLV-I immortalized cells, the role of βγ receptor signaling is an important area that requires further investigation. Two recent reports suggest that the switch to IL-2 independence in HTLV-I immortalized lines may be associated with the constitutive activation of Jak3, a tyrosine kinase associated with IL-2Rγ.[41,42] Thus, signaling by βγ receptor may contribute to the growth of these cell lines during the IL-2 independent phase. The development of intracellular antibodies directed against IL-2Rβ and γ_c should permit direct experimental testing of this hypothesis.

References

1. Smith KA. Interleukin-2: inception, impact, and implications. Science 1988; 240:1169-1176.
2. Waldmann TA. The interleukin-2 receptor. J Biol Chem 1991; 266:2681-2684.
3. Takeshita T, Asao H, Ohtana K. Cloning of the γ chain of the human IL-2 receptor. Science 1992; 257:379-382.
4. Taniguchi T and Minami Y. The IL-2/IL-2 receptor system: a current overview. Cell 1993; 73:5-8.
5. Bamford RN, Grant AJ, Burton JD et al. The interleukin (IL) 2 receptor beta chain is shared by IL-2 and a cytokine, provisionally designated IL-T, that stimulates T-cell proliferation and the induction of lymphokine-activated killer cells. Proc Natl Acad Sci USA 1994; 91:4940-4944.
6. Leonard WJ, Noguchi M, Russell SM et al. The molecular basis of X-linked severe combined immunodeficiency: the role of the interleukin-2 receptor γ chain as a common gamma chain. γ_c. Immunol Rev 1994; 138:61-86.

7. Siegel JP, Sharon M, Smith PL et al. The IL-2 receptor β chain (p70) role in mediating signals for LAK, NK, and proliferative activities. Science 1987; 238:75-78.

8. Tsudo M, Goldman CK, Bongivanni KF et al. The p75 peptide is the receptor for interleukin 2 expressed on large granular lymphocytes and is responsible for the interleukin 2 activation of these cells. Proc Natl Acad Sci USA 1987; 84:5394-5398.

9. Lê thi Bich-Thuy, Dukovich M, Peffer NJ et al. Direct activation of human resting T cells by IL 2: The role of an IL 2 receptor distinct from the Tac protein. J Immunol 1987; 139:1550-1556.

10. Leonard WJ, Depper JM, Robb RJ et al. Characterization of the human receptor for T-cell growth factor. Proc Natl Acad Sci USA 1983; 80:6957-6961.

11. Smith KA and Cantrell DA. Interleukin 2 regulates its own receptors. Proc Natl Acad Sci USA 1985; 82:864-868.

12. Meuer SC, Hussey RE, Cantrell DA et al. Triggering of the T3-Ti antigen-receptor complex results in clonal T-cell proliferation through an interleukin 2-dependent autocrine pathway. Proc Natl Acad Sci USA 1984; 81:1509-1513.

13. Wang R, Rogers AM, Rush BJ et al. Induction of sensitivity to activation-induced death in primary CD4⁺ cells: a role for interleukin-2 in the negative regulation of responses by mature CD4⁺ T cells. Eur J Immunol 1996; 26:2263-2270.

14. Willerford DM, Chen J, Ferry JA et al. Interleukin-2 receptor α chain regulates the size and content of the peripheral lymphoid compartment. Immunity 1995; 3:521-530.

15. Minami Y, Kono T, Miyazaki T et al. The IL-2 receptor complex: its structure, function, and target genes. Annu Rev Immunol 1993; 11:245-268.

16. Taniguchi T. Cytokine signaling through nonreceptor protein tyrosine kinases. Science 1995; 268:251-255.

17. Nelson BH, Lord JD and Greenberg PD. Cytoplasmic domains of the interleukin-2 receptor β and γ chains mediate the signal for T-cell proliferation. Nature 1994; 369:333-336.

18. Nakamura Y, Russell SM, Mess SA et al. Heterodimerization of the IL-2 receptor β- and γ-chain cytoplasmic domains is required for signaling. Nature 1994; 369:330-333.

19. Uchiyama T, Yodoi J, Sagawa K et al. Adult T cell leukemia: Clinical and hematologic features of 16 cases. Blood 1977; 50:481-492.

20. Takatsuki K, Matsuoka M, Yamaguchi K. Adult T-cell leukemia in Japan. J Acquir Immune Defic Syndr Hum Retrovirol 1996; 19:S15-S19.

21. Popovic M, Lange-Wantzin G, Sarin PS et al. Transformation of human umbilical cord blood T cells by human T-cell leukemia/lymphoma virus. Proc Natl Acad Sci USA 1983; 80:5402-5406.

22. Miyoshi I, Kubonishi I, Yoshimoto S et al. Type C-virus particles in a cord T-cell line derived by cocultivating normal human cord leukocytes and human leukemic T cells. Nature 1981; 294:770-771.

23. Salahuddin SZ, Markham PD, Wong-Staal F et al. Restricted expression of human T-cell leukemia-lymphoma virus (HTLV) in transformed human umbilical cord blood lines. Virology 1983; 129:51-64.

24. Hoshino H, Esumi H, Miwa M et al. Establishment and characterization of 10 cell lines derived from patients with adult T-cell leukemia. Proc Natl Acad Sci USA 1983; 80:6061-6065.

25. Yssel H, Malefyt R, Duc Dodon M et al. Human T cell leukemia/lymphoma virus type I infection of a CD4+ proliferative/cytotoxic T cell clone progresses in at least two distinct phases based on changes in function and phenotype of the infected cells. J Immunol 1989; 142:2279-2289.

26. Uchiyama T, Hori T, Tsudo M et al. Interleukin-2 receptor (Tac antigen) expressed on adult T cell leukemia cells. J Clin Invest 1985; 76:446-453.

27. Maeda M, Shimizu A, Ikuta K et al. Origin of human T-lymphotropic virus 1-positive T cell lines in adult T cell leukemia: analysis of T cell receptor gene rearrangement. J Exp Med 1985; 162:2169-2174.

28. Maeda M, Daitoku Y, Kashihara M et al. Evidence for the interleukin-2 dependent expansion of leukemic cells in adult T cell leukemia. Blood 1987; 70:1407-1411.

29. Yoshida M, Suzuki T, Fujisawa J et al. HTLV-1 oncoprotein tax and cellular transcription factors. Curr Top Microbiol Immunol 1995; 193:79-89.

30. Franchini G. Molecular mechanisms of human T-cell leukemia/lympho-tropic virus type 1 infection. Blood 1995; 86:3619-3639.

31. Siekevitz M, Feinberg MB, Holbrook N et al. Activation of interleukin 2 and interleukin 2 receptor (Tac) promoter expression by the trans-activator (tat) gene product of human T cell leukemia virus, type I. Proc Natl Acad Sci USA 1987; 84: 5389-5393.

32. Maruyama M, Shibuya N, Harada H et al. Evidence for aberrant activation of the interleukin-2 autocrine loop by HTLV-I-encoded p40x and T3/Ti complex triggering. Cell 1987; 48:343.

33. Franchini G, Wong-Staal F, Gallo RC. Human T-cell leukemia virus (HTLV-I) transcripts in fresh and cultured cells of patients with adult T-cell leukemia. Proc Natl Acad Sci USA 1984; 81:6207-6211.

34. Tendler CL, Greenberg SJ, Blattner WA et al. Transactivation of interleukin 2 and its receptor induces immune activation in human T-cell lymphotropic virus type 1-associated myelopathy: Pathogenic implications and a rationale for immunotherapy. Proc Natl Acad Sci USA 1990; 87:5218-5222.

35. Hattori T, Uchiyama T, Toibana T et al. Surface phenotype of Japanese adult T cell leukemia cells characterized by monoclonal antibodies. Blood 1981; 58:645-647.

36. Waldmann TA, Greene WC, Sarin PS et al. Functional and phenotypic comparison of human T cell leukemia/lymphoma virus positive adult T cell leukemia with human T cell leukemia/lymphoma virus negative Sezary leukemia, and their distinction using anti-Tac monoclonal antibody identifying the human receptor for T cell growth factor. J Clin Invest 1984; 73:1711-1718.

37. Kodaka T, Umadome H, Uchino H. Expression of cytokine mRNA in leukemia cells from adult T cell leukemia patients. Jpn J Cancer Res 1989; 80:531-536.

38. Arya SK, Wong-Staal F, Gallo RC. T-cell growth factor gene: lack of expression in human T-cell leukemia-lymphoma virus-infected cells. Science 1984; 223:1086-1087.

39. Yodoi J, Uchiyama T, Maeda M. T cell growth factor receptor in adult T cell leukemia (correspondence). Blood 1983; 62:509-511.

40. Sharon M, Gnarra JR and Leonard WJ. The beta-chain of the IL-2 receptor (p70) is tyrosine-phosphorylated on YT and HUT-102B2 cells. J Immunol 1989; 143:2530-2533.

41. Migone TS, Lin JX, Cereseto A et al. Constitutively activated Jak-STAT pathway in T cells transformed with HTLV-I. Science 1995; 269:79-81.

42. Xu X, Kang S, Heidenreich O et al. Constitutive activation of different Jak kinases in human T cell leukemia virus type 1 (HTLV-I) Tax protein or virus-transformed cells. J Clin Invest 1995; 96:1548-1555.

43. Uchiyama T, Broder SA, Waldmann TA. A monoclonal antibody reactive with activated and functionally mature human T-cells. J Immunol 1981; 126:1393.

44. Richardson JH, Sodroski JG, Waldmann TA et al. Phenotypic knockout of the high-affinity human interleukin 2 receptor by intracellular single-chain antibodies against the α subunit of the receptor. Proc Natl Acad Sci USA 1995; 92:3137-3141.

45. Marasco WA, Haseltine WA, Chen SY. Design, intracellular expression and activity of a human anti-HIV-1 gp120 single chain antibody. Proc Natl Acad Sci USA 1993; 90:7889-7893.

46. Beerli RR, Wels W and Hynes NE. Autocrine inhibition of the epidermal growth factor receptor by intracellular expression of a single-chain antibody. Biochem Biophys Res Commun 1994; 204:666-672.

47. Jost CR, Kurucz I, Jacobus CM et al. Mammalian expression and secretion of functional single-chain Fv molecules. J Biol Chem 1994; 269:26267-26273.

48. Greene WC, Robb RJ, Depper JM et al. Phorbol diester induces expression of Tac antigen on human acute T lymphocytic leukemic cells. J Immunol 1984; 133:1042-1047.

49. Wano Y, Uchiyama T, Fukui K et al. Characterization of human interleukin 2 receptor (Tac antigen) in normal and leukemic T cells: coexpression of normal and aberrant receptors on Hut-102 cells. J Immunol 1984; 132:3005-3010.

50. Doms RW, Russ G and Yewdell JW. Brefeldin A redistributes resident and itinerant Golgi proteins to the endoplasmic reticulum. J Cell Biol 1989; 109:61-72.

51. Emmanuel J, Wiertz HJ, Tortorella D et al. Sec61-mediated transfer of a membrane protein from the endoplasmic reticulum to the proteasome for destruction. Nature 1996; 384:432-438.

52. Richardson JH, Hofmann W, Sodroski JG et al. Intrabody-mediated knockout of the high-affinity IL-2 receptor in primary human T cells using a bicistronic HIV-1 based vector. Submitted

53. Rubin LA, Kurman CC, Biddison WE et al. A monoclonal antibody 7G7/B6, binds to an epitope on the human interleukin-2 (IL-2) receptor that is distinct from that recognized by IL-2 or anti-Tac. Hybridoma 1985; 4:91-102.

54. Grossen M and Bujard H. Tight control of gene expression in mammalian cells by tetracycline-responsive promoters. Proc Natl Acad Sci USA 1992; 89:5547-5551.

55. Hori T, Uchiyama T, Tsudo M et al. Establishment of an interleukin 2-dependent human T cell line from a patient with T cell chronic lymphocytic leukemia who is not infected with human T cell leukemia/lymphoma virus. Blood 1987; 70:1069-1072.

56. Richardson JH, Waldmann TA, Sodroski JG et al. Inducible knockout of the interleukin-2 receptor α chain: expression of the high-affinity IL-2 receptor is not required for the in vitro growth of HTLV-I transformed T cell lines. Virology (in press).

57. Hemar A, Subtil A, Lieb M et al. Endocytosis of interleukin 2 receptors in human T lymphocytes: Distinct intracellular localization and fate of the receptor α, β, γ chains. J Cell Biol 1995; 129:55-64.

CHAPTER 4

Intracellular Antibodies as Tools to Study ErbB Receptor Tyrosine Kinases

Roger R. Beerli, Diana Graus-Porta and Nancy E. Hynes

Introduction

Intracellular expression of antibodies provides a novel approach for the manipulation of biochemical processes in intact cells. We have used this approach to test two alternative strategies for the inactivation of individual members of the erbB family of receptor tyrosine kinases (RTKs). Epidermal growth factor (EGF) receptor and erbB-2, two receptors whose aberrant expression is frequently involved in human cancer, were chosen as targets. Genes encoding single-chain Fv domains (scFvs) directed to the extracellular portion of these receptors were derived from hybridoma cells producing the corresponding monoclonal antibodies. The scFvs were provided with a signal peptide to direct their synthesis to the secretory compartment of the cell. Intracellular expression of an scFv that competes with EGF was found to inhibit EGF receptor function in an autocrine manner. In contrast, scFvs provided with an additional endoplasmic reticulum (ER) retention signal were retained in the ER and inactivated their target receptor by preventing its appearance on the cell surface. In NIH/3T3 fibroblasts transformed by a mutated, constitutively active erbB-2, scFv-mediated intracellular retention of erbB-2 led to complete reversion of the transformed phenotype. The potential for the targeted inactivation of cell surface proteins by a gene therapy approach has implications for the treatment of many human diseases including cancer and viral infections. Apart from its potential for future medical applications, intracellular antibody expression has a wide range of immediate applications in basic research. For example, the phenotypic knock-out of growth factor receptors by scFv mediated intracellular retention provides an attractive tool for investigating complex receptor-ligand interactions. This is exemplified by our studies of the erbB signaling network with its four related receptors and multiple activating ligands. Analysis of signaling emanating from this receptor family is complicated by the extensive crosstalk which occurs among the individual receptors. We have downregulated cell surface expression of individual erbB receptors via intracellular antibody

Intrabodies: Basic Research and Clinical Gene Therapy Applications, edited by
Wayne A. Marasco. © 1998 Springer-Verlag and R.G. Landes Company.

expression prior to analyzing ligand induced signaling. The results from these experiments have allowed us to draw some important conclusions concerning erbB receptor interplay and to show that erbB-2 is the key member of this family.

The ErbB Family of Receptor Tyrosine Kinases

The erbB family is a subclass of receptor tyrosine kinases (RTKs), cell surface proteins with an extracellular ligand binding domain and an intracellular tyrosine kinase domain.[1] There are four structurally related erbB proteins, the epidermal growth factor (EGF) receptor/erbB-1, erbB-2, erbB-3 and erbB-4[2] (Fig. 4.1A). erbB receptors are widely expressed in epithelial, mesenchymal and neuronal tissues and play fundamental roles during development.[3-5] Aberrant expression of members of the erbB family has been observed in several human cancers,[6,7] making it crucial that their function be elucidated in detail.

The erbB receptors are activated following binding of specific ligands, members of the EGF-related family of peptides.[2,7] Ligand binding leads to receptor dimerization, stimulation of kinase activity and autophosphorylation on tyrosine residues. These phosphorylated residues in turn serve as binding sites for SH2-containing[8] and phosphotyrosine-binding domain (PTB)-containing[9] proteins

Fig. 4.1. The erbB receptor family and their activating ligands. In (A) the EGF-related peptides EGF, HB-EGF, TGFα, AR, BTC and NDF, as well as the erbB receptors to which they bind, are depicted. Ligand binding to erbB-1 and erbB-3/4 induces the formation of erbB-2 containing receptor heterodimers (B).

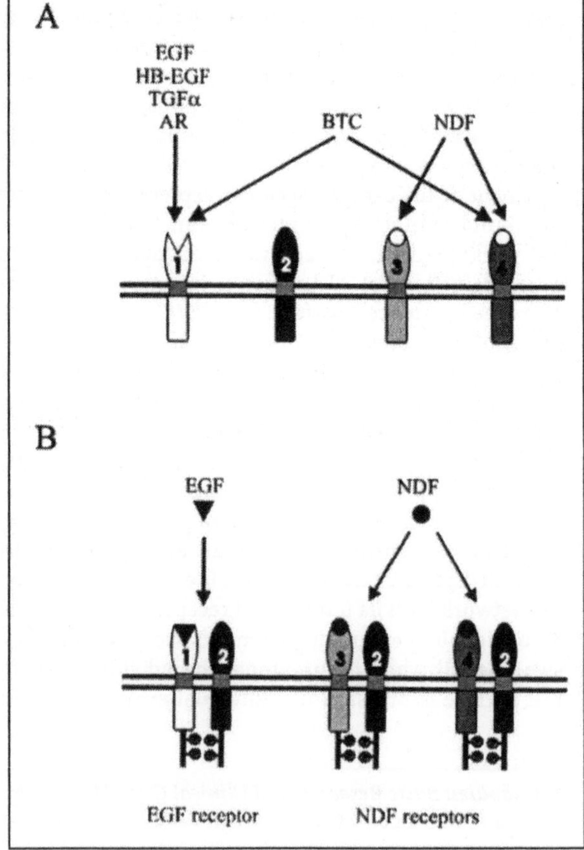

which couple the activated receptors to diverse intracellular signaling pathways. Regulation of erbB receptor activity is quite complex. First, multiple ligands have been identified and they can be classified into three groups based upon their target receptor: EGF, transforming growth factor-α (TGF-α), amphiregulin (AR) and heparin binding-EGF (HB-EGF) which bind EGF receptor[17]; betacellulin (BTC) which is a ligand for EGF receptor and erbB-4;[10-11] and the neu differentiation factors (NDFs), also referred to as heregulins, which bind erbB-3 and erbB-4[2] (Fig. 4.1A). Second, both receptor homodimers and receptor heterodimers form following ligand binding (Fig. 4.1B). Thus, although none of the EGF-related peptides directly bind erbB-2, all of them induce its tyrosine phosphorylation by triggering heterodimerization and cross phosphorylation.[12-14] Receptor heterodimerization serves to diversify the signals elicited by a specific ligand. The specificity of interactions of an intracellular binding protein is determined by the phosphorylated tyrosine residues as well as the surrounding amino acids in the intracellular domain of the receptor,[15] therefore, a heterodimer conceivably provides not only quantitatively but also qualitatively different docking sites compared to a homodimer.

The extensive crosstalk occurring among erbB receptors imposes significant complications for studying the function of individual receptors during ligand-induced signaling. Expressing pairs of erbB receptors in heterologous cells which have no endogenous erbB receptors is one viable approach to analyze receptor-receptor interactions. Such studies have shown that almost all the receptor homo- and heterodimers are formed in response to the appropriate ligand.[11,16,17] While analyses of this type have been important for probing receptor heterodimerization, they are difficult to interpret from a biological standpoint. The isolated expression of pairs of erbB receptors does not reflect the natural situation since most epithelial cells express all four erbB receptors. Thus, it is difficult to predict which of the receptor-receptor interactions observed in the heterologous cells are physiologically relevant. We investigated the function of erbB receptors in their natural cellular setting by targeted inactivation of individual receptors via intracellular expression of scFvs. This approach has yielded valuable information on the behavior of the erbB receptor family (see Study of erbB receptor function by intracellular antibody expression, later in this chapter).

Intracellular Retention or Autocrine Inhibition of ErbB Receptors by Expression of Single-Chain Antibodies

EGF receptor and erbB-2 were chosen as targets to test two alternative methods of receptor inactivation by intracellular antibody expression. Fv domains rather than whole antibodies were used for this purpose. ScFvs are especially suitable for intracellular expression due to their small size and their lack of assembly requirements in the cell. EGF receptor is particularly amenable for testing the concept of autocrine inhibition of receptor function since antagonistic antibodies which block ligand binding are available. This has allowed us to directly assay for effects of scFvs on EGF-induced cell growth. ScFvs targeted to, and retained in the endoplasmic reticulum (ER) were used to test the concept of intracellular retention of receptors (Fig. 4.2).

The EGF receptor specific mAbs EGFR1 and 225 recognize two different epitopes on the extracellular domain of the human EGF receptor, and the latter competes with EGF for binding to the receptor.[18,19] The erbB-2-specific mAbs FRP5 and FWP51 bind to different epitopes on the extracellular domain of human erbB-2.[20]

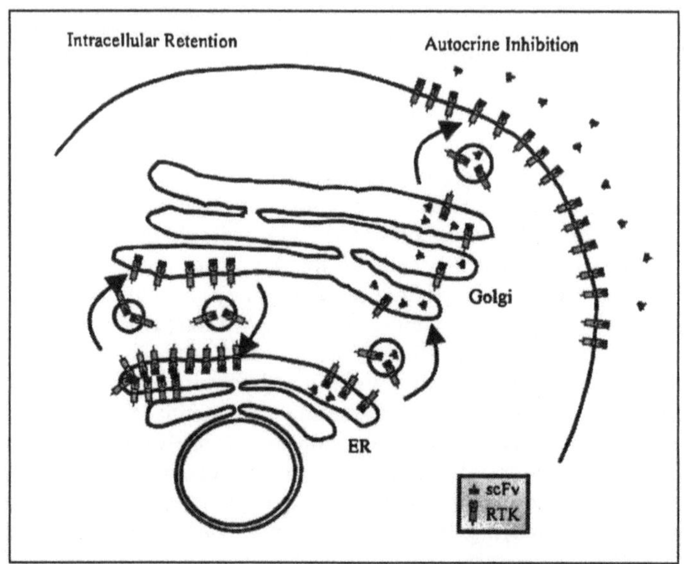

Fig. 4.2. Scheme of intracellular vs. autocrine inhibition of erbB receptors by intracellular antibody expression. For expression in eukaryotic
cells the scFvs were provided with an N-terminal signal peptide which
directs them to the secretory compartment of the cell, the same compartment through which the erbB receptors (RTK) pass on their way to
the plasma membrane. Two versions of each scFv were constructed.
The retained version has the KDEL ER retention signal whereas the
secreted version does not. The former inhibits receptor activation by
preventing its appearance on the plasma membrane. The latter functions in an autocrine fashion via competition with activating ligands.

Hybridoma cells producing these four mAbs were used to construct genes encoding scFvs.[21,22] For expression in eukaryotic cells the scFv proteins were provided
with an N-terminal Ig heavy chain-derived signal peptide which directs them to
the secretory compartment of the cell. Two versions of each cDNA were created.
The scFv-S cDNAs encode secreted proteins, whereas the scFv-R cDNAs encode
variants localized to the lumen of the ER. The latter have the KDEL ER retention
signal at their C-terminus. This peptide causes the retention of soluble proteins in
the lumen of the ER[23] by binding to a specific receptor present in this cellular
compartment.[24]

Expression of scFv cDNA constructs was first examined using transient transfection assays in COS-1 cells. Figure 4.3 shows a Western analysis of cellular extracts made from COS-1 cells transfected with the secreted (S) and the retained (R)
version of three different scFv-encoding cDNAs. ScFv protein was detected using a
specific polyclonal anti-scFv antiserum.[22] The highest levels of scFv protein were
found in cells expressing the retained versions (R) of the erbB-2 specific scFvs,
FRP5 and FWP51 (Fig. 4.3 lanes 4 and 6). The upper band seen in extracts from the
former corresponds to a N-glycosylated form of the scFv-FRP5.[25] The secreted versions of these two scFvs were also found in the triton extracts (Fig. 4.3 lanes 3 and

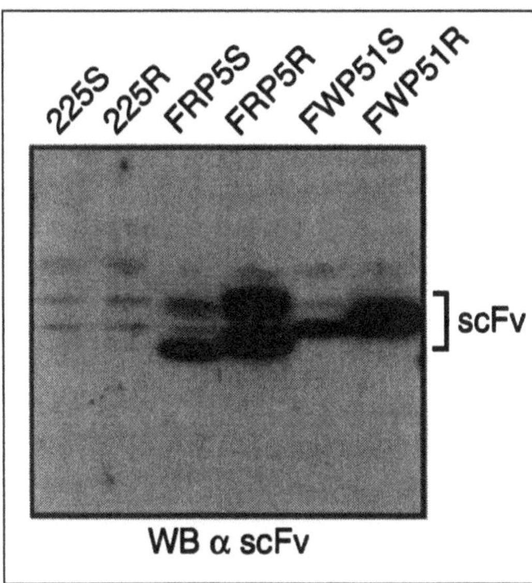

Fig. 4.3. Expression of different scFvs in eukaryotic cells. cDNAs encoding the secreted (S) or retained (R) version of the indicated scFvs were cloned into the pECE expression vector, transiently introduced into COS-1 cells via the lipofectamine technique and 48 hrs later, cellular lysates were prepared in a Triton extraction buffer. Equal aliquots of protein were analyzed by SDS-PAGE, blotted onto PVDF membranes, and the level of scFv protein was determined with a specific serum22 followed by the ECL detection technique. The faint bands in the lanes marked 225S and R represent nonspecific background proteins.

5), however, high levels of both proteins were also found in the conditioned medium.[26] In contrast, there was no detectable scFv-FRP5R or scFv-FWP51R in the conditioned medium.[26] The retained version of the scFv-EGFR1 was also expressed at high levels in various cells[27] (Table 4.1).

In contrast, the EGF receptor specific scFv-225S and scFv-225R were neither detectable in triton extracts (Fig. 4.3) nor in the conditioned medium.[28] The faint bands visible in lanes 1 and 2 of Figure 4.3 are nonspecific background bands. However, scFv-225 protein was detected in the triton-insoluble pellet of COS-1 cells.[28] The results shown in Figure 4.3 and in Table 4.1 are characteristic of the variability which we have observed in the expression of different scFvs in eukaryotic cells. It is conceivable that parameters such as hydrophobicity or solubility contribute to the expression level of a particular scFv. Interestingly, high amounts of a scFv-225-toxin fusion protein were produced in bacteria, suggesting that the addition of the toxin sequences increased the solubility of scFv-225.[29]

To test for biological activity, the erbB2 and EGFR directed scFvs were cloned into the pBabe-puromycin retroviral vector[30] and introduced into eukaryotic cells via viral infection, followed by selection in puromycin-containing medium. We have observed that intracellular expression of the KDEL-tagged, ER-lumenal variants of scFv-FRP5, scFv-FWP51, and scFv-EGFR1 inhibit the function of their target receptor by preventing its appearance on the cell surface. Expression of both erbB-2 directed scFvs led to complete intracellular retention of the receptor,[26,31,33] while scFv-EGFR1 retained approximately 70% of the EGF receptor in the cell.[27] Intracellular expression of scFv-225R did not lead to ER retention of the EGF receptor.[28] However, scFv-225S which competes with EGF for binding the receptor, functions in an autocrine fashion to inhibit EGF-dependent cell growth.[25,28]

Table 4.1. Intracellular expression of ErbB receptor specific scFvs

scFv	Target Protein	Expression Level	Intracellular Retention	Autocrine Inhibition
225	EGFR	+	–	+
EGFR1	EGFR	++++	+	ND
FRP5	ErbB-2	++++	+	+
FWP51	ErbB-2	++++	+	–
SGP1	ErbB-3	+	–	ND
105	ErbB-3	++	–	ND
252	ErbB-3	+	–	ND
72	ErbB-4	+	–	ND
77	ErbB-4	–	–	ND
179	ErbB-4	+	–	ND

cDNAs encoding the different scFvs were cloned into the pECE vector and expression levels were determined by a Western analysis following transient transfection into COS-1 cells. To test for intracellular retention of the target receptor the KDEL-containing version of each scFv cDNA was cloned into the pBabe-puromycin retroviral vector. Following infection and selection of the appropriate cell line (NE1 cells for 225; NE1 cells and T47 cells for EGFR1; T47D cells for the remaining scFvs), a FACS analysis was performed on pools of puromycin resistant cells to quantitate the level of the specific cell surface receptor.[30,31] The 225, FRP5 and FWP51 scFvs were tested for their ability to inhibit their target receptor in an autocrine fashion using the secreted version of each scFV.[26,28]

Two other laboratories have reported on the inhibition of cell surface expression of transmembrane proteins by ER-lumenal expression of scFvs. An scFv binding to the extracellular portion of the HIV envelope protein, scFv-F105, was used to inhibit its ER-transit and consequently the production of infectious virus.[33] In the same laboratory, a phenotypic knock out of the high-affinity interleukin-2 receptor was achieved by ER-lumenal expression of scFv-Tac which is directed against the a subunit of the receptor.[34] Finally, a second laboratory reported on the downregulation of cell surface erbB-2 following ER-lumenal expression of scFv-e23.[35]

It is worth noting that divergent results have been obtained with respect to the requirement for a C-terminal KDEL signal for intracellular retention of a particular scFv. The results obtained with scFv-FRP5, scFv-FWP51 and scFv-Tac indicate that the KDEL peptide is absolutely required for retention of scFvs in the ER-lumen, since variants lacking the peptide were secreted.[23,26,27,34] In contrast, scFv-F105 and scFv-e23 remained in the ER despite the lack of an ER-retention signal and effectively interfered with the ER-transit of their respective target protein.[33,35] Moreover, in the case of scFv-F105, the presence of a KDEL peptide was even undesirable since it caused the protein to become unstable, leading to its rapid degradation.[33] At present, no general rules can be given about the structural requirements for high expression levels, or the necessity of a KDEL-peptide for intracellular retention.

Variable Expression of Endoplasmic Reticulum-Targeted scFvs

The experiments showing the potency of ER-lumenal scFvs in inhibiting the transit of proteins to the plasma membrane suggest that this approach might be universally applicable. However, we have observed that the behavior of scFvs expressed in this compartment is variable (Table 4.1). For instance, while the scFvs -FRP5, -FWP51, and -EGFR1 were expressed at high enough levels to have an effect on the ER-transit of their respective target protein,[26,27,31,32] scFv-225 was only expressed at a very low level and a KDEL-tagged variant failed to have an appreciable effect on cell surface localization of the EGF receptor.[28] However, as previously mentioned, the secreted version of scFv-225 interfered with EGF stimulated cell growth.

In order to study erbB-3 and erbB-4 function we have also tested scFvs directed to these receptors. Unfortunately, the results have been very disappointing (Table 4.1). mAbs SGP1,[36] 105 and 252[37] recognize the extracellular domain of erbB-3. mAbs 72, 77 and 179 are specific for erbB-4.[37] mAbs 105 and 72 compete with NDF for binding to erbB-3 and erbB-4, respectively.[37] Expression of these six scFvs was examined COS-1 cells following transient transfection of the cDNA vectors. Low to intermediate levels of the erbB-3 specific scFv-SGP1, scFv-252 and scFv-105, as well as the erbB-4 specific scFv-72 and scFv-179 were observed, while scFv-77 was not detectable (Table 4.1). In order to test their ability to prevent the appearance of their target receptor on the plasma membrane, cDNAs encoding the retained versions of the erbB-3 and erbB-4 specific scFvs were stably introduced, via retroviral infection, into T47D human breast tumor cells. These cells express moderate levels of all four erbB receptors and we have routinely used them to test the ability of a particular scFv to downregulate a receptor.[31,32] Unfortunately, none of these scFvs caused an intracellular retention of the respective target receptor (R.R. Beerli and D. Graus-Porta, unpublished results) (Table 4.1). It may be possible that the affinity of these particular scFvs was too low to bind their receptor. Alternatively, the rather low expression level of the scFvs in COS-1 may account for their failure to retain the target receptor in the ER. Each of the scFvs has a different amino acid sequence which might explain the different level of expression. The potential of intracellularly expressed scFvs would be greatly enhanced if it were possible to consistently achieve high expression of these molecules in the ER, or in other compartments of the cell.

Effect of scFv on ErbB Receptor Turnover

Results obtained with metabolically labeled cells suggest that in the nonactivated state erbB receptors turn over slowly with the bulk plasma membrane.[38] In contrast, ligand activated receptors, particularly EGF receptor, are rapidly cleared from the cell surface by the endocytic pathway and show an enhanced turnover, leading to a lower level of expression.[39] We examined the effect of scFv-mediated ER retention of erbB proteins upon the turnover of the target receptor by analyzing the receptor levels. The most dramatic effect of scFv expression was observed in the NIH/erbB2* cells. These fibroblasts ectopically express a point mutated, oncogenically activated mutant of erbB-2.[40] The mutation, which is found in the transmembrane domain of the receptor, causes erbB-2 to spontaneously form kinase-active homodimers which have a short half life,[41] presumably because they are rapidly cleared from the plasma membrane. Figure 4.4 shows that the amount of erbB-2 protein is dramatically elevated in NIH/erbB2* cells expressing the retained

Fig. 4.4. erbB-2 protein level in scFv expressing NIH/erbB-2* cells. Equal aliquots of cellular lysates prepared from the control SKBR3 breast tumor cell line and the NIH/erbB-2* control and scFv-expressing cell lines were analyzed by SDS-PAGE, blotted onto a PVDF membrane and the level of erbB-2 protein was determined using the FRP5 mAb[20] and the ECL detection technique.

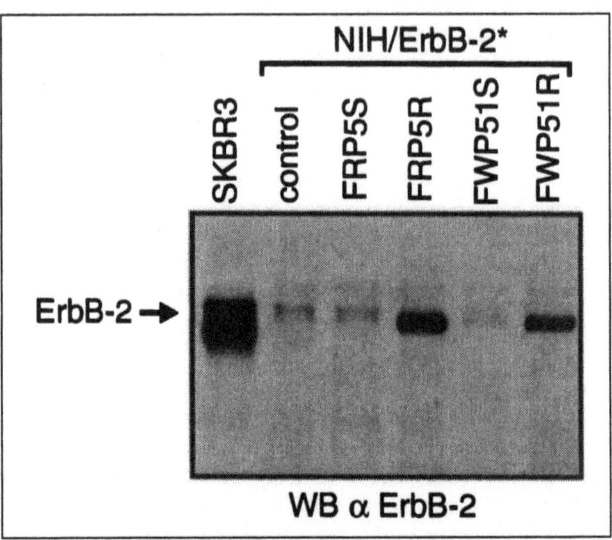

versions of the two scFvs, FRP5R and FWP51R (lanes 4 and 6) compared to the control cells (lane 2) or the cells expressing the secreted version of the two scFvs (lanes 3 and 5). In contrast, introduction of scFv-FRP5R into human tumor cell lines expressing normal wild type erbB-2 did not alter the level of erbB-2 expression.[31,32] These results suggest that the half life of the wild type receptor is the same in every compartment of the cell, i.e., it probably turns over with the other cellular membrane proteins.[42] In contrast, it appears that the point mutated, kinase-active erbB-2 must be localized to the plasma membrane in order for it to be rapidly degraded. When it is retained in the ER it turns over with the same rate as the other membrane proteins, leading to its accumulation in this compartment of the cell. Finally, we have observed slightly elevated levels of the intracellularly retained EGF receptor in cells expressing scFv-EGFR1R.[27,43] These experiments show first, that the ER localized scFv-receptor complex is not rapidly targeted for degradation since compared to the level of receptor in control cells, we have never observed less of the scFv-receptor complex in a cell. Second, the scFv-FRP5R and -EGFR1R each appears to have a distinctive effect on its target receptor since the former does not affect the turnover of erbB-2 while the latter retards the turnover of EGF receptor. This could be due to differences in the natural turnover of the target receptor as well as to the particular characteristics of each scFv.

How Many Receptors Can Be Retained in the ER

We have routinely observed high levels of scFv protein following infection of different eukaryotic cell lines with retroviruses encoding scFv-FRP5R, -FWP51R and -EGFR1R.[27,31,32] These levels are sufficient to cause ER retention of low (1×10^4) to moderately high (1-2×10^5) levels of erbB-2[26,31,32] or EGF receptor.[27] The results obtained with the two erbB-2 specific scFvs are shown in Figure 4.4. The SKBR3 breast tumor cells (lane 1) express 1-2×10^6 erbB2 molecules per cell.[27] Compared to SKBR3 cells the NIH/erbB-2* cell lines, scFv-FRP5R and -FWP51R (lanes 4 and 6) express approximately 1-2 x 10^5 molecules of erbB-2. Expression of both scFvs led to essentially complete retention of erbB-2 in the ER.[26,31,32]

Intracellular scFv Expression Leads to Inhibition of Transformed Cell Growth

c-erbB-2 gene amplification leading to overexpression of the protein occurs in many primary human tumors arising at various sites, including breast, ovaries and lung.[6] Amplification of the gene encoding EGF receptor occurs in a different spectrum of human tumors, including glioblastomas and head and neck cancers.[7] Since both receptors have been implicated in the development of so many different human cancers, they are under intense scrutiny as targets for cancer therapy. ScFvs could potentially be used as tumor growth inhibitory agents.

We have taken two approaches to examine the effects of scFv expression on the growth of transformed cells. In the first, we tested the scFvs for their effects using the model system of NIH/3T3 mouse fibroblasts transformed by ectopic expression of either erbB-2 or EGF receptor. NIH/3T3 NE1 cells overexpress human EGF receptor and are transformed in an EGF-dependent manner.[23] ER retention of the EGF receptor via scFv-EGFR1R expression caused a substantial decrease in the EGF-induced ability of the NE1 cells to grow anchorage independently in soft agar.[27] These results were anticipated since the receptor must be on the plasma membrane in order to bind its ligand which in turn leads to cellular transformation. The NIH/erbB-2* cells which were introduced above express a point-mutated, transforming erbB-2 which requires no ligand for activation. Introduction of scFv-FRP5R and -FWP51R into NIH/erbB-2* cells prevented erbB-2 transit through the ER, resulting in essentially complete reversion of the transformed phenotype.[26] It has been suggested that constitutively active receptor tyrosine kinases, such as erbB-2* or the platelet-derived growth factor receptor in *sis*-transformed cells, might at least partially exert their transforming activity intracellularly in the ER and/or Golgi compartment.[44] The results obtained with the scFv-FRP5R and -FWP51R into NIH/erbB-2* cells suggest that the activated receptor must be on the cell surface to cause transformation. While the exact mechanism leading to reversion is still unknown, one interesting possibility it that erbB-2* only contacts intracellular substrates important for transformation when present on the plasma membrane.

In the second approach we examined the effect of scFv expression on the growth of human tumor cells naturally overexpressing EGF receptor or erbB-2 due to gene amplification. Different tumor cell lines with high levels (1-2 x 10^6 molecules/cell) of erbB-2 or EGF receptor were infected with pBabe-puromycin retroviruses encoding scFv-FRP5R or scFv-EGFR1R and their colony forming ability following selection in puromycin-containing medium was determined. The same cell lines were also infected with the control puromycin encoding retrovirus. Infection with the control retrovirus gave rise to many stable colonies, whereas infection with the scFv encoding vectors gave variable results and, in some instances, we were unable to isolate stable cell lines following viral infection and selection.[27] This suggests that interfering with the transit of the receptor to the plasma membrane was deleterious for their long-term cell growth. In contrast, infection of tumor cells with low levels of erbB receptors, such as the T47D cells, with the scFv-FRP5R and scFv-EGFR1R retroviruses resulted in the appearance of many stable puromycin resistant colonies.[27,31,32,43] Thus, it appears that in some tumor cell lines the overexpressed, plasma membrane-localized erbB-2 or EGF receptor is essential for growth. Therefore, intracellular expression of scFv targeted to erbB receptors may be a suitable approach for inhibiting growth of some human tumor cells.

Study of erbB Receptor Function by Intracellular Antibody Expression

Intracellular antibody expression offers a novel experimental approach for the study of transmembrane signaling, especially for investigating complex receptor-ligand systems like that observed in the erbB receptor family. In addition, it has allowed us to uncover novel receptor-ligand interactions. As an example, intracellular retention of the EGF receptor via expression of scFv-EGFR1R has revealed that this receptor is not the only one which can bind HB-EGF. In T47D cells expressing scFv-EGFR1R the EGF receptor is not available for ligand binding and, as anticipated, no activation of EGF receptor was observed following treatment of these cells with HB-EGF.[43] However, following HB-EGF treatment we found an equivalent level of phosphotyrosine in erbB-4 immunoprecipitated from control T47D cells (puro) and from T47D/EGFR1R cells (Fig. 4.5 lanes 3 and 6), suggesting that there is another receptor for HB-EGF in these cells. Since it has recently been shown that BTC, another EGF receptor agonist, is also a ligand for erbB-4,[11] it is tempting to speculate that HB-EGF directly binds erbB-4. This could be tested in cells which ectopically express only erbB-4.

The functional inactivation of erbB-2 and EGF receptor via intracellular scFv expression has revealed a number of interesting characteristics of this ligand-receptor family:[31,32,45]

1) erbB-2 can physically interact with all other erbB receptors. All EGF receptor-binding ligands induce EGF receptor/erbB-2 dimers and NDF induces erbB-3/erbB-2 and erbB-4/erbB-2 dimers (Fig. 4.1).[16,17,31,32,45] It is likely that the physiologically important, high-affinity receptors for NDF and EGF agonists are erbB-2-containing heterodimers. Thus, the function of erbB-2 is analogous to gp130, the nonbinding subunit shared by several cytokine receptors.[46] This could explain why no direct erbB-2 ligand has yet been isolated. If erbB-2 is an integral part of the NDF receptors and EGF receptor, a direct ligand may not exist.

2) erbB-2 is the preferred heterodimerization partner for all other erbB receptors.[16,43] Activation of erbB2 can be readily detected following treatment of cells with all the EGF receptor agonists.[26] If no erbB-2 is available on the plasma membrane, erbB-3 and/or erbB-4 heterodimerize with erbB-1 in response to NDF. However, if all four erbB proteins are present, the NDF receptors preferentially dimerize with erbB-2 (Fig. 4.6).[43]

3) The presence of erbB-2 dramatically increases the affinity of both NDF and EGF to their respective receptors, predominantly by decelerating the ligand dissociation rates.[45]

4) erbB-2 actively participates in signaling and potentiates the extent of NDF- and EGF-induced activation of various intracellular signaling pathways, as well as enhances the tyrosine phosphorylation of the other erbB receptors in response to their ligands.[31,32,43]

5) EGF-induced activation of erbB-3 and erbB-4 is impaired in the absence of erbB-2, suggesting that erbB-2 has a role in the lateral transmission of signals between other erbB receptors (Fig. 4.6).[43] All of these attributes may help explain the importance of erbB-2 in cancer development.

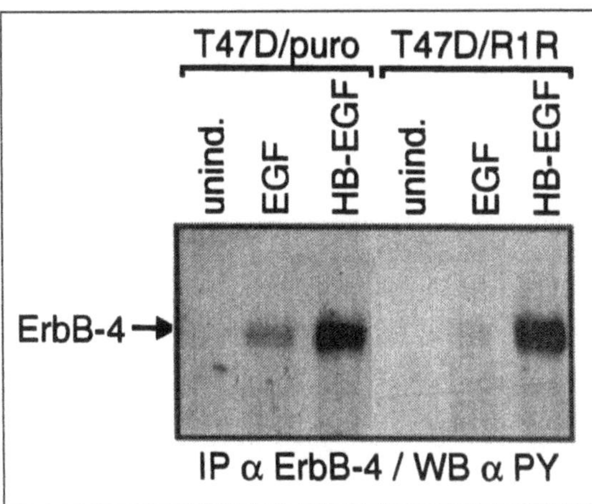

Fig. 4.5. Ligand induced tyrosine phosphorylation of erbB-4. T47D breast tumor cells resulting from infection with either the control puromycin encoding retrovirus (puro) or the scFv-EGFR1R-expressing retrovirus (R1R) were treated for 10 min with 4 nM EGF or with 4 nM HB-EGF or left untreated (unind.). erbB-4 was immunoprecipitated from 2 mg of cellular lysates with a specific antiserum (C18, Santa Cruz) and its phosphotyrosine content was determined with a specific mAb[32] following SDS-PAGE analysis and blotting onto a PVDF membrane.

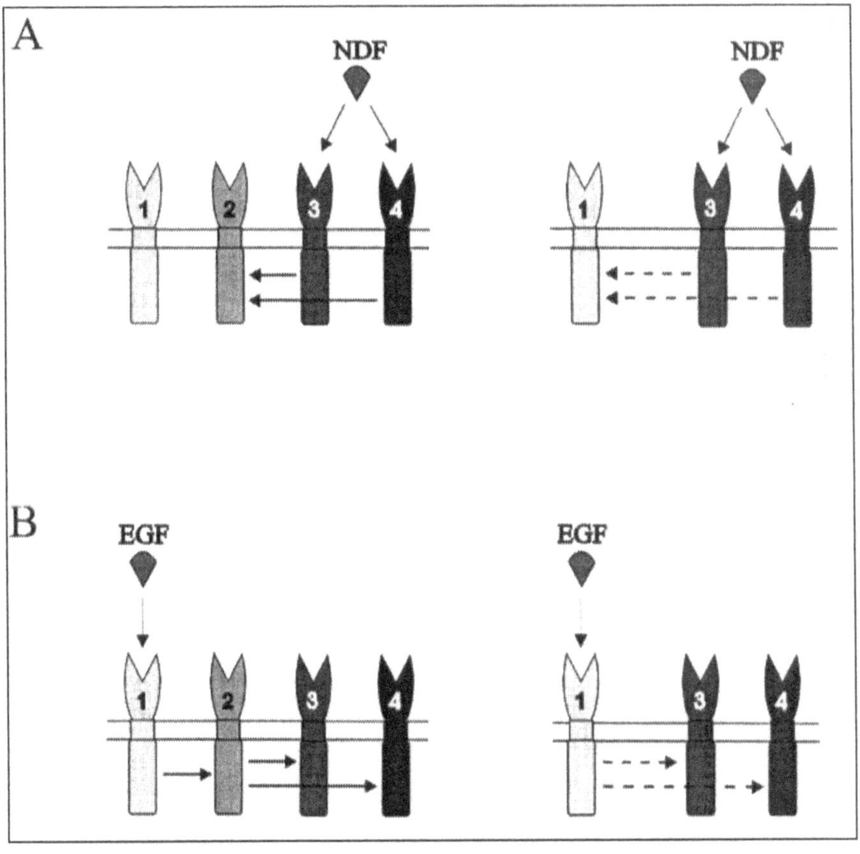

Fig. 4.6. Model of erbB receptor interactions. erbB receptor interactions induced by EGF-related factors in the presence (left) or absence (right) of erbB-2. erbB receptor dimerization is a hierarchical process and erbB-2 is the preferred heterodimerization partner of the receptors for NDF (A) and EGF (B). Activation of erbB-2 in response to EGF (B), but not to NDF (A), allows for lateral transmission of signals.

Overexpression of erbB-2, leading to constitutive activation of its kinase is observed in many human tumors.[47,48] In some of these tumors activation of other erbB receptors, as evidenced by the presence of phosphotyrosine, is also observed.[49] This probably reflects the fact that erbB-2 mediates lateral transmission of signals between other erbB receptors. Moreover, cancer cells often express EGF-family ligands.[7] The remarkable transforming potency of erbB-2 could be due to a combination of all these characteristics: its ability to heterodimerize with other erbB receptors, transmit signals between these receptors, as well as to potentiate EGF-related growth factor signaling.

Acknowledgments

D. Graus-Porta was supported by a grant from the Beider Basel Krebsliga. We thank Dr. Y. Yarden (Weizmann Institute, Rehovot) for providing us with RNA from the hybridoma cells producing mAbs 105, 252, 72, 77 and 179. We thank Dr. W. Gullick (ICRF, London) for making the hybridoma cells producing mAb SGP1 available to us.

References

1. Fantl WJ, Johnson DE, Williams LT. Signaling by receptor tyrosine kinases. Annu Rev Biochem 1993; 62:453-481.
2. Peles E, Yarden Y. Neu and its ligands: from an oncogene to neural factors. Bioassay 1993; 15:815-824.
3. Gassmann M, Casagrands F, Orioli D et al. Aberrant neural and cardiac development in mice lacking the erbB4 neuregulin receptor. Nature 1995; 378:390-394.
4. Lee KF, Simon H, Chen H, et al. Requirement for neuregulin receptor erbB2 in neural and cardiac development. Nature 1995; 378:394-398.
5. Sibilia M, Wagner EF. Strain-dependent epithelial defects in mice lacking the EGF receptor. Science 1995; 269:234-238.
6. Hynes NE, Stern DF. The biology of erbB-2/neu/HER2 and its role in cancer. Biochim Biophys Acta 1994; 1198:165-184.
7. Salomon DS, Brandt R, Ciardiello F et al. Epidermal growth factor-related peptides and their receptors in human malignancies. Crit Rev Oncol Hematol 1995; 19:182-232.
8. Cohen GB, Ren R, Baltimore D. Modular binding domains in signal transduction proteins. Cell 1995; 80:237-248.
9. Kavanaugh WM, Williams LT. An alternative to SH2 domains for binding tyrosine-phosphorylated proteins. Science 1994; 266:1862-1865.
10. Beerli RR, Hynes NE. Epidermal growth factor-related peptides activate distinct subsets of erbB receptors and differ in their biological activities. J Biol Chem 1996; 271:6071-6076.
11. Riese DJ II, Bermingham Y, van Raaij TM et al. Betacellulin activates the epidermal growth factor receptor and erbB-4, and induces cellular response patterns distinct from those stimulated by epidermal growth factor or neuregulin. Oncogene 1996; 12:345-353.
12. King CR, Borrello I, Bellot F et al. EGF binding to its receptor triggers a rapid tyrosine phosphorylation of the erbB-2 protein in the mammary tumor cell line SK-BR-3. EMBO J 1988; 7:1647-1651.
13. Plowman GD, Green JM, Culouscou JM et al. Heregulin induces tyrosine phosphorylation of HER4/p180erbB4. Nature 1993; 366:473-475.
14. Sliwkowski MX, Schaefer G, Akita RW et al. Coexpression of erbB2 and erbB3 proteins reconstitutes a high affinity receptor for heregulin. J Biol Chem 1994; 269:14661-14665.

15. Songyang Z, Shoelson SE, Chaudhuri M et al. SH2 domains recognize specific phosphopeptide sequences. Cell 1993; 72:767-778.
16. Tzahar E, Waterman H, Chen X et al. A Hierarchial network of interreceptor interactions determines signal transduction by neu differentiation factor/neuregulin and epidermal growth factor. Mol Cell Biol 1996; 16:5276-5287.
17. Riese DJ II, van Raaij TM, Plowman GD et al. The cellular response to neuregulins is governed by complex interactions of the erbB receptor family. Mol Cell Biol 1995; 15:5770-5776.
18. Waterfield MD, Mayes ELV, Stroobant T et al. A monoclonal antibody to the human epidermal growth factor receptor. J Cell Biochem 1982; 20:149-161.
19. Sunada H, Magun BE, Mendelsohn J et al. Monoclonal antibody against epidermal growth factor receptor is internalized without stimulating receptor phosphorylation. Proc Natl Acad Sci USA 1986; 83:3825-3829.
20. Harwerth IM, Wels W, Marte BM et al. Monoclonal antibodies against the extracellular domain of the erbB-2 receptor function as partial ligand agonists. J Biol Chem 1992; 267:15160-15167.
21. Wels W, Harwerth IM, Mueller M et al. Selective inhibition of tumor cell growth by a recombinant single-chain antibody-toxin specific for the erbB-2 receptor. Cancer Res 1992; 52:6310-6317.
22. Wels W, Harwerth IM, Zwickl M et al. Construction, bacterial expression and characterization of a bi-functional single-chain antibody-phosphatase fusion protein targeted to the human erbB-2 receptor. BioTechnology 1992; 10:1128-1132.
23. Munro S, Pelham HRB. A C-terminal signal prevents secretion of luminal ER proteins. Cell 1987; 48:899-907.
24. Lewis MJ, Pelham HRB. A human homologue of the yeast HDEL receptor. Nature 1990; 348:162-163.
25. Beerli RR, Wels W, Hynes NE. Inhibition of signaling from Type 1 receptor tyrosine kinases via intracellular expression of single-chain antibodies. Breast Cancer Res Tr 1996; 38:11-17.
26. Beerli RR, Wels W, Hynes NE. Intracellular expression of single chain antibodies reverts erbB-2 transformation. J Biol Chem 1994; 269:23931-23936.
27. Jannot CB, Beerli RR, Mason S et al. Intracellular expression of a single-chain antibody directed to the EGFR leads to growth inhibition of tumor cells. Oncogene 1996; 13:275-282.
28. Beerli RR, Wels W, Hynes NE. Autocrine inhibition of the epidermal growth factor receptor by intracellular expression of a single-chain antibody. Biochem Biophys Res Comm 1994; 204:666-672.
29. Wels W, Beerli RR, Hellmann P et al. EGF receptor and p185^{erbB-2}-specific single-chain antibody toxins differ in their cell-killing activity on tumor cells expressing both receptor proteins. Int J Cancer 1995; 60:137-144.
30. Morgenstern JP, Land H. Advanced mammalian gene transfer: high titre retroviral vectors with multiple drug selection markers and a complementary helper-free packaging cell line. Nucl Acids Res 1990; 18:3587-3596.
31. Beerli RR, Graus-Porta D, Woods-Cook K et al. Neu differentiation factor activation of erbB-3 and erbB-4 is cell specific and displays a differential requirement for erbB-2. Mol Cell Biol 1995; 15:6496-6505.
32. Graus-Porta D, Beerli RR, Hynes NE. Single-chain antibody-mediated intracellular retention of erbB-2 impairs neu differentiation factor and epidermal growth factor signaling. Mol Cell Biol 1995; 15:1182-1191.
33. Marasco WA, Haseltine WA, Chen S. Design, intracellular expression, and activity of a human anti-human immunodeficiency virus type 1 gp120 single-chain antibody. Proc Natl Acad Sci USA 1993; 90:7889-7893.

34. Richardson JH, Sodroski JG, Waldmann TA et al. Phenotypic knockout of the high-affinity human interleukin 2 receptor by intracellular single-chain antibodies against the a subunit of the receptor. Proc Natl Acad Sci USA 1995; 92:3137-3141.

35. Deshane J, Loechel F, Conry RM et al. Intracellular single-chain antibody directed against erbB-2 downregulates cell surface erbB2 and exhibits a selective anti-proliferative effect in erbB2 overexpressing cancer cell lines. Gene Therapy 1994; 1:332-337.

36. Rajkumar T, Gullick WJ. A Monoclonal antibody to the human c-erbB3 protein stimulates the anchorage-independent growth of breast cancer cell lines. Br J Cancer 1994; 70:459-465.

37. Chen X, Levkowitz G, Tzahar E et al. An immunological approach reveals biological differences between the two NDF/heregulin receptors, erbB-3 and erbB-4. J Biol Chem 1996; 271:7620-7629.

38. Lamaze C, Schmid SL. The emergence of clathrin-independent pinocytic pathways. Curr Opin Cell Biol 1995; 7:573-580.

39. Sorkin A, Waters CM. Endocytosis of growth factor receptors. Bioassay 1993; 15:375-382.

40. Bargmann CI, Hung M-C, Weinberg RA. Multiple independent activations of the neu oncogene by a point mutation altering the transmembrane doemin of P185. Cell 1986; 45:649-657.

41. Stern DF, Kamps MP, Cao H. Oncogenic activation of p185^{neu} stimulates tyrosine phosphorylation in vivo. Mol Cell Biol 1988; 8:3969-3973.

42. Damke H, Baba T, van der Bliek AM et al. Clathrin-independent pinocytosis is induced in cells overexpressing a temperature-sensitive mutant of dynamin. J Cell Biol 1995; 131:69-80.

43. Graus-Porta D, Beerli RR, Daly J et al. erbB-2, the preferred heterodimerization partner of all erbB receptors, is a mediator of lateral signaling. EMBO J 1997; in press.

44. Huang SS, Koh HA, Konish Y et al. Differential processing and turnover of the oncogenically activated neu/erbB2 gene product and its normal cellular counterpart. J Biol Chem 1990; 265:3340-3346.

45. Karunagaran D, Tzahar E, Beerli RR et al. erbB-2 is a common auxiliary subunit of NDF and EGF receptors: implications for breast cancer. EMBO J 1996; 15:254-264.

46. Kishimoto T, Taga T, Akira S. Cytokine signal transduction. Cell 1994; 76:253-262.

47. DiGiovanna MP, Stern DF. Activation state-specific monoclonal antibody detects tyrosine phosphorylated p185neu/erbB-2 in a subset of human breast tumors overexpressing this receptor. Cancer Res 1995; 55:1946-1955.

48. Wildenhain Y, Pawson T, Blackstein ME et al. p185^{neu} is phosphorylated on tyrosine in human primary breast tumors which overexpress neu/erbB-2. Oncogene 1990; 5:879-833.

49. Alimandi M, Romano A, Curia MC et al. Cooperative signaling of erbB3 and erbB2 in neoplastic transformation and human mammary carcinomas. Oncogene 1995; 10:1813-1821.

CHAPTER 5

Neuroantibodies: The Use of Recombinant Antibody Expression in the Central Nervous System

Antonino Cattaneo, Patrizia Piccioli and Francesca Ruberti

Introduction

In this chapter we shall describe how ectopic antibody expression can be used to interfere with the function of molecules that are located in the extracellular environment of cells of the nervous system.

The concept of ectopic antibody expression is based on the exploitation of antibody genes, rather than antibody proteins. This concept was initially introduced to direct antibodies against molecules acting extracellularly in the nervous system,[1] but with the demonstration that intracellular targeting of antibodies can be achieved in mammalian cells,[2,3] the principle of ectopic antibody expression can be (and has been) extended to intracellular target antigens as well. This chapter will not cover this aspect of the technology, however, as it is extensively covered in the rest of the volume, as well as elsewhere.[4]

Antibodies are normally secreted by plasma cells. It is now well established that virtually all of the non-lymphoid cells tested, both of animal and of plant origin, can support the secretion of functional immunoglobulins. However, as already demonstrated in the initial study,[1] the efficiency of antibody secretion by different cell types varies dramatically, by orders of magnitude. In that study it was found, in particular, that the efficiency of antibody secretion by cells related to the nervous system, both of neuronal and of glial origin, is very high and is comparable to that of lymphoid cells transfected with the same antibody genes. Thus, it was proposed[1] that the local secretion (by cells of the nervous system) of specific monoclonal antibodies, cloned from the corresponding hybridoma cell lines, could be utilized to perform functional and developmental studies in the otherwise intact nervous system of transgenic mice (the so called neuroantibody approach).[1,5,6] The advent of phage technology, with new ways of isolating, selecting and engineering cloned recombinant antibodies, greatly enriches the potential of this experimental approach.[7]

Intrabodies: Basic Research and Clinical Gene Therapy Applications, edited by Wayne A. Marasco. © 1998 Springer-Verlag and R.G. Landes Company.

How to get antibodies to be produced across the blood-brain barrier by cells of the nervous system? Different strategies for tackling the problem can be pursued, depending on the application envisaged, and including the production of transgenic mice, the grafting of cells engineered to secrete antibody genes and the direct infection of neural cells with viral vectors harboring the antibody genes.

While the transgenic approach may find applications in a research context or in the creation of experimental models for neurological pathologies, the local expression of recombinant antibodies by cells of the nervous system may provide an experimental scenario of some therapeutical potential, as it would circumvent the problem that the blood-brain barrier poses to the accessibility of a circulating antibody to the central nervous system (CNS). This represents a major hurdle limiting the application of therapeutic antibodies to CNS diseases as a therapeutic antibody, potentially useful for some CNS disorder, would not have access to its target in the CNS, if delivered systemically. Strategies to achieve the delivery of the corresponding gene for local expression across in the CNS, such as the grafting of cells engineered to secrete antibodies or the use of viral vectors, would therefore be useful.

In conclusion, the basis of the neuroantibody technique is to harness the efficient secretion of antibodies by glial and neuronal cells, and to use antibodies as a local immunological "knife", to create neurological lesions in the CNS.

Neuroantibodies: Studies with Transgenic Mice

Selective lesioning techniques are at the heart of functional neuroscience research. The availability of gene transfer techniques, and in particular the ability to create lines of transgenic mice, has opened new possibilities to study the physiology of the nervous system. In particular, "loss-of-function" mutations can be created in mice by different strategies and study of the response of the perturbed system should provide insights into the function of the complex system. The ectopic expression of antibodies is a recent addition to the different methods presently being used to inhibit the function of selected genes in mammalian organisms. The first transgenic studies with antibodies are now appearing,[8,9] and therefore their merits and problems may be evaluated.

The CNS, because of its complexity and its highly organized cytoarchitecture, represents an attractive target for the use of antibodies as perturbing agents, in interfering with the action of extracellularly acting molecules. Indeed, the importance of extracellular signaling molecules is paramount in the CNS, as is the geometry of the system, and the possibility of exploiting the richness of the antibody repertoire to modulate in a controlled way the function of these molecules would be very fruitful. Functional and developmental studies on the mammalian nervous system would greatly benefit from the ability to specifically interfere with selected neuronal sub-populations or pathways, by perturbing the action of extracellular or extracellularly exposed molecules, to produce functional lesions in a controlled fashion. Synaptic pathways may conceivably be facilitated, for instance by antibodies against molecules such as acetyl-cholinesterase, which limit the synaptic action of neurotransmitters.

The first attempts to pursue this strategy have relied on the creation of transgenic mice lines carrying a transgene encoding for the desired antibody, under the control of a suitable promoter.

Different choices in terms of promoters and enhancers will allow different questions to be addressed and different experimental models to be produced; the possible choices include:

1) strong promoters, without a particular tissue specificity, but permissive for expression in the CNS; one example is the cytomegalovirus early region promoter (CMV promoter);[10,11]

2) promoters and/or enhancers with a broad specificity for the CNS, and, in particular, for terminally differentiated neuronal or glial cells; these include the promoters of neuronally expressed genes such as neuronal specific enolase (NSE),[12] α-tubulin,[13] synapsin (R. Heumann, personal communication), neurofilament proteins,[14] Thy-1,[15] and PDGF-B chain.[16] Expression in glial cells may be achieved by using the promoters of glial specific genes such as the myelin basic protein (MBP)[17] or the glial fibrillar acidic protein (GFAP). All these have been used to direct the transcription of transgenes in the CNS, with varying degrees of efficacy and stringency;

3) promoters with a restricted regional specificity for CNS subregions, or cell types; expression in postmitotic neurons of the forebrain (and the hippocampus in particular) has been achieved using the promoter from the α-CaMKII gene,[18] while more restricted expression patterns of transgenes, such as in the retina or the cerebellum,[19] can be achieved with other promoters;

4) inducible promoters such as the tetracycline regulated[20] or the ecdysone regulated[21] systems;

5) the transcription regulatory sequences of the gene, whose product one might block with antibodies, could conceivably be used to drive the transcription of the corresponding antibody, so that all cells expressing a particular gene would also express an antibody directed against its gene product; and

6) activity-dependent transcription is an important, emerging theme in neurobiological studies, and promoters of genes transcriptionally activated by neuronal electrical activity could be exploited.

As for the antibody forms utilized, the transgenic experiments performed so far[8,9] have utilized full length antibodies. This is partly because the secretion of whole antibody molecules is much more efficient and certainly more predictable than that of scFv fragments. Indeed, the ability of individual scFv fragments to be efficiently secreted depends in a crucial but still unknown way on the primary sequence of the variable regions. The presence of two separate chains contributing to the antigen binding site may add some flexibility to the approach (see below), but there is no reason why other antibody forms could not be considered for transgenic studies in the future.

Transgenic Mice Expressing Anti NGF Antibodies

The neuroantibody approach is illustrated by a study in progress, in which transgenic mice expressing antibodies against the neurotrophic factor NGF (nerve growth factor) were produced.[9,22]

NGF, a member of the so-called neurotrophin gene family, regulates survival, differentiation and maintenance of specific neuronal populations, both in the peripheral and in the central nervous system, and was first characterized as a target-derived survival factor for developing sympathetic and sensory neurons.[23] Competition for limited amounts of target derived survival signals is thought to underlie the death of many central (CNS) and peripheral (PNS) neurons during development, and NGF is one of these.[23] Minute amounts of NGF control neuronal survival of specific cell populations during the critical period of development, when cellular death occurs in the nervous system. This classical paradigm of neurotrophic action by NGF has been extended to the central nervous system,[24] where striatal and basal forebrain cholinergic neurons have been identified as target cells for NGF. Moreover, the action of NGF and other neurotrophins, in the CNS, is broader than in the PNS, being related not just to neuronal survival during development, but also to the modulation of activity-dependent synaptic plasticity, in the adult, after the termination of development.[25] A very important role of NGF and of other neurotrophins as neuroprotective agents towards a variety of brain insults has also been reported. Moreover, NGF regulates the function of nociception in the postnatal life as well as in adulthood, and when administered to aged, cognitively impaired rats, NGF improves memory and other behavioral responses. These results suggested that NGF may have beneficial roles in a variety of neurodegenerative diseases, or neuropathological situations, and have prompted investigations into its possible therapeutical use.[26] Clinical trials using NGF, as well as other neurotrophic factors, for the treatment of different neurological disorders are underway,[27] but they mostly involve peripheral neuropathies, not only because these are more accessible to systemic administration of the factor, but also because the disorders are better characterized. On the contrary, the translation of animal experiments, involving NGF and other neurotrophins, to the treatment of human CNS diseases is still in its infancy and much more evidence is needed to strengthen the case for a therapeutic use of NGF. The lack of animal models, in which the actions of endogenous levels are competed in a chronic way in the adult CNS, is probably the main hindrance to progress in the field and there is therefore a great need of reproducible models to approach the problem. The knock-out by homologous recombination of the genes for NGF[28] and for its receptor TrkA[29] have been reported; however, in both cases the mice die after the first postnatal weeks, because of severe developmental defects arising from failure of NGF to act on its well-known target cells. Thus, these knock-out models do not allow us to study questions related to the role of NGF in the adult nervous system.

We decided, therefore, to exploit the neuroantibody approach, and to produce transgenic mice expressing a neutralizing antibody against NGF, that would be amenable to study the wide spectrum of activities of NGF. The aims of this work were, at first, to demonstrate that the expression of transgenic antiNGF antibodies could successfully compete with endogenous NGF on classical targets such as sympathetic neurons, and second, to also achieve a ubiquitous expression in adult mice. For this reason, the early region promoter of the human cytomegalovirus was used to direct the expression of the recombinant antiNGF.

The starting point was the rat monoclonal antibody α-D11,[30] which neutralizes very efficiently the biological action of NGF, both in vitro[30] and in vivo,[31,32] by a direct competition between NGF itself and the TrkA receptor. αD11 recognizes[33] the NGF loop region from residues 41-49, which forms part of the interaction surface of NGF with its receptor TrkA, and distinguishes NGF from the other members of the neurotrophin family. Consistently, mAb αD11 does not crossreact with the other members of the neurotrophin family BDNF, NT3 and NT4.[33] These characteristics of the mAb αD11 make this antibody an ideal reagent for in vivo studies.

The cloned variable regions of the α-D11 antibody[34] were linked to human γ1 constant regions, to facilitate the detection of the transgenic antibody against the back-ground of the endogenous mouse immunoglobulins, and placed under the transcriptional control of the early region promoter of the human cytomegalovirus. Mice transgenic for both transgenes (heavy and light chain genes) or for each individual transgene were derived.

Double transgenic founder mice express functional anti-NGF chimeric antibodies in the serum. Moreover, analysis of their superior cervical ganglia showed a marked immunosympathectomy, thus proving that the transgenic antibodies are effective in competing with endogenous NGF, at least in one classical target, sympathetic neurons. However, these mice did not mate and reproduce and a double transgenic line could not be obtained. On the other hand, transgenic lines expressing only the heavy or the light chain of the αD11 could be derived and were intercrossed to produce, in a two-tiered approach, transgenic mice which express functional anti-NGF antibodies. The advantage in this approach derives from the possibility of obtaining stable lines of mice in which the expression of the transgene should not produce an NGF-deprived phenotype. In this way, two families of αD11 transgenic mice were derived (CMV-αD11-A and CMV-αD11-B), both expressing functional anti-NGF antibodies in the serum. The amount of circulating anti NGF transgenic antibodies was in the range of 5-10 ng/ml for mice of family A, and of 50-100 ng/ml for family B. The difference in antibody levels, between family A and family B, is due to a difference in the levels of light chain mRNA (and hence protein) expression. High levels of antiNGF antibodies could be demonstrated in the brain of family B mice (but not of family A mice), with a widespread spatial distribution throughout the CNS. This demonstrates that the heavy and light chain are coexpressed in the same set of cells. The expression of the αD11 antibody chains in the CNS of transgenic mice, studied by immunocytochemistry, is illustrated in Figure 5.1. The overall picture is that of an abundant staining of many discrete areas throughout the nervous system. The staining is mainly neuronal and is distributed throughout the cell extensions, including their cellular processes and arborizations.

Interestingly, the αD11 transgenic antibody is expressed in a developmentally regulated manner—the expression of the mRNAs for both the heavy and the light αD11 chains being about 5-fold higher at postnatal day 90 than at birth, in different tissues. At the protein level, the αD11 antibody levels in the brain are one or two orders of magnitude higher in the adult than at birth (for mice of Family B). This pattern of expression is very convenient for an anti NGF transgenic model.

The transgenic antibodies compete successfully with the action of endogenous NGF. Even in Family A transgenic mice, which express lower circulating αD11 antibody levels, the Superior Cervical ganglion (SCG) was found to be visibly smaller than that of controls, with a greatly reduced number of sympathetic neurons, reflecting cell death induced by the neutralization of endogenous NGF.

In conclusion, the production of transgenic mice which express functional anti NGF antibody was achieved by a two-tiered approach, the intercrossing of transgenic mice which express the heavy and the light chain of the αD11 antibody. This transgenic model provides a proof of principle for the activity of transgenic antibodies, and is now amenable to study different aspects of the biological roles of NGF in the adult PNS and CNS, which cannot presently be easily addressed with other transgenic models (Ref. K.O.). This model, however, represents only a first

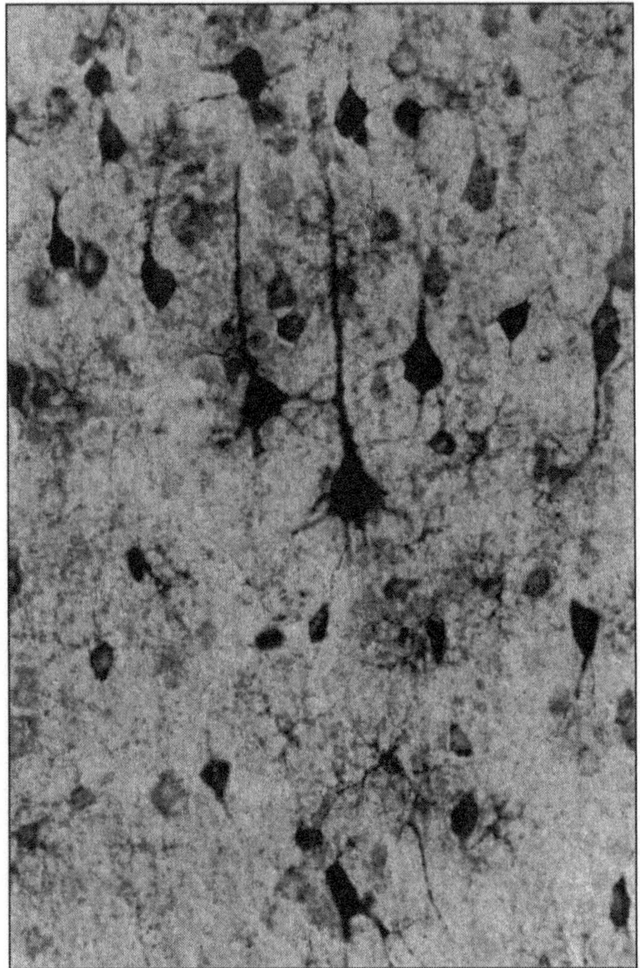

Fig. 5.1. Neuronal expression of αD11 heavy chain in the cortex of transgenic mice. Brain sections were stained with antibodies against human γ1 constant region.

step, and lends itself to further refinements, by restricting the spatiotemporal expression of the antibodies. The lines of transgenic mice, expressing heavy or light α-D11 chains in isolation, could be further exploited to obtain a more refined spatial and/or temporal control of αD11 expression. This will be achieved by mating mice transgenic for one of the two chains to new transgenic lines expressing the cognate chain under the control of a more restricted tissue specific or of an inducible promoter. Also, the cognate antibody chain may be delivered by the use of a viral vector (Fig. 5.2). This will provide a spatially and/or temporally more restricted expression of functional antibody. Expression of one or both antibody chains driven by a promoter specific for expression in mammary glands (such as for instance the promoter of the goat β-casein gene) will allow a high expression of the recombi-

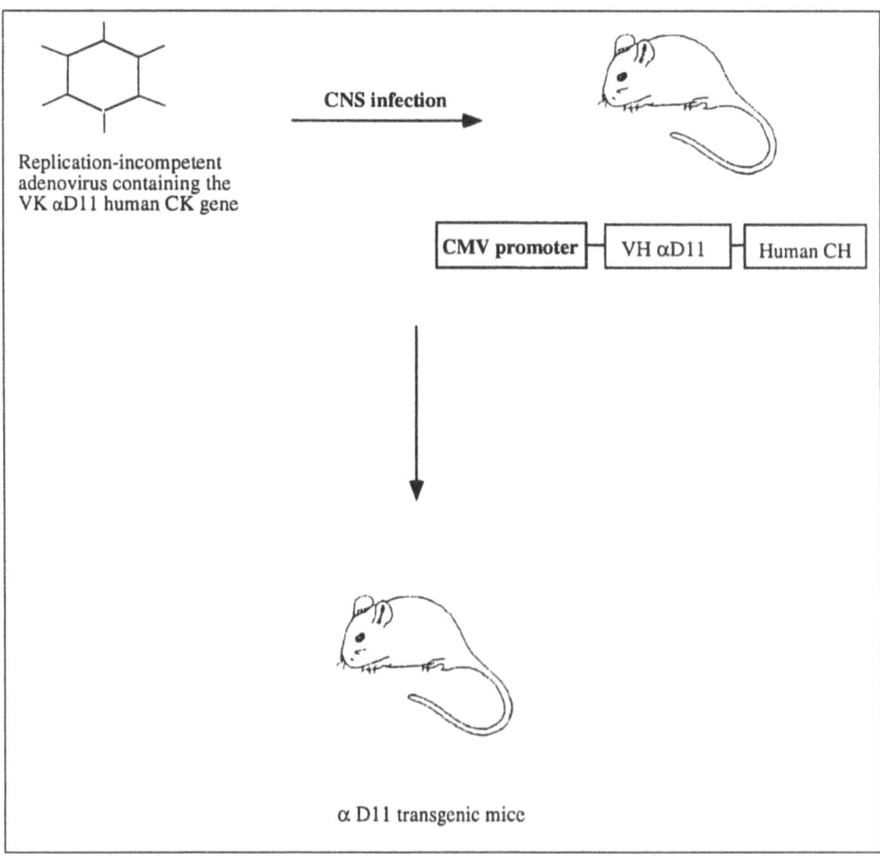

Fig. 5.2. Spatially restricted functional αD11 antibody could be achieved by viral-mediated delivery of the αD11 light chain into the CNS of mice expressing the heavy chain.

nant antibody in the milk of lactating transgenic mice. This would allow us to achieve a time controlled delivery of antibodies to neonate mice, through a time controlled exposure of mice pups to transgenic lactating mothers. During this period, the blood-brain barrier is not yet fully developed in mice and the antibodies would have access to the developing nervous system.

Transgenic Mice Expressing Antibodies Against the Neuropeptide Substance P (SP)

Another application of the neuroantibody experimental strategy concerns the production and characterization of transgenic mice expressing neutralizing antibodies against the neuropeptide substance P (SP).[8] This study provided the first demonstration of a transgenic model in which antibodies expressed in the CNS interfere functionally with the corresponding target molecule.[8]

SP is a peptide belonging to the tachykinin family, that has been associated with the transmission of sensory and nociceptive information in the spinal cord (see ref. 35 for a review). Antidromic release of SP from sensory afferents is responsible for vasodilatation and plasma extravasation in neurogenic inflamma-tion. SP is also present in CNS areas such as the substantia nigra and hypothala-mus, where its function is unknown, and in peripheral neuronal structures, where it is involved in regulating smooth muscle motility. The actions of SP during devel-opment are still largely unknown. Recent data suggest a role for SP in the innerva-tion of the developing spinal cord floor plate.[36]

The rat monoclonal antibody NC1/34HL[37] binds to the amidated C-terminal portion of the SP peptide, responsible for receptor binding, and does not crossreact with the related tachykinin peptides NKA and NKB. To engineer the expression of the chimeric NC1/34HL antibody in transgenic mice, the cloned variable regions of the NC1 mAb were placed under the control of the promoter of the *neuroendo-crine vgf* gene.[38] The *vgf* gene is expressed in cells of neuronal and endocrine ori-gin, in regions such as cerebral cortex, thalamus, hypothalamus, adenohypophysis, spinal cord and dorsal root ganglia, as well as in the adrenal medulla, and this promoter region has been characterized.[39] Many of these regions show SP and/or NK1 receptor expression as well.

Transgenic mice expressing the NC1 chimeric antibody under the transcrip-tional control of the *vgf* promoter are viable and have a normal life span. Their overall phenotype is rather mild (see below). The expression of the heavy and light NC1 antibody chains parallels that of the endogenous *vgf* gene, and peaks in the second postnatal week, after which it remains at fairly high levels into adulthood. Both antibody chains are abundantly present in many discrete areas throughout the nervous system, with a predominantly neuronal localization. Most important, from the point of view of the neutralization of SP action(s), the antibodies are present in many CNS regions where the SP peptide is also present. Functional an-tibodies were present both in the serum and in the brain of transgenic mice, dem-onstrating assembly (and therefore coexpression) of the two antibody chains. The adrenal medulla is the probable source for the transgenic antibodies found in se-rum. The levels of SP binding transgenic antibodies, in serum and brain, increase after birth up to a maximum around the second to third postnatal week, with an average value of 60 ng/ml in serum and of 700 ng/100mg in brain. These values are comparable to those obtained for the antiNGF mice.

In order to verify, from a functional point of view, whether the transgenic anti-bodies affect some of the different systems in which the SP peptide is involved, we tested in these mice acute nociceptive behavior (tail immersion and hot plate test), neurogenic inflammation upon mustard application on the skin, and motor activ-ity and exploratory behavior in the open field test. For practical reasons, the be-havioral tests were performed on adult mice, even if the levels of the antibody are lower in adult than in younger animals.

The reaction time in the hot plate test and the tail withdrawal latencies did not differ in anti-SP transgenic mice with respect to age matched control mice. On the contrary, mustard-oil induced plasma protein extravasation was greatly reduced in mice expressing the substance P antibody. Tests for motor activity showed a marked inhibition of locomotory activity in the transgenic mice.

The lack of inhibition in the acute nociceptive behavioral test is in line with the postulated role of neurokinin peptides in chronic, rather than in acute pain.[40] On the other hand, the primary role of SP in neurogenic inflammation and in smooth muscle contraction and vasodilation is well established,[41] and the observed inhibi-tion of neurogenic inflammation in the transgenic mice confirms the validity of the transgenic model. As for the central nervous system, where the SP peptide and its receptor are widely distributed, their functional role(s) are far from clear. It has been previously shown that central injections of substance P induce locomotion activity.[42,43] Consistently, transgenic mice expressing anti-substance P antibodies show a decreased locomotion activity and exploration behavior. The precise site of action of the antibodies in causing these motor deficits remains to be determined. Answers to this question will be provided by the study of lines of transgenic mice with a different transcriptional control of the transgenic antibodies.

In conclusion, the local expression in the CNS of recombinant antibodies directed against the neuropeptide substance P, under the transcriptional control of the promoter of the neuronal *vgf* gene, is functionally effective in neutralizing the actions of the neuropeptide.

The neuroantibody approach allows the creation of different experimental models for the same target antigen of interest, by changing the spatiotemporal expression patterns of the transgenic antibody, according to the promoter utilized for the transcriptional control of the transgene.

The expression of the transgenic antibodies in the CNS of *vgf* anti-SP transgenic mice starts postnatally. In order to investigate the role of SP earlier in development, we compared the phenotype in the *vgf* anti-SP mice with that of mice in which the same antibody is expressed earlier in development. Toward this aim, we constructed another set of transgenic mice[44] in which the expression of the anti-SP antibodies is controlled by the promoter and transcriptional enhancer of the immunoglobulin heavy chain gene, which is transcriptionally active already during embryonic development. Even if the antibody is expressed in cells of the immune system, the circulating antibodies should have access also to the nervous system, since the blood-brain barrier only forms postnatally.

Of the double transgenic founder mice born after coinjection of the heavy and light chain plasmids, five founders were selected for further analysis (giving rise to families #1, #2, #6, #8 and #9).

In these mice, the transgenic antibody chain mRNAs are expressed in a tissue-specific fashion, being found in spleen and thymus, but not in brain, kidney or other tissues. Chimeric anti-SP antibodies are found in the serum at all the postnatal ages tested, and their levels are similar to those of the *vgf* anti-SP mice.

Unlike the *vgf* anti-SP mice, Ig anti-SP mice show a very high level of perinatal mortality, as summarized in Figure 5.3. This was particularly severe for all mice, except for those of family #8, which show a longer life span: 75% of family #1 mice die before postnatal day 10 (P10), while this value is around 50% for mice from families #2 and #9, and only 5% for mice from family #8. Thus, a very small percentage of Ig anti-SP transgenic mice live for more than two months. This is in sharp contrast with the mild phenotype observed with the *vgf* anti-SP mice.[8] The reasons for the mortality are presently under investigation. Transgenic mice which do survive after the first postnatal weeks are much smaller than age-matched controls.

The question arises as to why the Ig anti-SP mice show such a dramatic lethal phenotype, while the *vgf* anti-SP do not, even if, in the latter, the transgenic antibodies are effective in blocking some of the actions of the SP neuropeptide in adults.[8] The levels of transgenic anti-SP antibodies in the serum of *vgf* anti-SP mice are very similar to those of the Ig anti-SP mice, at all the postnatal ages tested. Thus, the different phenotypes observed in the two transgenic lines are not due to

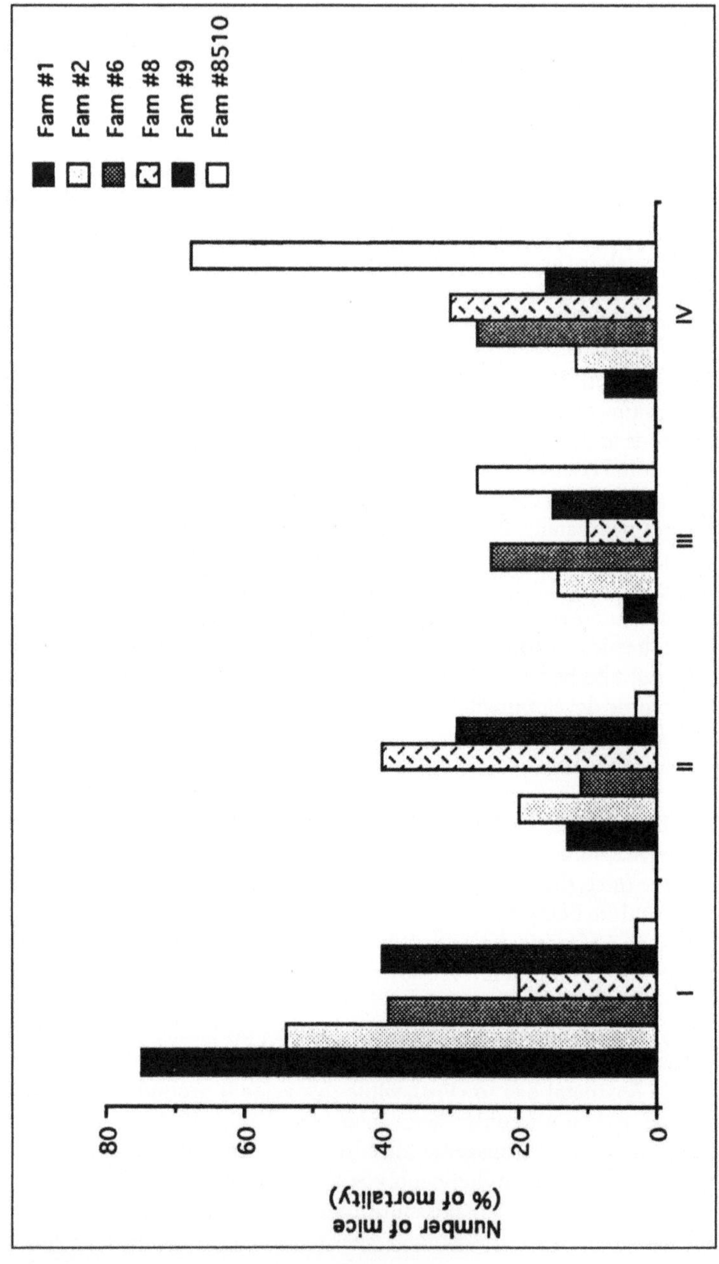

Fig. 5.3. Postnatal lethality in Ig anti-SP transgenic mice. The percentage of mice dying within the indicated time periods (I=p1-p10; II=p10-2 months; III=2 months-1 year; IV= > 1 year) is reported for different transgenic families. Fam 1: X=19, N= 93; Fam 2: X= 11, N=70; Fam 9; X=7, N=74; Fam 6: X=7, N=54; Fam 8: X= 5, N= 50; #8510 (*vgf* anti-SP mice[8]): X= 5, N=48. p0 (postnatal day 0) is the day of birth, X = number of crossings, N= number of mice at p0.

differences in the postnatal expression levels of the circulating antibodies. On the other hand, the developmental onset of expression of the transgenic antibodies is different in the two lines of mice. Very high levels of anti-SP antibodies are found in embryos (embryonic age E16-E19) of all Ig anti-SP mice (except for those of family #8), while no significant amount of anti-SP chimeric antibody can be detected in *vgf* anti-SP embryos, nor in family #8 Ig anti-SP embryos (Fig. 5.4). Thus, there appears to be a direct correlation between the presence of high levels of transgenic antibodies between embryonic days E16 and E19 and the subsequent fate of the litters: high perinatal lethality for family #9 Ig anti-SP mice, less than 5% perinatal lethality for family #8 Ig anti-SP and for *vgf* anti-SP mice.

In adult *vgf* anti-SP mice, the expression of the transgenic antibodies induces a marked inhibition of neurogenic inflammation and of general motor behavior.[8] In mice from the family #8 Ig anti-SP, which do not show a lethal phenotype, neurogenic inflammation is also inhibited, showing that the antibodies are effective in

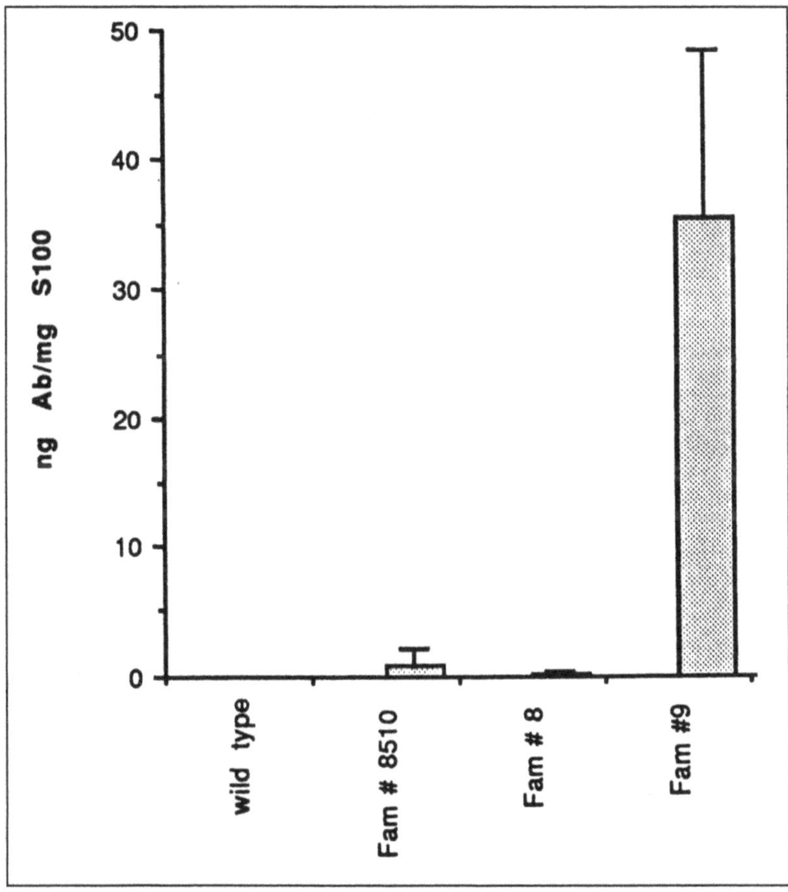

Fig. 5.4. Embryonic expression of the recombinant anti-Substance P antibody in transgenic mice. The levels of recombinant anti-SP antibodies in transgenic embryos were determined by RIA. Embryos were collected at E16-E17.

counteracting the actions of the neuropeptide in adult mice from this family as well. On the other hand, motor activity is not influenced in family #8 Ig anti-SP mice, suggesting that the inhibition observed for the *vgf* anti-SP mice is due to antibodies acting in the CNS.

These results point to a fundamental, yet still unknown, role of the neuropeptide SP during embryonic development. In principle, the lethal phenotype observed in the Ig anti-SP mice may be caused by the antibody-mediated disruption of the interactions of SP with its receptor, the so-called NK1-R receptor, or with alternative receptors. Interestingly, recently derived transgenic mice in which the gene for the substance P receptor NK1-R has been disrupted by gene targeting[45] show a normal development, are healthy and do not show the lethal phenotype displayed by the Ig anti-SP transgenic mice. Therefore, it may be inferred that the lethal phenotype in these mice is not due to block of the interaction of SP with the NK1-R receptor. While many, well-characterized actions of SP are mediated by the NK1-R receptors, actions not mediated by this receptor have been described, both in neuronal[46-48] and non neuronal[49] cells. It thus appears that the Ig anti-SP transgenic model may have uncovered an important developmental role of the SP neuropeptide, not mediated by the NK1-R receptor.

To summarize, the results obtained with the transgenic mice expressing antibodies against the neuropeptide SP, as well as with mice expressing antibodies against NGF, provide proofs of principle of the validity of the neuroantibody approach. Changing the spatiotemporal expression of a recombinant antibody, by the use of different promoters, can lead to very different transgenic models. These may lend themselves to study different physiological aspects of one given target molecule.

These transgenic studies also provide the background information necessary to evaluate the potential of locally secreted antibodies to interfere with their corresponding antigens active in the brain parenchyma, upon grafting of antibody secreting cells (see below).

Refining the Mode of Action of Neuroantibodies

In the transgenic studies performed so far, antibodies are constitutively secreted from neuronal cells and act as dominant inhibitors, by competing with the extracellular pool of their target antigens. An intracellular interaction between antigen and antibody is possible in principle, if the two molecules are coexpressed in the same cell. This would make the inhibition more stringent, in not relying only on extracellular interactions, but also on intracellular ones. For NGF and SP, an intracellular interaction with the constitutively secreted antibodies is, however, unlikely, since both NGF and SP are processed into their mature form from larger precursor pre-pro-proteins. This processing occurs in specialized compartments of the secretory pathway, where the constitutively secreted antibodies have no access, and the unprocessed antigen is not necessarily recognized by the antibody. Both NGF and SP are secreted by neurons through the regulated secretory pathway. Secretory antibodies could be re-routed from the constitutive to the regulated secretory pathway, by appropriate targeting signals such as, for instance, that of pro-opiomelanocortin.[50] The signal for targeting to the regulated pathway appears to be dominant over that for the constitutive secretion. This was shown[51] by injecting in PC12 cells the mRNA from a hybridoma-producing an antibody against secretogranin I. The secretogranin I antibody was packaged into secretory granules and was released by regulated exocytosis.[51] Thus, a constitutive secretory antibody can be diverted to the regulated pathway of secretion by its protein-protein interaction with a regulated secretory protein.

Engineering the regulated secretion of antibodies in combination with expression controlled by inducible promoters should allow tight control of antibody secretion, if needed. This could apply, for instance, to the regulated secretion of immunotoxins (see below), in response to external stimuli of various nature.

Competition with a secreted antigen could, therefore, be approached by a combination of an "extracellular" mode of action, with an "intracellular" one (intracellular anchors), to block the antigen from within the cell.

If the strategy of blocking the secretion of the target molecule is pursued, this should exploit anchoring or retention sequences. This approach has been used to block the appearance of membrane proteins to the plasma membrane,[52-54] but has not yet been applied to blocking the secretion of a soluble protein. We have shown that SEKDEL-tagged antibodies can successfully prevent the secretion of various proteins, including the 7B2 chaperonin (A. Braks and AC, unpublished). Of note, in the "intracellular anchor" mode of action, the antibody does not have to be a neutralizing one. The only binding prerequisite is that the target antigen is expressed, in the ER, in a form recognizable by the antibody.

The morphological complexity of neuronal cells and, in particular, the spatial separation between functional compartments of these cells, opens a further challenge for the cell biologist interested in carefully controlling the secretion of antibodies from neurons: is it possible to achieve a localized secretion of antibodies by a neuronal cell, for instance at sites close to the synaptic cleft? In principle, secretion into the synaptic cleft could be achieved presynaptically or postsynaptically. We could envisage implementing the first possibility (presynaptic secretion) by engineering the targeting of antibodies to synaptic vesicles, possibly by constructing antibody fusions with proteins of the synaptic vesicles. The secretion from the postsynaptic end of a synapse (dendrites) could exploit the fact that some neural mRNAs are targeted to dendrites and that the machinery for local protein synthesis is present in dendrites.[55] Incorporating mRNA targeting sequences in the antibody expression vector, to direct the mRNA for the antibody to dendrites, could provide a further level of localization control and help obtaining a localized secretion of antibodies by neurons.

In more refined schemes, antibodies would be not just inhibitors, but could be engineered with effector functions, coupled to their antigen binding moiety. This would provide the antibodies with effector functions triggered by the binding to the corresponding antigen, as in the normal immune response. Thus, the logic of the immune system, according to which antibodies have a binding function, provided by the variable regions, coupled to an effector function that is activated upon binding the antigen, could, in principle, be adapted to the particular needs of ectopic expression (see below). One example of effector function, inducing antibody-mediated cell killing, is discussed below.

Bypassing the Blood-Brain Barrier: Grafting of Engineered Cells or Direct Gene Transfer

Besides the production of transgenic models, another approach that could be used to achieve the local production of antibodies across the blood-brain barrier, by cells of the nervous system, involves the transplantation of genetically engineered cells into the CNS or a direct delivery of antibody genes. Although this has not been done with antibody genes yet, it is in principle feasible, since it is an approach that is being pursued for the delivery to the CNS of other gene products. In fact, while the CNS still represents a difficult target for gene therapy, practical paradigms are being developed for a genetic approach toward therapy for a variety of CNS pathologies, that allow us to circumvent the physiological barrier posed by the blood-brain barrier.[56,57]

The compartmentalization and relative inaccessibility of the CNS has greatly limited the use of antibodies as prospective therapeutic reagents for CNS disorders. Facilitated access of antibodies to the CNS has been achieved by engineering antibody fusions to transferrin,[58] but it is clear that methods to achieve the local expression of antibodies within the CNS[1,2] across the blood-brain barrier, would greatly increase the prospective use of antibodies as candidate genes for CNS therapy. We have successfully used the transplant of antibody secreting hybridoma cells in the CNS to block the action of NGF in the visual system[31] and in the basal forebrain,[32] over a period of several weeks, with a single injection. This shows that the approach of transplanting antibody secreting cells can be very effective. Provided that more suitable cellular hosts are engineered to secrete recombinant antibodies, this represents a method with great potential for research and therapy.

Brain tissue grafts have been employed for many years as a tool to study aspects of developmental and functional neurobiology, to create experimental models or, more recently, to develop prospective therapies, for instance, as a replacement therapy for Parkinson's disease.[59] Grafted cells can exert a functional influence over the host brain by various mechanisms, including:

1) the release of growth and trophic factors, of deficient neurochemicals or hormones (paracrine model); and
2) the precise reinnervation and reformation of synaptic connections between the neurons of the graft and host brain.

More recently, grafted cells have been engineered to express and produce molecules of interest, such as a neurotrophic factor or an enzyme responsible for the synthesis of a missing neurotransmitter. Along the same lines, it is proposed that cells engineered to secrete a recombinant antibody could be grafted in the brain to achieve a localized but continuous infusion of antibodies.

This ex vivo approach is not different, in principle, to ex vivo approaches that are being pursued for various human diseases, including tumor therapy:[60] the appropriate (autologous or heterologous) cells are transduced in vitro, to express the therapeutic gene of interest, and transduced cells are (stereotaxically, in the case of the CNS) reintroduced in vivo, into the appropriate site of the nervous system.[61,62] The choice of the engineered donor cell to be grafted back into the CNS is obviously an important issue. In rat and primate models, genetically modified autologous fibroblasts or myoblasts, and genetically engineered xenogenic cells isolated within a permeability-selective polymer capsule, that restricts cell growth, have been successfully grafted, to achieve high and stable levels of foreign gene expression in the CNS. In many cases, experimental neurodegenerative models have been successfully cured by the transplant of engineered cells, and clinical trials involving the grafting of encapsulated genetically modified cells are underway.[63]

Recent advances in the biology of neural stem cells[64,65] open new exciting possibilities, that may have profound consequences for the treatment of neurological diseases, through ex vivo manipulation of such cells, before transplantation. Stem cells from the CNS of mouse, rat or human can be isolated from various regions of the developing CNS. These cells have the potential to differentiate into neurons, astrocytes and oligodendrocytes and to self-renew. In vitro, these cells divide in response to epidermal growth factor or other mitogens and, following removal of the mitogen and exposure to a suitable substrate, differentiate into neurons, astrocytes and oligodendrocytes.[66-69] The intermediate filament nestin is a major cytoskeletal protein in neuronal precursors in the mammalian CNS. Coincident with their exit from the cell cycle, neurons downregulate nestin. Also the adult nervous system contains multipotential precursors for neurons, astrocytes and oligodendrocytes. The current view on the lineal relationship between the major CNS cell types is illustrated in Figure 5.5. Both embryonic and adult CNS stem cells can proliferate extensively and differentiate in vitro and both types of stem cells have been transplanted into the developing brain, where they switch to the locally appropriate fate and generate region-specific neurons in response to local cues. However, adult stem cells, which differentiate efficiently in vitro, do so inefficiently when transplanted in the adult brain. This discrepancy is a critical but unresolved question for the field.

Neural stem cells could, in principle, represent suitable hosts for the secretion of gene products of interest, including antibodies. Stem cells immortalized with oncogenes have been shown to differentiate into neurons when grafted into the developing or adult brain. More recently, the field has shifted away from the use of oncogene-immortalized cells towards the grafting of primary cells expanded in vitro. Perhaps one of the greatest challenges at present is to characterize the various types of CNS stem cells driven by different mitogens in vitro, with regard to their proliferative capacity, differentiation potential and possibility of being transduced by (retro)viral infection or by transfection in vitro, prior to their grafting. Once these questions receive adequate answers, CNS stem cells could provide the preferred source of neural tissue for transplantation. Engineering neural stem cells to secrete antibodies is likely to become a very exciting possibility.

Direct delivery to the CNS, with viral vectors, is another possible strategy to achieve local secretion of antibodies, in a gene therapy perspective. This is a problem in common with the delivery of other therapeutic genes, potentially useful for CNS disorders.[56] The ideal vector system is not yet available and, anyway, will depend on the application envisaged. Retroviruses are ineffective for introducing genes in the mitotically quiescent cells of the CNS, but they are suitable for gene delivery into brain tumors.[60] A lentiviral vector, based on vesicular stomatitis virus G protein (VSV-G)-pseudotyped HIV-1 vector, has recently been described,[70] which is capable of stable gene transfer to CNS postmitotic neurons in vivo in adult rats, but its potential needs to be further evaluated. Gene delivery to peripheral and central neurons has been achieved with herpes simplex virus-type 1 and adenovirus vectors.[56,57,62] Recombinant adeno-associated virus (rAAV) vectors are presently being actively considered,[71] and may have some advantages over other systems currently used to deliver genes to the CNS. Among these, AAV vectors are capable of transducing and eliciting long-term stable expression from postmitotic cells and produce little inflammatory response. Any of these vectors could be used to express antibody genes in the CNS, with the same indications and caveats as for other genes.

In conclusion, the local expression of antibodies within the CNS, obtained by the grafting of engineered cells or by direct delivery of the corresponding genes, would allow us to circumvent the obstacle that the blood-brain barrier poses to

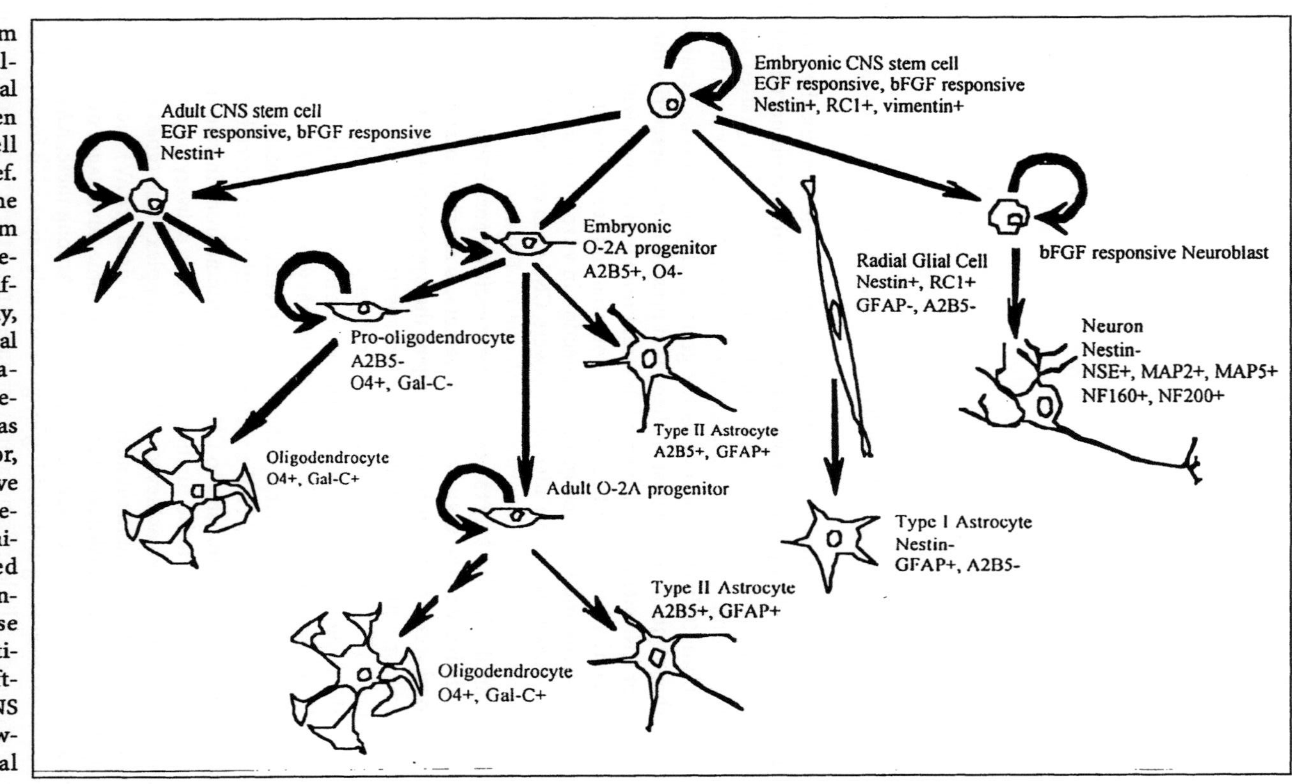

Fig. 5.5. Neural stem cells. Current idealized view of the lineal relationships between the major CNS cell types (taken from ref. 65). At the top is the embryonic CNS stem cell, which can self-renew and produce differentiated progeny, perhaps by a gradual restriction mechanism, where intermediate cell types such as the O-2A progenitor, or neuroblasts serve as the immediate precursors of the terminally differentiated cell types. Ex vivo engineering of these cells, to express antibodies, before grafting back into the CNS could provide a powerful experimental tool.

systemically delivered useful antibodies. The potential use of neural stem cells as cellular hosts for the secretion of antibodies is an exciting possibility. This may increase the range of targets for therapeutic antibodies for CNS disorders.

Antibody Expression in the CNS for Intercellular Immunization

Local Expression of Immunotoxins

As mentioned above, antibodies engineered for schemes of intercellular immunization can be equipped with effector functions that, combined with their binding properties, may transform them into more powerful reagents. Particularly attractive are effector functions that, following the binding event mediated by the antibody, lead to the death of the cell recognized by the antibody. The best example is that of immunotoxins,[72,73] fusion proteins in which antibodies are fused, or coupled, to toxic subunits. Thus, the specificity provided by the antibody variable regions can be harnessed to direct antibody-fusion toxins towards target cells and to achieve an antibody-mediated cell ablation. This type of strategy has been developed in the context of cancer therapy, and immunotoxins are normally delivered systemically, as recombinant proteins, purified from *Escherichia coli*. Due to the blood-brain barrier problem, immunotoxins have therefore not been considered so far for brain tumors. Two recent papers[74,75] describe how mammalian cells can produce and secrete targeted toxin proteins, while remaining viable. This was done by fusing a scFv fragment or a light chain (to be expressed with the corresponding heavy chain) to truncated *Pseudomonas* exotoxin A (PEA), encompassing the translocation and the toxic domains. The PEA toxin kills cells by inactivation of the elongation factor-2 in the cytosol, thereby inhibiting protein synthesis. These authors demonstrate that the antibody-toxin fusion can be secreted by mammalian cells, which remain viable and therefore show that it is feasible to produce, in vivo, targeted toxins to destroy specific cells. In one application of this strategy,[74] HIV-specific killer cells were generated by genetically modifying lymphocytes, and shown to have potent and selective cytotoxicity to the targeted HIV-infected cells in vitro. In another application,[75] tumoral cells were selectively killed by cocultivating them with cells expressing the toxin fused to an antibody directed against a membrane antigen of tumor cells.

Using antibodies secreted from one cell type, be it a grafted cell or a resident cell of the CNS, to deliver a toxic gene to another cell population, in an intercellular mode of communication, is a very powerful application of intercellular immunization in the CNS. This approach, which is based on the local ectopic expression of recombinant antibodies within the CNS, is likely to have many applications, not only in a prospective therapeutic settings but also in research. Immunotoxin proteins have been used as research tools in neurobiology, for immunolesioning studies.[73] Recombinant immunotoxins, locally expressed in the CNS, add a new dimension to these tools. The production of the recombinant immunotoxin may be potentially controlled at different levels, including regulated transcription and regulated secretion. In a more hypothetical scenario, engineering a highly localized secretion and/or uptake of the recombinant immunotoxin could be exploited to block local protein synthesis in dendrites, without affecting the global protein synthesis process at the level of the cell soma. If this was possible, one would have a very important research tool for studies of synaptic plasticity.

Linking Antibodies to Signal Transduction Pathways

Antibodies targeted to the plasma membrane, by fusion of the scFv fragment with transmembrane and cytoplasmic domains of receptors and signaling membrane proteins, have been shown to be functionally linked to the corresponding signal transduction pathways, activated by cell-bound[76-79] or a soluble[80] ligands. In this case, the effector function provided by the chimeric construction is the ability to participate in a signal transduction event triggered upon the binding of the corresponding antigen. The studies reported have not been developed with the nervous system in mind, but, again, the principle may lend itself to interesting applications to CNS studies.

In one approach[76-79] antibody variable regions have been combined to the constant regions of T cell receptor chains, resulting in chimeric genes endowing T lymphocytes expressing them with antibody-type recognition (i.e., non-MHC restricted). In this case, the effector function is provided by the activation of the T cell, following the binding event. For instance, antigen specific scFv fragments were linked to gamma or zeta chains, the signal-transducing subunits of T cell receptor and the chimeric proteins were expressed on the membrane of a cytolytic T cell. Upon antigen binding, these cells produced interleukin-2 and were able to lyse target cells in an antigen-dependent way. Such chimeric antibody fusions can be exploited to provide T cells and other effector lymphocytes with antibody-type recognition directly coupled to cellular activation. In principle, this type of scheme could be extended to the antigen-antibody mediated activation of different cell types, providing effector functions that may not be limited to target cell lysis. In fact, the principle of linking antibodies targeted to the plasma membrane to signal transduction pathways has been extended to the signal transduction pathway activated by the Met tyrosine kinase receptor.[80] A scFv-Met chimeric receptor was expressed on the surface of MDCK cells and upon antigen binding, cells responded by activation of the downstream signal transduction pathway of the Met receptor, i.e., tyrosine phosphorylation of the receptor, colony scattering and cell proliferation in soft agarose. In conclusion, engineering the membrane expression of antibodies may provide a way to control signal transduction events in target cells by endogenous or exogenous soluble or cell-bound ligands.

Conclusions

As discussed, prospective applications of therapeutic antibodies to the treatment of CNS disorders have been severely hampered by accessibility problems of systemically delivered antibodies. The efficient secretion of antibodies by glial and neuronal cells,[1,6] together with the demonstration that transgenic antibodies are effective in neutralizing the activity of the corresponding antigens,[8,9,22] provide the background for proposing the local production of antibodies within the CNS as a promising therapeutic strategy. Studies exploiting the transplant of antibody secreting cells have demonstrated that locally produced antibodies can successfully compete with endogenous molecules active in the extracellular environment of the CNS.[31,32] CNS stem cells could provide the preferred source of neural tissue for transplantation in the future and engineering neural stem cells to secrete antibodies is likely to become a very exciting possibility, adding antibody genes to the repertoire of candidate genes for neural gene therapy.

The local expression of antibodies in the form of immunotoxins may find applications for brain tumors, but other potentially very promising areas of application are represented by neurodegenerative diseases, such as Alzheimer's disease

(AD). The hallmarks of AD are neuritic plaques (extracellular deposits of the Aβ peptide) and neurofibrillary tangles (intracellular and extracellular structures consisting of twisted filaments of abnormally modified microtubule associated protein tau). Interfering with the formation or with the neurotoxic action(s) of these structures represents a major therapeutic target, in the AD field, and the neuroantibody technique, described here, may provide a new approach to the problem.

Acknowledgments

We would like to thank all our friends and colleagues in the lab for their support, help and discussions. This chapter was written while AC was on sabbatical leave at the MRC Laboratory of Molecular Biology (Cambridge, UK). AC is very grateful to Terry Rabbitts and Greg Winter, Joint Heads of the PNAC Division of the LMB, for this exciting opportunity.

Work in the authors' laboratory was funded by MURST, CNR and HFSPO (Grant 93/93).

References

1. Cattaneo A, Neuberger MS. Polymeric immunoglobulin M is secreted by transfectants of nonlymphoid cells in the absence of immunoglobulin J chain. EMBO J 1987; 6:2753-2758.
2. Biocca S, Neuberger MS, Cattaneo A. Expression, targeting of intracellular antibodies in mammalian cells. EMBO J 1990; 1:101-108.
3. Biocca S, Cattaneo A. Intracellular immunization: antibody targeting to sub-cellular compartments. Trends Cell Biol 1995; 5:248-252.
4. Cattaneo A, Biocca S (eds.). Intracellular Antibodies. Ectopic Antibody Expression for Intracellular and Intercellular Immunization. Austin TX: Landes Bioscience, 1997.
5. Cattaneo A. Neuroantibodies: expression of specific monoclonal antibodies in cells of the nervous system. Ann Ist Super. Sanita' 1988; 24:531-535.
6. Piccioli P, Ruberti F, Biocca S et al. Neuroantibodies: molecular cloning of a monoclonal antibody against substance P for expression in the central nervous system. Proc Natl Acad Sci USA (1991); 88:5611-5615.
7. Winter G, Griffiths AD, Hawkins RE et al. Making antibodies by phage display technology. Ann Rev Immunol 1994; 12:433-455.
8. Piccioli P, Di Luzio A, Amann R et al. Neuroantibodies: ectopic expression of a recombinant anti-substance P antibody in the central nervous system of transgenic mice. Neuron 1995; 15:373-384.
9. Cattaneo A, Ruberti F. Transgenic mice expressing NGF antibodies during postnatal life. Soc Neurosci Abstracts 1996; 22, Part I: 301.18.
10. Boshart MF, Weber G, Jahn K et al. A very strong enhancer is located upstream of an immediate early gene of human cytomegalovirus. Cell 1985; 41:521-530.
11. Baskar JF, Smith PP, Nilaver G. The enhancer domain of the human cytomegalovirus major immediate early promoter determines cell type specific expression in transgenic mice. J Virol 1996; 70:3207-3214.
12. Forss-Petter S, Danielson PE, Catsicas S et al. Transgenic mice expression of β-galactosidase in mature neurons under neuron-specific enolase promoter control. Neuron 1990; 5:187-197.
13. Gloster A, Wu W, Speelman A et al. The T alpha-tubulin promoter specifies gene expression as a function of neuronal growth and regeneration in transgenic mice. J Neurosci 1994; 14:7319-7330.

14. Reeben M, Halmekyto M, Alhonen L et al. Tissue-specific expression of rat light neurofilament promoter-driven reporter gene in transgenic mice. Biochem Biophys Res Comm 1993; 192:465-470.

15. Vidal M, Morris R, Grosveld F et al. Tissue specific control elements of the Thy 1 gene. EMBO J 1990; 9:833-840.

16. Sasahara M, Fries JW, Fujita T et al. PDGF B-chain in neurons of the central nervous system, posterior pituitary and in a transgenic model. Cell 1991; 64:217-227.

17. Readhead C, Popko B, Takahashi N et al. Expression of a myelin basic protein gene in transgenic sheeverer mice: correction of the dysmyelinating phenotype. Cell 1987; 48:703-712.

18. Mayford M, Wang J, Kandel E et al. CaMKII regulates the frequency-response function of hippocampal synapses for the production of both LTD and LTP. Cell 1995; 81:891-904.

19. Oberdick J, Smeyne RJ, Mann JR et al. A promoter that drives transgene expression in cerebellar Purkinje and retinal bipolar neurons. Science 1990; 248: 223-226.

20. Furth PA, St Onge L, Boger H et al. Temporal control of gene expression in transgenic mice by a tetracycline-responsive promoter. Proc Natl Acad Sci USA 1994; 91:9302-9306.

21. No D, Yao TP, Evans RM. Ecdysone-inducible gene expression in mammalian cells and transgenic mice. Proc Natl Acad Sci USA 1996; 93:3346-3351.

22. Ruberti F. Use of anti-NGF monoclonal antibodies to study synaptic plasticity in the developing and adult hippocampus. PhD thesis, SISSA 1966.

23. Levi-Montalcini R. The nerve growth factor 35 years later. Science 1987; 237:1154-1162.

24. Thoenen H. The changing scene of neurotrophic factors. TINS 1991; 14:165-170.

25. Thoenen H. Neurotrophins, neuronal plasticity. Science 1995; 270:593-598.

26. Lindsay RM, Wiegand SJ, Altar CA et al. Neurotrophic factors: from molecule to man. TINS 1994; 17:182-190.

27. Schaetzl HM. Neurotrophic factors: ready to go? TINS 1995; 18:463-464.

28. Crowley C, Spencer SD, Nishimura MC et al. Mice lacking nerve growth factor display perinatal loss of sensory and sympathetic neurons yet develop basal forebrain cholinergic neurons. Cell 1994; 76:1001-1011.

29. Smeyne RJ, Klein R, Schnapp A et al. Severe sensory and sympathetic neuropathies in mice carrying a disrupted Trk/NGF receptor gene. Nature 1994; 368:246-251.

30. Cattaneo A, Rapposelli B, Calissano P. Three distinct types of monoclonal antibodies after long term immunization of rats with mouse NGF. J Neurochem 1988; 50:1003-1010.

31. Berardi N, Cellerino A, Domenici L et al. Monoclonal antibodies to nerve growth factor affect the postnatal development of the visual system. Proc Natl Acad Sci USA 1994; 91:684-688.

32. Molnar M, Ruberti F, Domenici L et al. A critical period in the sensitivity of basal forebrain cholinergic neurons to NGF deprivation. Neuroreport 1997, in press.

33. Gonfloni S. Recombinant antibodies as structural probes for neurotrophins. PhD thesis, SISSA 1995.

34. Ruberti F, Bradbury A, Cattaneo A. Cloning and expression of an anti-NGF antibody for studies using the neuroantibody approach. Cell Mol Neurobiol 1993; 13:559-568.

35. Otsuka M, Yoshioka K. Neurotransmitter functions of mammalian tachykinins. Physiol Rev 1993; 73:229-307.

36. De Felipe C, Pinnock RD, Hunt SP. Modulation of chemotropism in the developing spinal cord by substance P. Science 1995; 267:899-902.

37. Cuello C, Galfre G, Milstein C. Detection of substance P in the central nervous system by a monoclonal antibody. Proc Natl Acad Sci USA 1979; 76:3532-3536.

38. Levi A, Eldridge J, Paterson BM. Molecular cloning of a gene sequence regulated by nerve growth factor. Science 1985; 229:393-395.

39. Possenti R, Di Rocco G, Nasi S et al. Regulatory elements in the promoter region of vgf, a nerve growth factor-inducible gene. Proc Natl Acad Sci USA 1992; 89: 3815-3819.

40. Dray A, Urban L, Dickenson A. The pharmacology of chronic pain. TiPS 1994; 15:190-197.

41. Lembeck F, Donnerer J, Tsuchiya M et al. The nonpeptide tachykinin antagonist CP-96,345, is a potent inhibitor of neurogenic inflammation. Br J Pharmacol 1992; 105:527-530.

42. Kelley AE, Iversen SD. Substance P infusion into the substantia nigra of the rat: behavioral analysis and involvement of striataldopamine. Eur J Pharmacol 1979; 60:171-179.

43. Naranjo JR, Del Rio J. Locomotor activity induced in rodents by substance P and analogues. Neuropharmacol 1984; 23:1167-1171.

44. Piccioli P, Cattaneo A. Lethal phenotype induced by recombinant anti-SP antibodies in transgenic mice. Submitted for publication, 1997.

45. Bozic CR, Lu B, Hoepken UE et al. Neurogenic amplification of immune complex inflammation. Science 1996; 273:1722-1725.

46. Clapham DE, Neher E. Substance P reduces acetylcholine-induced currents in isolated bovine chromaffin cells. J Physiol 1984; 347:255-277.

47. Simasko SM, Durkin JA, Wieland GA. Effect of substance P on nicotinic acetylcholine receptor function in PC12 cells. J Neurochem 1987; 49:253-260.

48. Stafford GA, Oswald RE, Wieland GA. The beta subunit of neuronal nicotinic acetylcholine receptors is a determinant of the affinity for substance P inhibition. Mol Pharmacol 1994; 45:758-762.

49. Kavelaars A, Broeke D, Jeurissen F et al. Activation of human monocytes via a nonneurokinin substance P receptor that is coupled to Gi protein, calcium, phospholipase D, MAP kinase and IL-6 production. J Immunol 1994; 153:3691-3699.

50. Cool DR, Fenger M, Snell CR et al. Identification of the sorting signal motif within pro-opiomelanocortin for the regulated secretory pathway. J Biol Chem 1995; 270: 8723-8729.

51. Rosa P, Weiss U, Pepperkok R et al. An antibody against secretogranin I (chromo-granin B) is packaged into secretory granules. J Cell Biol 1989; 109:17-34.

52. Marasco WA, Haseltine WA, Chen SY. Design, intracellular expression, and activity of a human anti-human immunodeficiency virus type 1 gp120 single-chain antibody. Proc Natl Acad Sci USA 1993; 90:7889-7893.

53. Beerli RR, Wels W, Hynes NE. Autocrine inhibition of the epidermal growth factor receptor by intracellular expression of a single-chain antibody. Biochem Biophys Res Comm 1994; 204:666-672.

54. Richardson JH, Sodroski JG, Waldmann TA et al. Phenotypic knockout of the high-affinity human interleukin 2 receptor by intracellular single-chain antibodies against the α subunit of the receptor. Proc Natl Acad Sci USA 1995; 92:3137-3141.

55. Steward O. mRNA localization in neurons: a multipurpose mechanism? Neuron 1997; 18:9-12.

56. Karpati G, Lochmuller H, Nalbantoglu J et al. The principles of gene therapy for the nervous system. TINS 1996; 19:49-54.

57. Fisher LJ, Ray J. In vivo and ex vivo gene transfer to the brain. Curr Opin Neurobiol 1994; 4:735-741.

58. Shin SU, Friden P, Moran M et al. Transferrin-antibody fusion proteins are effec-

tive in brain targeting. Proc Natl Acad Sci USA 1995; 92:2820-2824.

59. Lindvall O, Sawle G, Widner H et al. Evidence for long-term survival and function of dopaminergic grafts in progressive Parkinson's disease. Ann Neurol 1994; 35:172-180.

60. Culver KW, Blaese RM. Gene therapy for cancer. TIG 1994: 10:174-178.

61. Friedmann T. Gene therapy for neurological disorders. TIG 1994; 10:210-214.

62. Wolff JA. Postnatal gene transfer into the central nervous system. Current Opin Neurobiol 1993; 3:743-748.

63. Aebischer P, Schluep M, Deglon N et al. Intrathecal delivery of CNTF using encapsulated genetically modified xenogeneic cells in amyotrophic lateral sclerosis patients. Nature Med 1996; 2:696-699.

64. McKay R. Stem cells in the central nervous system. Science 1997; 276:66-71.

65. Stemple DL, Mahanhappa NK. Neural stem cells are blasting off. Neuron 1997; 18:1-4.

66. Cattaneo E, McKay R. Proliferation and differentiation of neuronal stem cells regulated by nerve growth factor. Nature 1990; 347:762-765.

67. Gage FH, Ray J, Fisher LJ. Isolation, characterization and use of stem cells from the CNS. Annu Rev Neurosci 1995; 18:159-192.

68. Weiss S, Reynolds BA, Vescovi AL et al. Is there a neural stem cell in the mammalian forebrain? TINS 1996; 19:387-393.

69. Svendsen CN, Rosser AE. Neurones from stem cells? TINS 1995; 18:465-467.

70. Naldini L, Blomer U, Gallay P et al. In vivo gene delivery and stable transduction of nondividing cells by a lentiviral vector. Science 1996; 272:263-267.

71. Peel AL, Zolotukhin S, Schrimsher GW et al. Efficient transduction of green fluorescent protein in spinal cord neurons using adeno-associated virus vectors containing cell type specific promoters. Gene Therapy 1997; 4:16-24.

72. Pastan I, FitzGerald D. Recombinant toxins for cancer treatment. Science 1992; 254:1173-1177.

73. Wiley RG. Neural lesioning with ribosome-inactivating proteins: suicide transport and immunolesioning. TINS 1992; 15:285-290.

74. Yang AG, Chen SY. A new class of antigen-specific killer cells. Nature Biotech 1997; 15:46-51.

75. Chen SY, Yang AG, Chen JD et al. Potent antitumour activity of a new class of tumour-specific killer cells. Nature 1997; 385:78-81.

76. Eshhar Z, Waks T, Gross G et al. Specific activation and targeting of cytotoxic lymphocytes through chimeric single chains consisting of antibody-binding domains and the gamma or zeta subunits of the immunoglobulin and T cell receptors. Proc Natl Acad Sci USA 1993; 90:720-724.

77. Lustgarten J, Eshhar Z. Specific elimination of IgE production using T cell lines expressing chimeric T cell receptor genes. Eur J Immunol 1995; 25:2985-2991.

78. Hwu P, Yang JC, Cowherd R et al. In vivo antitumor activity of T cells redirected with chimeric antibody/T cell receptor genes. Cancer Res 1995; 55:3369-3373.

79. Stancovski I, Schindler DG, Waks T et al. Targeting of T lymphocytes to Neu/HER2-expressing cells using chimeric single chain Fv receptors. J Immunol 1993; 151:6577-6582.

80. Heinrichs AAJ. Antibody display and selection on mammalian cells. Ph.D. thesis, University of Cambridge 1997.

CHAPTER 6

Intracellular Antibody-Mediated Knockout of the ErbB-2 Oncoprotein as a Cancer Gene Therapy Approach

David T. Curiel

Introduction

There is increasing recognition that cancer results from a series of accumulated, acquired genetic lesions. To a larger and larger extent, the genetic lesions associated with malignant transformation and progression are being identified. The recognition of, and definition of, the molecular basis of carcinogenesis makes it rational to consider genetic approaches to therapy. In this regard, a number of strategies have been developed to accomplish cancer gene therapy.[1-8] These approaches include: 1) mutation compensation; 2) molecular chemotherapy and 3) genetic immunopotentiation. For mutation compensation, gene therapy techniques are designed to rectify the molecular lesions in the cell having undergone malignant transformation. For molecular chemotherapy, methods have been developed to achieve selective delivery or expression of a toxin gene in cancer cells to achieve their eradication. Genetic immunopotentiation strategies attempt to achieve active immunization against tumor-associated antigens by gene transfer methodologies. Whereas the biology of each malignant disease target will likely dictate the approach taken, the overwhelming majority of human clinical gene therapy trials to date have been based upon the genetic immunopotentiation approach. For most tumor types, however, the absence of clear clinical evidence of anti-tumor effect has suggested the need for alternative approaches to achieve results. In addition, the lack of clear efficacy of any of the human studies to date argue for the development of additional, novel approaches for anti-cancer gene therapy.

The genetic lesions etiologic of malignant transformation may be thought of as a critical compilation of two general types: aberrant expression of "dominant" oncogenes or loss of expression of "tumor suppressor" genes. Gene therapy strategies have been proposed to achieve correction of each of these lesions. For approaching the loss of function of a tumor suppressor gene, the logical intervention

Intrabodies: Basic Research and Clinical Gene Therapy Applications, edited by Wayne A. Marasco. © 1998 Springer-Verlag and R.G. Landes Company.

is replacement of the deficient function with a wild-type tumor suppressor gene counterpart. This general strategy has been shown to allow phenotypic correction in vitro. For instance, Vogelstein et al have demonstrated that delivery of the wild-type p53 gene to transformed p53-deficient colonic carcinoma cells can abrogate the malignant phenotype.[9] In addition, similar studies carried out for other tumor suppressor genes and other tumor targets have further demonstrated the potential therapeutic effects achievable with re-establishment of wild-type tumor-suppressor gene function.[10-12] The concept has also been establishment in in vivo models. Roth et al have shown that delivery of the wild-type p53 gene via recombinant retrovirus or recombinant adenovirus, by direct in vivo delivery, can have a therapeutic effect in a murine model employing human lung cancer xenografts.[13] Importantly, it has been shown that despite the presence of multiple genetic lesions, the targeted rectification of only one of these is, in many instances, sufficient to revert the neoplastic phenotype.[10,11] This work has established the rationale for human clinical gene therapy trials designed to achieve mutation compensation in epithelial carcinomas of multiple sites, including lung, liver, and the head and neck.

For dominant oncogenes, it is the aberrant expression of the corresponding gene product which elicits the associated neoplastic transformation. In this context, molecular therapeutic interventions are designed to ablate expression of the dominant oncogene. The most universally employed methodology to achieve this is the utilization of "antisense" molecules (DNA or RNA oligonucleotides).[14-17] These molecules are designed to specifically target sequences to achieve blockade of the encoded genetic informational flow. Approaches have included the use of "triplex" DNA to achieve functional ablation of transcriptional activation through blockade of transcription factor binding sites. This has been used in in vitro model systems for targeting the c-myc,[18] ras,[19] and erbB-2 oncogenes.[20] Targeting has also been achieved at levels of gene expression distal to transcription. Specific antisense binding to transcribed RNA sequences may interrupt the flow of genetic information through several mechanisms, including degradation, impaired transport, and translational arrest. These interventions may be accomplished by simple antisense oligonucleotides, as well as by antisense molecules which possess catalytic activity to accomplish cleavage of target "sense" sequences.[19,21,22] A variety of experimental models have demonstrated the utility of the antisense approach as an anti-cancer therapeutic. Importantly, several studies have shown the ability to selectively ablate a dominant oncogene with reversion of the malignant phenotype.[18,19,22] In selected instances, the in vivo demonstration of this effect could also be accomplished by direct, in vivo delivery of the antisense molecules.[23,24] Thus, the antisense approach offers the potential to achieve targeted disruption of specific genes in anti-cancer therapy models.

Despite the potentially novel therapeutic strategies offered by the antisense approach, this methodology has in practice been associated with severe limitations. These practical constraints have limited wide employment for this technology in human gene therapy anti-cancer protocols. In this regard, there do not exist universal rules dictating the efficacy of a given antisense oligonucleotide for achieving specific gene inhibition.[14-16] Thus, despite the utility of antisense inhibition, there are a great many cancer-related genes which have resisted attempts to achieve their antisense ablation. In addition, delivery of antisense molecules has been highly problematic.[14-16] It is often difficult to achieve effective sustained intracellular levels of the antisense molecules sufficient for a therapeutic effect. To circumvent this problem, a number of design modifications have been developed to enhance their stability.[14-16,25] In addition, a number of vector approaches have been explored for effective cellular delivery.[24,26,27] Despite these maneuvers, the overriding limitations to the employment of this therapeutic modality remains the idiosyncratic efficacy of specific antisense for a given target gene and the sub-optimal delivery of antisense molecules.

Based upon these recognized limitations of antisense methodologies, we endeavored to develop alternative approaches, to achieve functional "knock-out' of oncoproteins. To this end, a large number of specific monoclonal antibodies (mAbs) have been developed to target these oncoproteins. Further, techniques have been developed to allow the derivation of recombinant molecules which possess antigen binding specificities expropriated from immunoglobins. In this regard, single-chain immunoglobin (scFv) molecules retain the antigen-binding specificity of the immunoglobulin from which they were derived; however, they lack other functional domains characterizing the parent molecule. The basis of constructing scFvs has been established. Pastan et al have developed methods to derive cDNAs which encode the variable regions of specific immunoglobins.[28,29] Specifically, a single-chain antibody gene is derived which contains the coding sequences for variable regions from the heavy chain (V_H) and the light chain (V_L) of the immunoglobulin separated by a short linker (L) of hydrophilic amino acids. The resultant recombinant molecule, when expressed in prokaryotic systems, is a single-chain antibody which retains the antigen recognition and binding profile of the parent. The development of recombinant immunotoxins employing scFv moieties achieves cell-specific binding of the toxin to the exterior of the target cell, allowing receptor-mediated endocytosis to accomplish toxin internalization. A variety of strategies employing the recombinant scFv-directed immunotoxins have been developed by a number of investigators.[28-33] In addition, it has recently been shown that scFv molecules may be expressed intracellularly in eucaryotic cells by gene transfer of scFv cDNAs. The encoded scFv may be expressed in the target cell and localized to specific, targeted subcellular compartments by appropriate signal molecules. Importantly, these intracellular scFvs may recognize and bind antigen within the target cell. Marasco et al have shown that intracellular antibodies against HIV can abrogate production of progeny virion in HIV-infected cells.[34,35] Thus, intracellular scFvs appeared to offer the means to achieve intracellular "knock-out" of specific oncoproteins for therapeutic purposes.

As an initial proof of principle, we endeavored to achieve functional knock-out of the erbB-2 oncoprotein. In this regard, erbB-2 is a 185-kDa transmembrane protein kinase receptor with extensive homology to the family of epithelial growth factor receptors.[36] Several lines of evidence suggest that aberrant expression of the erbB-2 gene may play an important role in neoplastic transformation and progression. Specifically, ectopic expression of erbB-2 has been shown to be capable of transforming rodent fibroblasts in vitro.[37] In addition, transgenic mice carrying either normal or mutant erbB-2 develop a variety of tumors, including neoplasms of mammary origin.[38] Importantly, it has been shown that amplification and/or overexpression of the erbB-2 gene occurs in a variety of human epithelial carcinomas, including malignancies of the ovary, breast, gastrointestinal tract, salivary gland, and lung.[39] In the instance of ovarian carcinoma, a direct correlation has been noted between overexpression of erbB-2 and aggressive tumor growth with reduced overall patient survival.[40,41] As erbB-2 overexpression may be a key event in malignant transformation and progression, strategies to ablate its expression would be rational as a therapeutic modality.

Use of Anti-erbB2 Monoclonal Antibodies in Cancer Therapy

To this end, monoclonal antibodies (mAbs) have been developed which exhibit high affinity binding to the extracellular domains of the erbB-2 protein.[42,43] A number of studies have demonstrated that a subset of these mAbs can elicit growth inhibition of erbB-2 overexpressing tumor cells, both in vitro and in vivo.[44,45] Based upon these observations, clinical trials in humans have been undertaken which exploit the direct anti-proliferative effect of anti-erbB-2 mAbs.[44] Antibody-based tumor targeting has also been utilized with radiolabeled anti-erbB-2 mAbs.[46] In addition, anti-tumor therapies directed at erbB-2 have been developed for targeted immunotoxins.[47] These experimental strategies have employed recombinant fusion proteins consisting of various bacterial toxins selectively targeted to tumor by virtue of single-chain anti-erbB-2 antibody moieties. Clinical trials to date, however, have not duplicated the impressive results obtained with in vitro or in vivo model systems.

It is important to note that in the antibody-based therapies described herein, the conceptual basis of immunologic employment is based upon exploiting cell surface erbB-2 expression as a marker of malignant cells. The molecular basis of erbB-2 mediated transformation, however, would predict that downmodulation of the oncoprotein per se would have an anti-neoplastic effect. To this end, methods have been proposed to downmodulate erbB-2 in overexpressing tumor cells by including antisense strategies targeted to the transcriptional and post-transcriptional levels of gene expression. In the former instance, triplex forming oligonucleotides have been shown to be capable of binding the erbB-2 promoter region to inhibit transcription of the erbB-2 gene.[20] In addition, antisense oligonucleotides targeted to the erbB-2 transcript have accomplished phenotypic alterations in erbB-2 overexpressing tumor cells, including downregulation of cell surface expression and inhibition of proliferation.[48,49] Thus, the conceptual basis of tumor targeting based upon modulation of erbB-2 levels has been established.

Construction and Anti-Proliferative Activity of Anti-erbB2 scFv Intrabody in an Ovarian Carcinoma Cell Line, SKOV3 Cells

Based upon these considerations, we hypothesized that we could employ scFv technology to downmodulate erbB-2 in overexpressing tumor cells to therapeutic end. As a method to effect targeted ablation of the erbB-2 oncoprotein, we have thus developed a strategy based upon intracellular localization of a single-chain antibody directed against erbB-2. We hypothesized that if an anti-erbB-2 scFv were localized to the endoplasmic reticulum (ER) of SKOV3 cells (an ovarian carcinoma cell line which overexpresses erbB-2), the nascent, newly synthesized erbB-2 protein would be entrapped within the ER of the cells, and unable to achieve its normal cell surface localization. It was further hypothesized that this intracellular entrapment would prevent erbB-2, a transmembrane tyrosine kinase receptor, from interacting with its ligand, thus abrogating the autocrine growth factor loop thought to be driving malignant transformation in erbB-2 overexpressing cell lines. To prevent maturational processing of the nascent erbB-2 protein during synthesis, a gene construct was thus designed which encoded an ER directed form of the anti-erbB-2 scFv (pGT21) (Fig. 6.1). As a control, a similar anti-erbB-2 scFv was designed which lacked a signal sequence dictating its localization to the ER (pGT20).

These scFv constructs were cloned into the eucaryotic expression vector pCDNA3, which directs high level gene expression from the cytomegalovirus (CMV) early intermediate promoter/enhancer. For this analysis, the plasmid DNAs pCDNA3, pGT20, and pGT21 were transfected into the erbB-2 overexpressing ovarian carcinoma cell line SKOV3 using the adenovirus-polylysine (AdpL) method.[50] Preliminary experiments had demonstrated that the adenovirus-polylysine-DNA complexes containing a β-galactosidase reporter gene (pCMVβ) effected detectable levels of reporter gene expression in > 95% of targeted cells. At various times after transfection, the cells were evaluated for cell surface expression of erbB-2 using an anti-human erbB-2 monoclonal antibody. Cells transfected with the irrelevant plasmid DNA, pCDNA3, exhibited high levels of cell surface erbB-2, as (cytosolic) form of the anti-erbB-2 (pGT20) exhibited levels of cell surface erbB-2 similar to the control (Fig. 6.2). In marked contrast, SKOV3 cells transfected with pGT21, which encodes an ER form of the anti-erbB-2 scFv, demonstrated marked downregulation of cell surface erbB-2 expression. This downregulation appeared to be time-dependent with cell surface erbB-2 levels progressively declining from 48 to 96 hours post-transfection. At 96 hours post-transfection, fewer than 10% of the pGT21 transfected cells exhibited detectable levels of cell surface erbB-2 protein. The cells otherwise appeared morphologically indistinguishable from the control groups (Fig. 6.2).

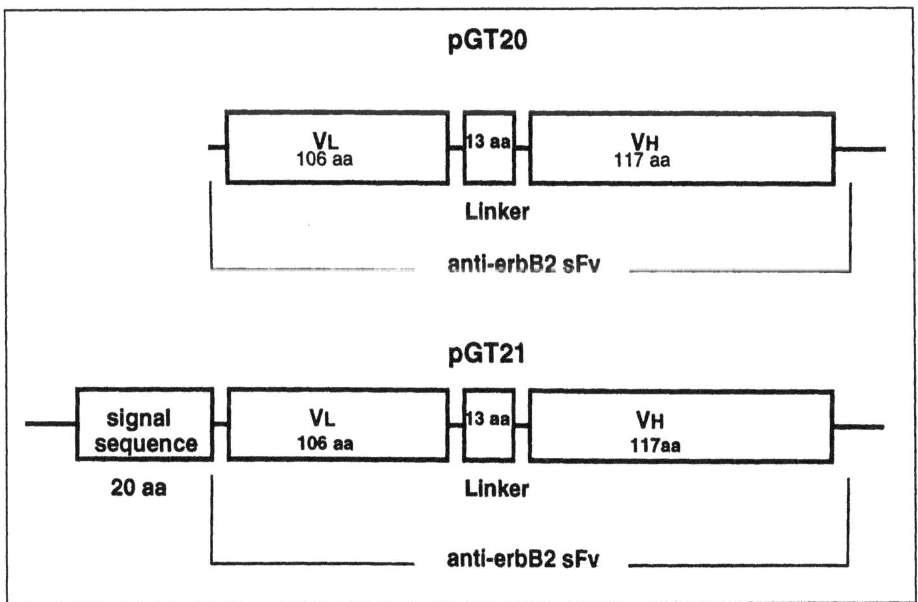

Fig. 6.1. Construction of anti-erbB-2 scFv expression plasmids. Gene constructs encoding anti-erbB-2s were derived from the plasmid e23scFv by PCR methods. The scFv gene constructs were cloned into the eucaryotic expression vector pCDNA3. The plasmid pGT20 was predicted to express a non ER-form (cytosolic) of the scFv. The plasmid pGT21 was predicted to express an ER-form of the scFv.

Fig. 6.2. Effect of intracellular anti-erbB-2 scFv on cell surface expression of erbB-2 protein. The human ovarian carcinoma cell line SKOV3 was transfected by the AdpL methods with the described plasmid constructs and analyzed for cell surface erbB-2 at 96 hours posttransfection using an anti-human erbB-2 monoclonal antibody. Original magnification 400X. A) Transfection with control plasmid pCDNA3. B) Transfection with nonER form of anti-erbB-2 scFv plasmid pGT20. C) Transfection with ER form of anti-erbB-2 scFv plasmid pGT21.

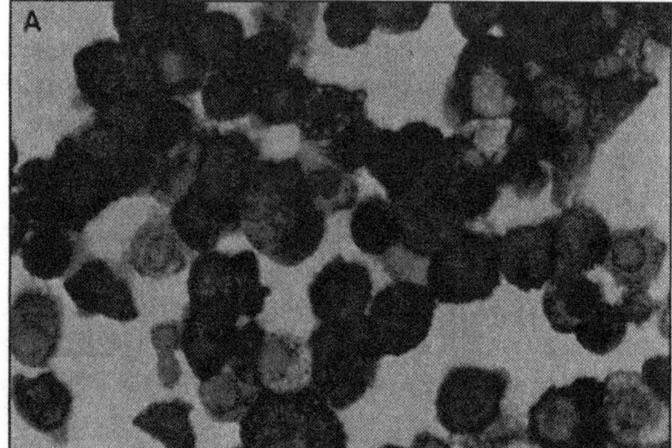

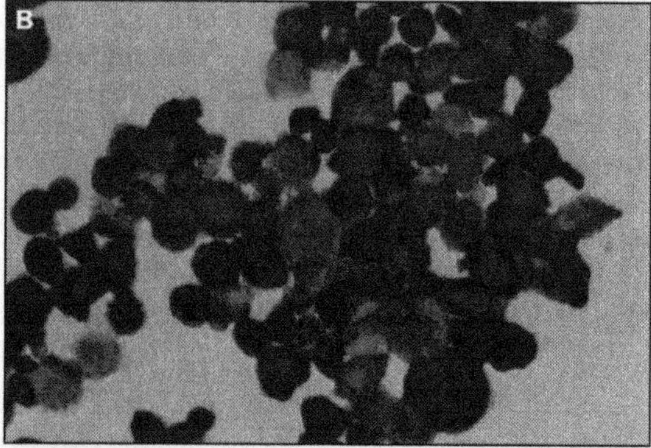

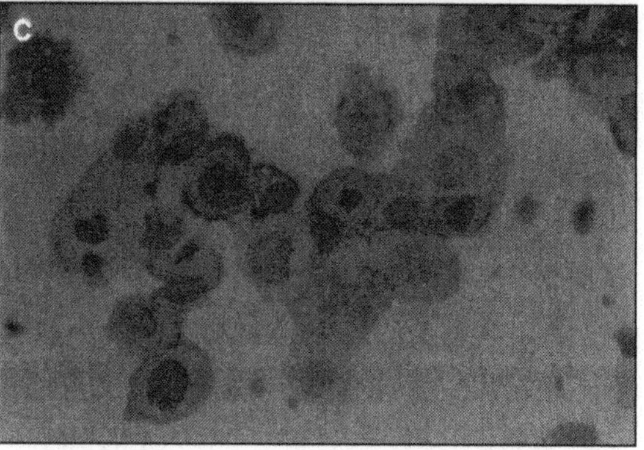

To determine whether modulation of cell surface expression of erbB-2 levels affected cellular proliferation in the SKOV3 cells, the various gene constructs were transfected using the AdpL vector as before. For this analysis, immunohistochemistry for the proliferation associated antigen Ki-67 resulted in the immunohistochemical detection of active cellular proliferation as indicated by intense nuclear staining. In addition, transfection with the nonER form of the anti-erbB-2 did not result in any net change in cell proliferation. In marked contrast, transfection of the erbB-2 overexpressing cell line SKOV3 with the ER form of the anti-erbB-2 scFv resulted in a dramatic inhibition of cellular proliferation (Fig. 6.3). Because

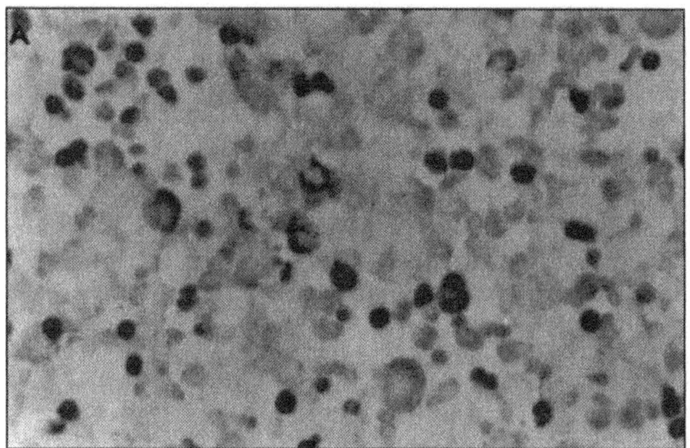
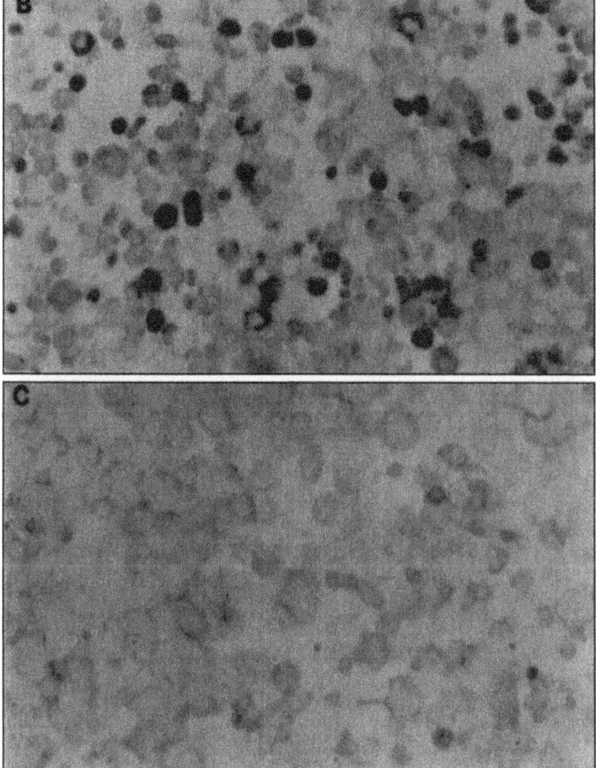

Fig. 6.3. Effect of intracellular anti-erbB-2 scFv expression on nuclear expression of proliferation-associated marker Ki-67. Transfected SKOV3 cells were analyzed for Ki-67 expression 96 hours post-transfection using a mouse monoclonal antibody. Original magnification 200X. A) Control plasmid pCDNA3. B) NonER form of anti-erbB-2 scFv-plasmid, pGT20. C) ER-form of anti-erbB-2 scFv plasmid, pGT21.

the ER anti-erbB-2 scFv exhibited a prominent anti-proliferative effect, it was hypothesized that it might also exhibit a tumoricidal effect in cells stably modified to express this gene construct. As the plasmids pCDNA3, pGT20, and pGT21 contained neomycin selectable markers, they were used to derive stable clones. In a preliminary experiment, the various plasmid constructs were used to derive G418 resistant clones in HeLa cells, a human cervical carcinoma cell line not characterized by overexpression of erbB-2. After selection, the number of clones derived from transfection with pGT20 and pGT21 was not significantly different (Table 6.1). Further, the number of clones did not differ after transfection with the control plasmid pCDNA3. A similar analysis was then carried out with the erbB-2 overexpressing tumor line SKOV3 as the target. In this study, the number of clones derived with pGT20 did not differ from the number derived with the control plasmid pCDNA3 (Table 6.1). Transfection with pGT21, however, resulted in a dramatic reduction in the number of stable clones derived (p < 0.001). Thus, it appeared that the expression of the ER form of the anti-erbB-2 scFv was incompatible with long-term viability of transfected SKOV3 cells. Further, this effect appeared specific for erbB-2 overexpressing cells as this differential clone survival was not noted in the HeLa cells. Thus, as would be predicted, nonerbB-2 expressing tumor cells did not appear to be affected by this specific anti-erbB-2 intervention.

Selective Cytocidal Activity of Anti-erbB2 scFv Intrabodies

As the expression of the anti-erbB-2 scFv appeared incompatible with the derivation of stable clones from SKOV3, we sought to determine whether the effect of expression of the single-chain antibody might be cytocidal. To test this concept, plasmid DNAs which encoded either the cytosolic form or the ER form of the anti-erbB-2 scFv, as well as the control plasmid pCDNA3 were delivered to SKOV3 cells.

Table 6.1. Derivation of stable colonies after transfection with anti-erbB2 scFv expression plasmids

| | G418-Resistant Colonies[a] | |
CELL LINES	Anti-erbB2 scFv NonER Form (pGT20)	Anti-erbB2 scFv ER Form (pGT21)
SKOV3	36	5
	28	5
	23	3
	26	3
	27	3
SW626	21	18
	24	16
	21	16
	28	21
	20	19
HeLa	68	77
	84	83
	91	93
	77	69
	88	89

[a]G418 resistant colonies were counted after staining with crystal violet at day 21 post-transfection

Transfected cells were evaluated for growth rates by analysis of the time-dependent increase in cell number. Cells transfected with the irrelevant plasmid DNA, pCDNA3, as well as the cytosolic pGT20 showed a time-dependent increase in cell number (Fig. 6.4). In marked contrast, transfection with the ER form of the anti-erbB-2 scFv, pGT21, resulted in a significant inhibition of cell growth. Analysis of cell growth kinetics suggested that intracellular expression of the ER form of the anti-erbB-2 scFv was cytocidal to erbB-2 overexpressing tumor cells, and not cytostatic. To more accurately determine the precise biologic effect of anti-erbB-2 scFv expression in the erbB-2 overexpressing tumor target, the XTT assay was employed as a direct determination of cell viability by quantifying a specific cellular enzymatic reaction.[52] For this analysis, the ovarian carcinoma line SKOV3 was transfected, as before, with the plasmid DNAs pCDNA3, pGT20 and pGT21. Transfection with pGT21 resulted in a time-dependent decrease in cell viability, with a > 95% decrement in the number of viable cells by 72 hours posttransfection (Fig. 6.5). Transfection with the control plasmid pCDNA3 and pGT20, however, did not exert any significant effect on cell viability. As an additional control, the ER form of the anti-erbB-2 scFv had no observable effect on the nonerbB-2 expressing human cervical carcinoma line HeLa. We have also transduced nonerbB-2 expressing human cell lines from a variety of different tissues (bladder, liver, mesothelium, kidney), and have not noted any significant cytotoxicity. These results confirmed our previous study demonstrating that the ER form of the scFv uniquely elicits phenotypic alterations in erbB-2 expressing cells.

Induction of Apoptosis by Anti-erbB2 scFv Intrabodies

The foregoing studies are consistent with the concept that entrapment of erbB-2 within the ER of erbB-2 overexpressing tumor cells elicits a selective cytotoxicity. However, this effect may not simply be based on erbB-2 downregulation, as antisense inhibition of erbB-2 gene expression elicited proliferative arrest of erbB-2 overexpressing cells, but not their death.[48,49] To further delineate the mechanistic basis of this effect, studies were carried out to determine if programmed cell death, i.e., apoptosis, was occurring. As before, the plasmid DNA constructs pCDNA3, pGT20, and pGT21 were delivered to the erbB-2 overexpressing SKOV3 cells and the nonerbB-2 expressing tumor cell line HeLa. At specific time points post-transfection, cells were harvested and evaluated for evidence of nuclear DNA fragmentation, a hallmark of programmed cell death.[53] In the HeLa cells, transfection with the various constructs did not demonstrate any evidence of apoptotic cellular events as determined by morphologic appearance or alterations in DNA as measured by gel electrophoresis (Fig. 6.6A). Transfection of the SKOV3 cells with the control plasmid pCDNA3 or the cytosolic anti-erbB-2 scFv pGT20 similarly did not elicit any evidence of cellular apoptosis. When the SKOV3 cells were transfected with the ER form of the anti-erbB-2 scFv, however, marked changes in chromosomal DNA were noted. These changes were first detected at 48 hours post-transfection and revealed on 2% agarose gel as a characteristic 200 bp apoptotic ladder (Fig. 6.6B). As independent confirmation, the presence of apoptotic nuclei was evaluated employing differential nuclear uptake of DNA-binding dyes.[53] In this analysis, SKOV3 cells transfected with the plasmid DNA pGT21 showed intense nuclear staining characteristic of cellular apoptosis. These alterations were not seen in cells transfected with the control plasmids pCDNA3 and pGT20 (Fig. 6.7A-C).

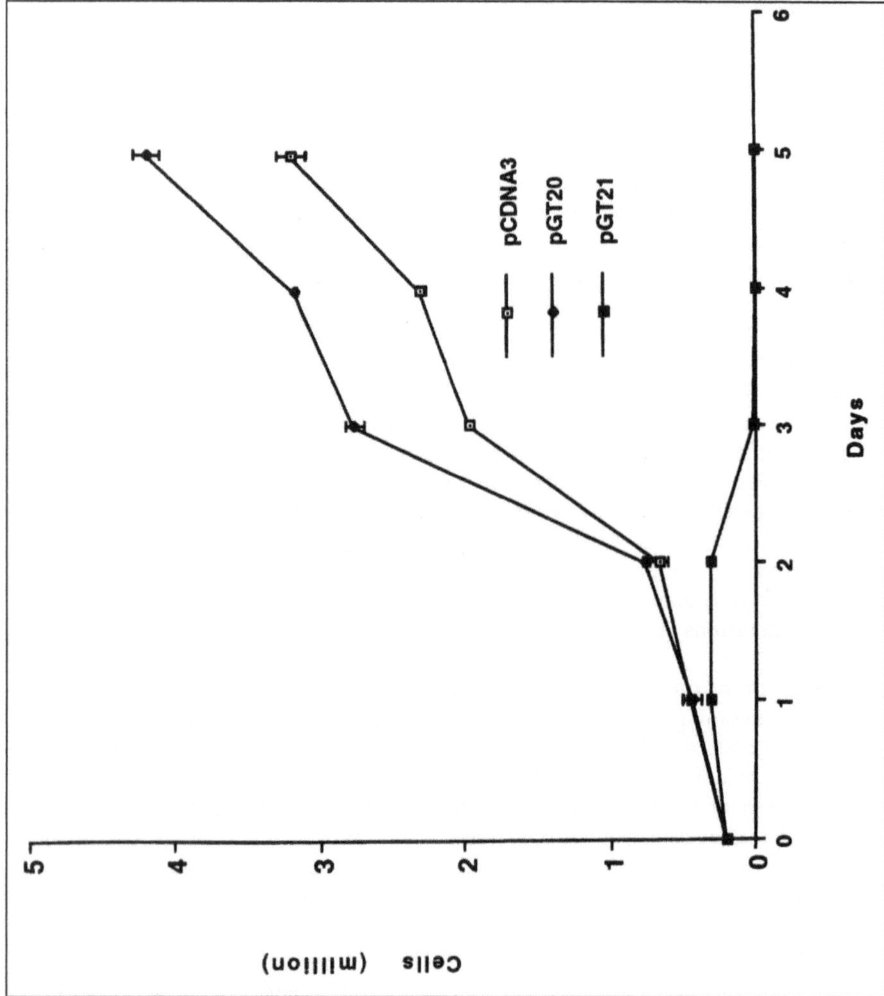

Fig. 6.4. Growth rate measurements of SKOV3 cells transfected with anti-erbB-2 scFv encoding plasmids. The erbB-2 overexpressing ovarian carcinoma cell line SKOV3 was transfected with a control plasmid (pCDNA3), a plasmid encoding a cytosolic form of the anti-erbB-2 scFv gene (pGT20), or a plasmid encoding an endoplasmic reticulum form of the anti-erbB-2 scFv gene (pGT21). Cell numbers were counted in triplicate by trypan blue exclusion of viable cells at indicated times post-transfection.

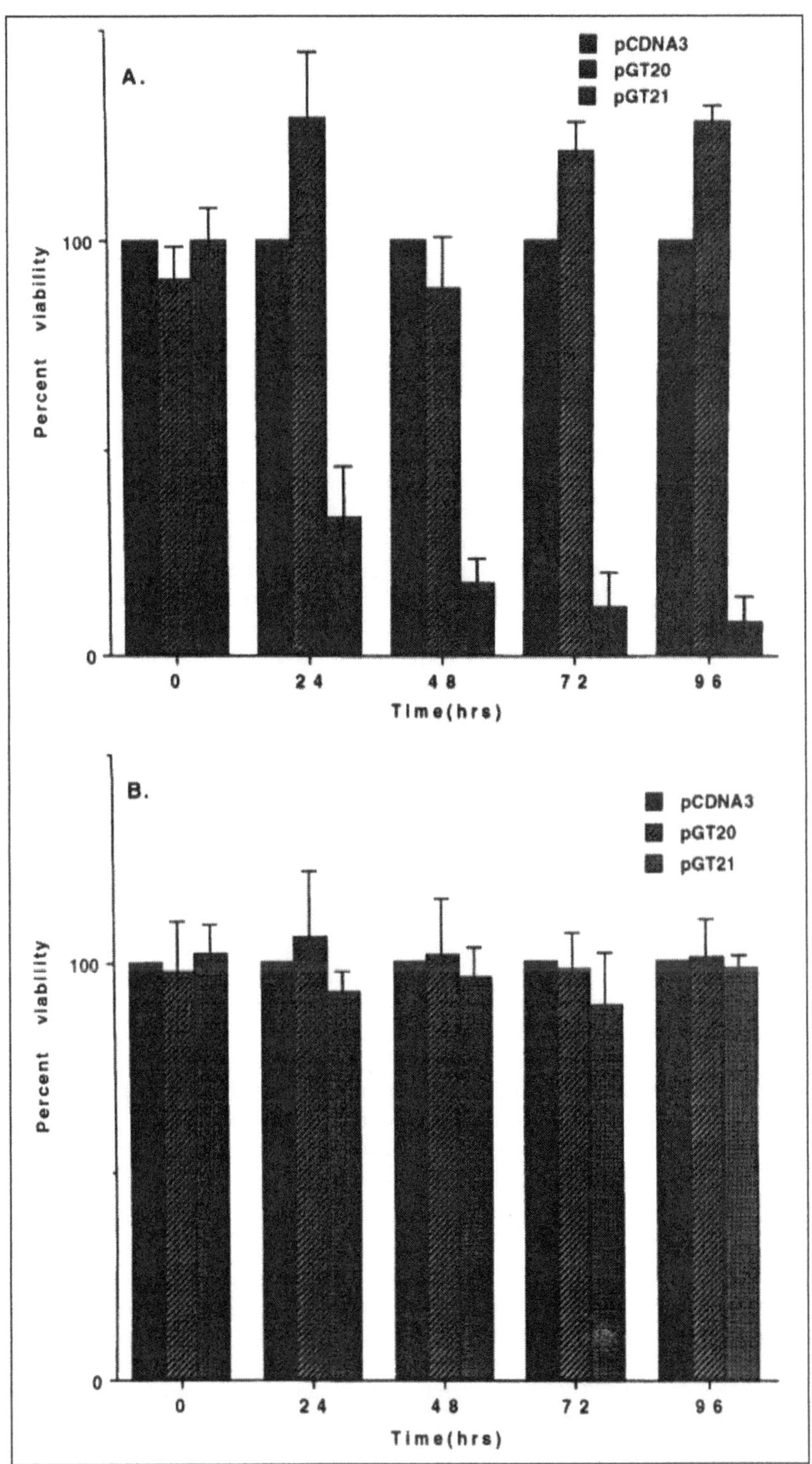

pCDNA3
pGT20
pGT21
A.
Percent viability
100
0
0
2 4
4 8
7 2
9 6
Time(hrs)
B.
pCDNA3
pGT20
pGT21
Percent viability
100
0
0
2 4
4 8
7 2
9 6
Time(hrs)

Quantitative analysis demonstrated that > 90% of the transfected SKOV3 cells exhibited apoptotic nuclear changes, whereas cells transfected with pCDNA3 and pGT20 did not exhibit levels of apoptosis different from untransfected controls. Thus, the basis of the cytocidal effect of the ER anti-erbB-2 scFv in the erbB-2 overexpressing cells was the induction of apoptosis. In the context of dominant oncogene induced tumorigenesis, downregulation of overexpressed immortalizing growth factor receptors may induce cellular apoptosis.[54,55] This suggests that the abrogation of the immortalizing stimulus allows cells to re-engage the previously over-ridden apoptotic program. Alternatively, ablation of dominant oncogene function may result in proliferative arrest, without induction of programmed cell death.[56,57] The precise mechanism distinguishing these alternate responses to oncogene ablation is not presently clear. It is interesting to note that erbB-2 downregulation mediated by antisense oligonucleotides induces proliferative arrest, but not apoptosis in erbB-2 overexpressing tumor targets.[48,49] In contrast, we have induced apoptosis by virtue of an alternate mechanism of erbB-2 down-regulation not inductive of apoptosis per se.

To determine the basis whereby the anti-erbB-2 scFv induced apoptosis, we attempted to reproduce this phenomenon in a heterologous system. For this analysis, ectopic localization of erbB-2, in nonerbB-2 transformed tumor cells, was accomplished by cotransfection of HeLa cells with wild-type human erbB-2 cDNA and the cDNA for the ER form of the anti-erbB-2 scFv. Transfection of the nonerbB-2 expressing HeLa cell line with the erbB-2 cDNA did not result in any change in cell viability, identical to that observed employing the irrelevant plasmid DNA control pCDNA3. In contrast, cotransfection of the erbB-2 cDNA with the anti-erbB-2 scFv construct caused a marked cytocidal effect (Fig. 6.8).

This cytotoxicity could also be shown to be on the basis of induction of apoptosis, as was observed in SKOV3 cells transfected with the anti-erbB-2 scFv (data not shown). Thus, where erbB-2 does not contribute to the transformed phenotype, ectopic localization of erbB-2 within the ER still induced apoptosis. Consistent with this concept, the cytocidal effect of ER entrapment of erbB-2 could be reversed by overexpression of bcl-2, a gene which encodes a mitochondrial protein that functions to promote cell survival through interference with the apoptosis program. In this regard, overexpression of the bcl-2 gene has been shown to revert apoptotic cell death induced by a variety of stimuli.[58-61] Whereas the ER-form of the anti-erbB-2 scFv-induced apoptotic cell death in the erbB-2 overexpressing ovarian carcinoma cell line SKOV3, this effect was abrogated by cotransduction of these cells with the bcl-2 gene (Fig. 6.9). These findings corroborate the concept that the ectopic localization of the erbB-2 oncoprotein specifically induces apoptosis. Further, this indicates that the induced apoptotic pathway is analogous to described mechanisms converging through the bcl-2 proto-oncogene. These data suggested

Fig. 6.5 (previous page). Effect of expression of intracellular anti-erbB-2 scFv genes on tumor cell viability. Tumor cell targets were transfected with the plasmids pCDNA3, pGT20, and pGT21. At indicated times post-transfection, cell viability was determined employing the XTT assay. A) Transfection of the erbB-2 overexpressing human ovarian carcinoma cell line SKOV3. B) Transfection of the nonerbB-2 overexpressing cervical carcinoma cell line, HeLa.

Fig. 6.6. Determination of apoptotic DNA fragmentation induced by ER anti-erbB-2 scFv. Tumor cells were transfected with the plasmids pCDNA3, pGT20, and pGT21. At indicated time points post-transfection, cells were harvested and chromosomal DNA analyzed by gel electrophoresis. A) Transfection of the nonerbB-2 overexpressing human cervical cell line HeLa. B) Transfection of the erbB-2 overexpressing human ovarian carcinoma cell line SKOV3.

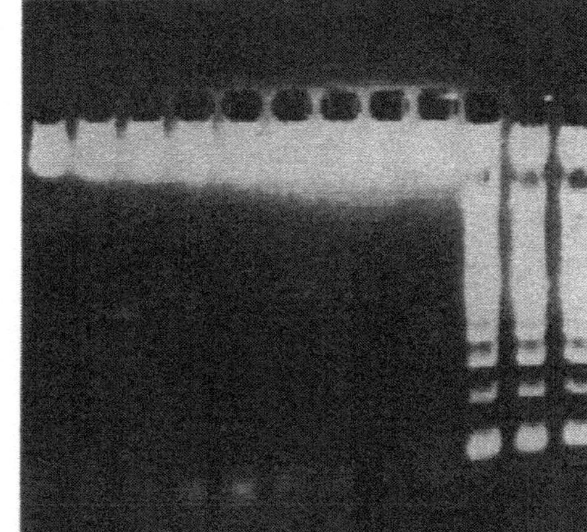

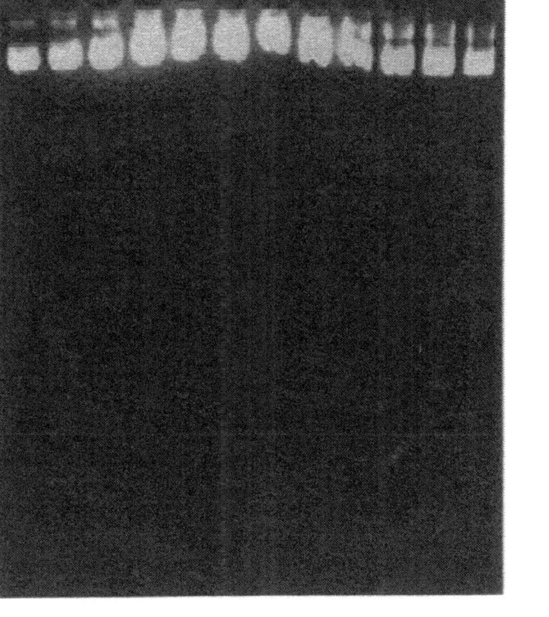

Fig. 6.7. Determination of apoptotic nuclear staining induced by ER anti-erbB-2 scFv. Tumor cell targets were transfected with the plasmids pCDNA3, pGT20, and pGT21. At 24 hours post-transfection, cells were harvested and nuclear uptake of fluorescent DNA-binding dyes were determined. A) SKOV3 cells transfected with pCDNA3. B) SKOV3 cells transfected with pGT20. C) SKOV3 cells transfected with pGT21. Original magnification 400X.

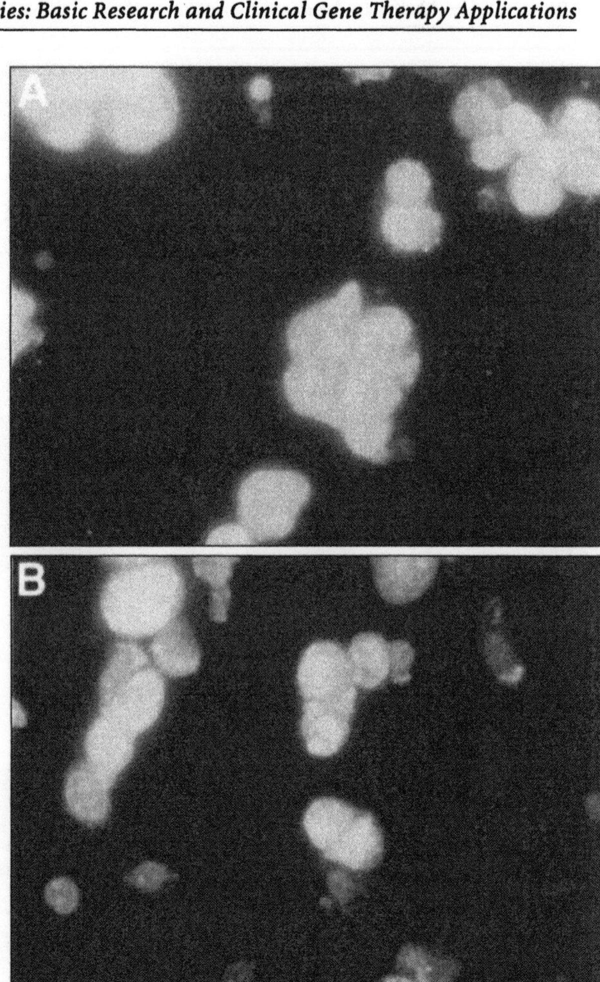
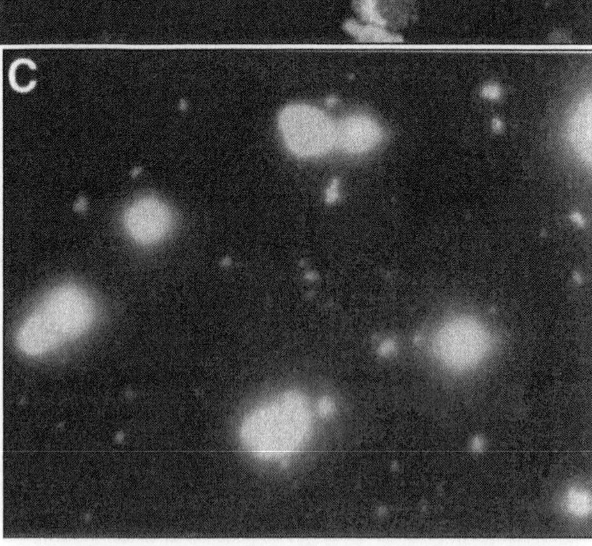

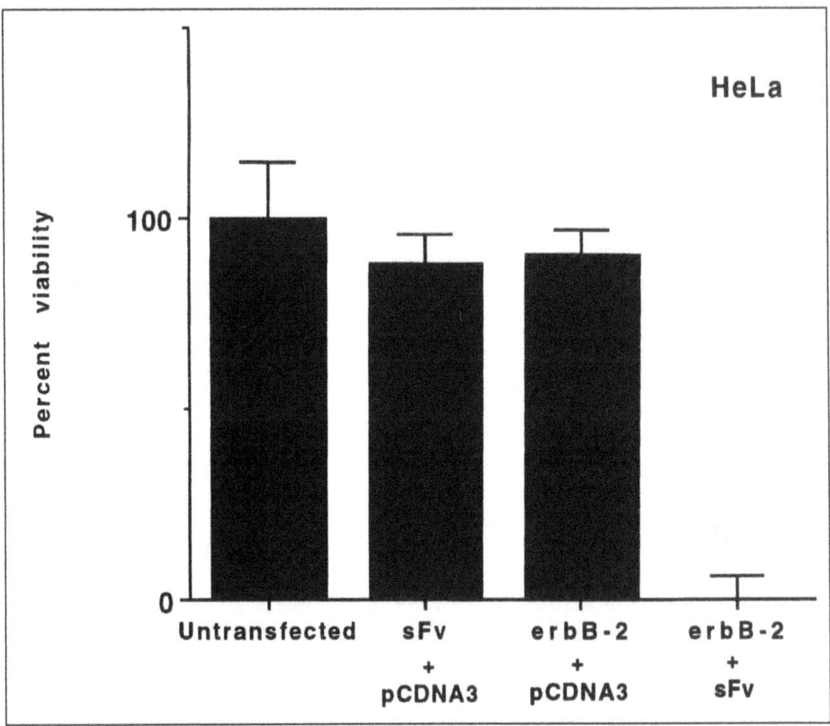

Fig. 6.8. Effect of coexpression of erbB-2 and the anti-erbB-2 scFv on HeLa cell viability. The nonerbB-2 expressing human tumor cell line HeLa was transfected with plasmids encoding the ER form of the anti-erbB-2 scFv (pGT21) and/or the human erbB-2 expression vector LTR-2/erbB-2. At 96 hours post-transfection, cell viability was determined employing the XTT assay. The means of 8 assays are shown.

that abrogation of a transforming oncogene, per se, is not necessarily the basis for apoptosis induction. The fact that mislocalization of erbB-2, in a nonerbB-2 transformed cell, caused apoptosis suggests that mislocalization may be a general means to induce cellular apoptosis. It is thus reasonable to speculate that other endogenous or xenotropic oncogenes, when abrogated in this manner, may trigger apoptotic cell death.

Cytocidal Effects of Anti-erbB2 scFv Intrabodies on Primary Ovarian Carcinoma Cells

We then sought to examine the effects of the anti-erbB-2 scFv in human tumor material isolated from a patient with primary ovarian carcinoma of epithelial origin. For this analysis, we developed methods to isolate primary ovarian tumor cells which maintain their viability and proliferation capacity in vitro for approximately 7-10 days. In addition, the amount of cell surface erbB-2 in these tumor explants had been rapidly determined employing a sensitive ELISA assay. To establish the biologic effects of intracellular single-chain antibody knockout of erbB-2 in these primary ovarian carcinoma cells, the various anti-erbB-2 constructs were delivered

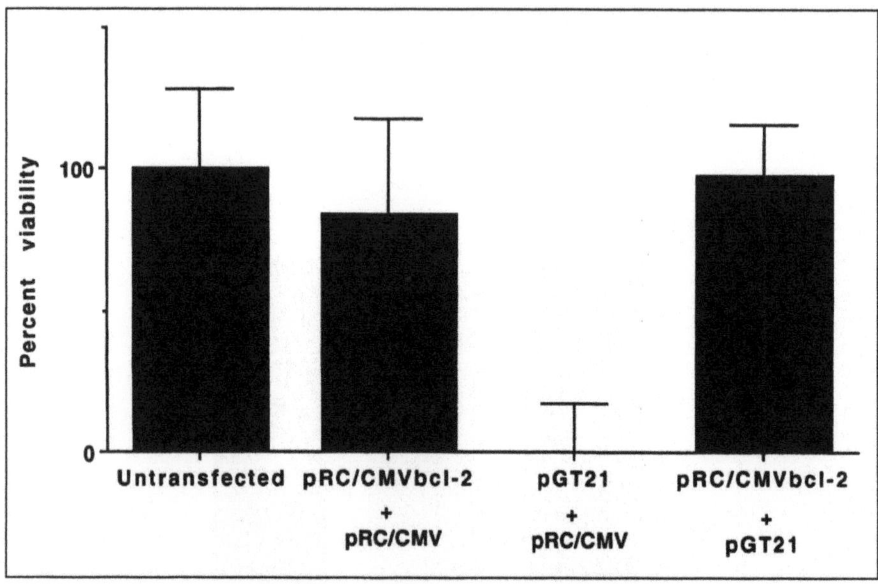

Fig. 6.9. Effect of bcl-2 expression of erbB-2 sFv-mediated apoptosis. The erbB-2 overexpressing ovarian carcinoma cell line SKOV3 was cotransfected with the plasmid pGT21, pRC/CMV control plasmid, and pRC/CMV bcl-2 encoding the wild-type bcl-2 cDNA. Cells were evaluated for viability at 72 hours post-transfection employing the XTT assay. The means of 8 assays are shown.

to cells employing the AdpL vector system, followed by the XTT assay for determination of cell viability. Control experiments employing LacZ reporter gene had demonstrated that > 95% of the isolated human primary ovarian carcinoma cells could be transduced in this manner. The human ovarian carcinoma cell line SKOV3 was employed as an additional control for these experiments. In this analysis, the ER form of the anti-erbB-2 scFv exhibited a specific cytotoxic effect in the human primary tumor cells at 96 hours post-transfection (Fig. 6.10). Interestingly, the magnitude of the scFv-mediated cell killing observed in the primary tumor material was as great as that observed in the erbB-2 overexpressing cell line SKOV3. We have expanded this analysis with additional samples of primary ovarian tumor cells (> 5) freshly isolated from malignant ascites fluid. In each instance, a selective cytotoxicity could be observed uniquely in the ER form of the anti-erbB-2 scFv. These findings strongly suggest that ovarian cancer cell lines represent appropriate models of the operative mechanisms utilized in actual patient tumor cells. These results thus exclude the possibility that the observed scFv-mediated cytotoxicity represent only an in vitro phenomenon. We have demonstrated here the utility of selective cytotoxicity in fresh tumor material derived from humans.

In Vivo Gene Therapy of Ovarian Carcinoma with Anti-erbB2 scFv Antibodies in a Murine Model

Having demonstrated the biologic utility of the anti-erbB-2 scFv strategy in human tumor material, we next sought to determine the therapeutic potential of such an approach in an appropriate disease context. To this end, we sought to de-

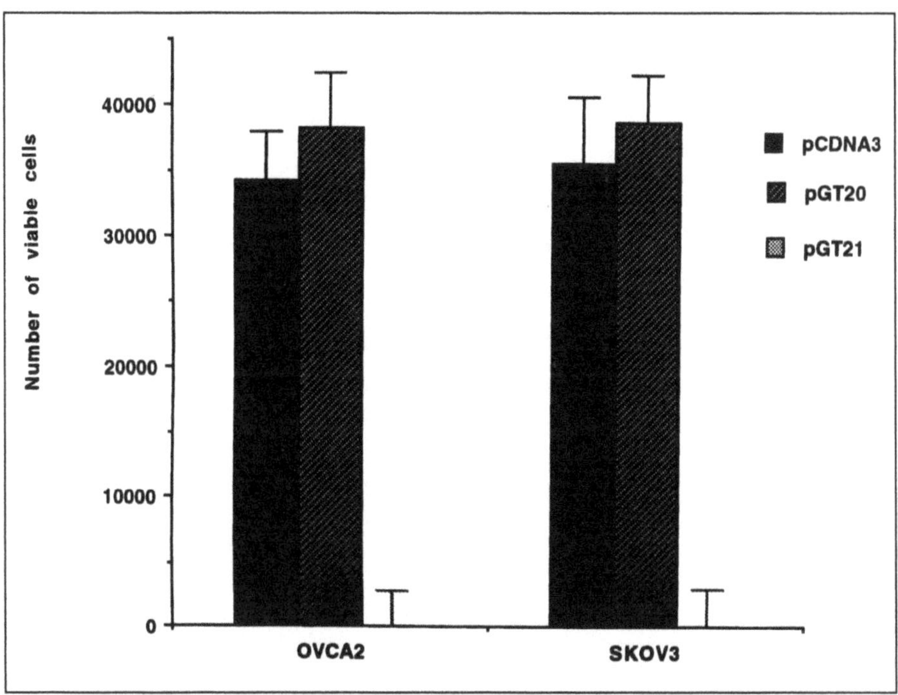

Fig. 6.10. Effect of expression of the anti-erbB-2 scFv gene on human ovarian tumor cell viability. ErbB-2 expressing human primary ovarian carcinoma cells isolated from malignant ascites were transfected with pCDNA3, pGT20, or pGT21. ErbB-2 overexpressing ovarian carcinoma cells (SKOV3) were used as additional controls. Cells were assayed for viability by the XTT assay at 96 hours post-transfection. This experiment was replicated 10x. Data represents Mean +/- SEM.

termine if human ovarian cancer cells could be selectively killed in a murine model of malignant ascites. For these studies, we engrafted athymic nude mice with the erbB-2 overexpressing human ovarian carcinoma line SKOV3. This model allows for the development of malignant ascites and peritoneal implants of neoplastic cells in a manner which parallels the human disease. For gene therapy to be of practical utility in human ovarian carcinoma, vector strategies must be capable of accomplishing direct, in situ delivery of the anti-erbB-2 gene scFv to tumor in vivo. We thus undertook a preliminary analysis to determine which of a series of vector systems could accomplish efficiency in situ transduction of the mobile tumor cells found in ovarian carcinoma malignant ascites fluid. We chose for this analysis, vector systems which had been previously reported to be capable of achieving a reasonable level of in vivo gene transfer. For this analysis, athymic nude mice (Balb/c) were transplanted intraperitoneally with 1×10^7 SKOV3 cells. After 48 hours, vectors were administered by the intraperitoneal route to accomplish delivery of an *E. coli* β-galactosidase reporter gene construct (LacZ) to target the mobile neoplastic cells. Evaluated vector systems included adenovirus-polylysine-DNA-complexes (AdpL), liposomes (DOTAP), and a recombinant adenovirus encoding lacZ (AdCMVLacZ). Forty-eight hours after vector administration, mobile tumor cells were harvested by peritoneal lavage and analyzed for expression of the LacZ reporter gene. This was accomplished by a FACS double-sorting procedure (Fig. 6.11). In this analysis, the highest level of gene transfer was accomplished with the recombinant adenoviral vector. These initial studies do not imply that the adenovirus will ultimately be the optimal vector for in vivo use in human ovarian carcinoma. In this regard, issues related to vector safety, toxicity, immunogenicity, and efficacy in the context of more advanced disease will need to be considered. This vector does, however, give us the means to ask additional questions at present, related to the potential efficacy of the anti-erbB-2 scFv approach as a gene therapy strategy in these model systems.

As the recombinant adenovirus proved of utility for in situ transduction of mobile neoplastic cells in vivo, we asked whether it was possible to accomplish anti-erbB-2 scFv-mediated selective toxicity in this setting. We therefore, constructed a recombinant adenovirus encoding the ER form of the anti-erbB-2 scFv

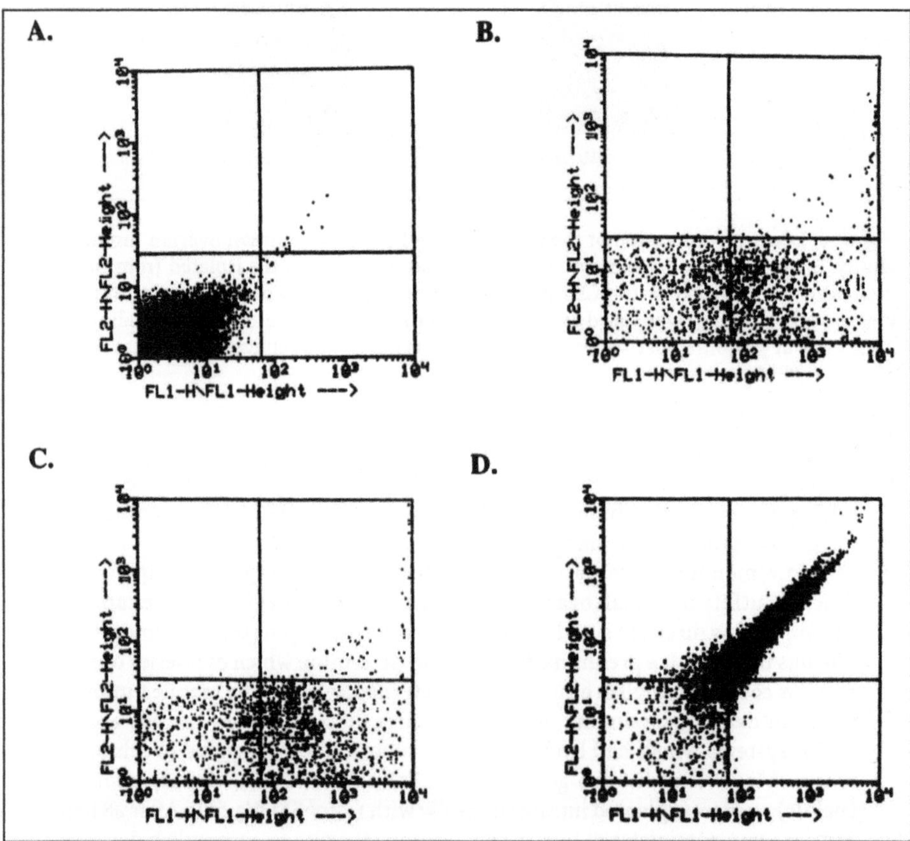

Fig. 6.11. The efficacy of various vectors in accomplishing in vivo gene delivery. Intraperitoneally transplanted SKOV3 cells were challenged with different vector systems delivering the LacZ reporter gene. Peritoneal lavage contents were subjected to FACS analysis for LacZ expression in erbB-2 overexpressing tumor cells. A) no vector transduction; B) DOTAP-DNA complexes; C) adenovirus-polylysine-DNA complexes; D) recombinant adenovirus.

(Ad21) using standard methods of homologous recombination. The resultant recombinant virus is E1A/B deleted and, thus, replication incompetent. Preliminary studies confirmed the structural integrity of the recombinant adenovirus genome. To establish that the anti-erbB-2 scFv gene functioned in this vector configuration, in vitro analysis was carried out employing the SKOV3 cells as the target. Cells were analyzed for viability employing the XTT assay. In this analysis, it could be seen that the anti-erbB-2 scFv encoding adenovirus accomplished the same selective cytotoxicity in the erbB-2 overexpressing targets as observed with AdpL-mediated delivery (Fig. 6.12). Notably, the adenovirus encoding an irrelevant gene (LacZ) had no effects on cell viability, even when delivered at an identical multiple of infection. Thus, a replication-defective adenovirus encoding the anti-erbB-2 scFv has been constructed which retains the capacity to express an ER-anti-erbB-2 scFv. This vector can achieve selective cytotoxicity based on the encoded scFv in human ovarian carcinoma cell lines.

To determine the feasibility of employing the adenoviral vector for in situ tumor cell killing via anti-erbB-2 scFv gene delivery, we undertook treatment experiments employing an orthotopic murine model. As before, SKOV3 cells were xenotransplanted into SCID mice. Forty-eight hours after engraftment with SKOV3 cells, the SCID mice were challenged intraperitoneally with the E1A/B-deleted recombinant adenovirus encoding the irrelevant reporter gene LacZ (AdCMVLac). Ninety-six hours after treatment, the animals underwent peritoneal lavage for analy-

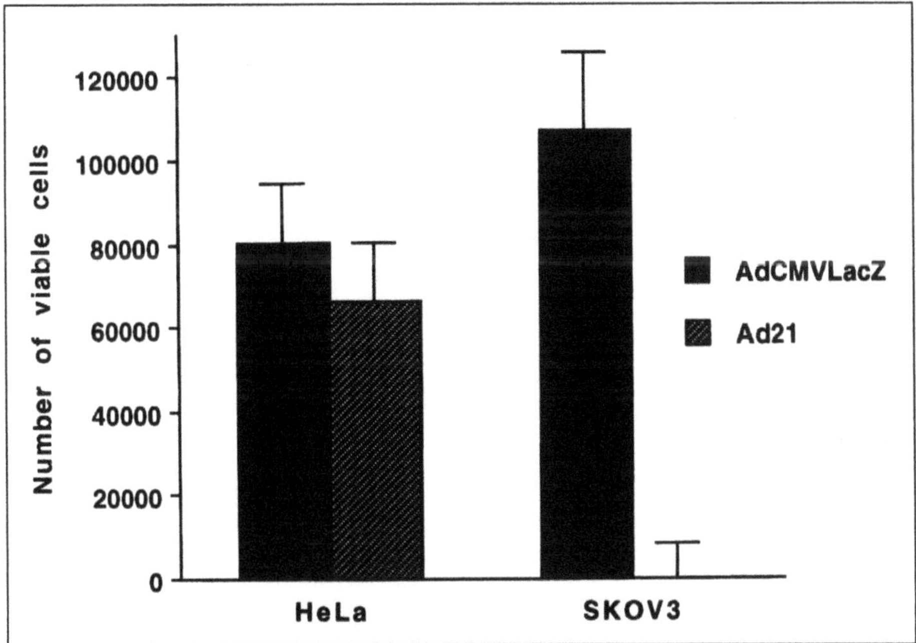

Fig. 6.12. Analysis of cell viability. Transduced cells were evaluated for viability using the XTT assay at 96 hours after transduction. Target cells included the erbB-2-negative cervical carcinoma cell line, HeLa, and the erbB-2 positive ovarian carcinoma cell line, SKOV3.ip1.

sis of harvested mobile tumor cells. Cells were analyzed for cell viability employing the XTT assay. It could be seen that the number of viable cells was dramatically decreased in the Ad21 group compared to the AdCMVLacZ group (Fig. 6.13). This cytotoxicity appeared to be specifically associated with the anti-erbB-2 scFv encoding adenovirus. Analysis of the mechanism of cell death demonstrated that the Ad21 virus induced cellular apoptosis (data not shown). Thus, the recombinant adenovirus encoding the anti-erbB-2 scFv accomplished a specific cytotoxicity in mobile neoplastic cells in an orthotopic murine model of human ovarian cancer. Based upon these findings, we have developed a human clinical gene therapy trial to evaluate the safety toxicity and biologic efficacy of this approach.[62]

Anti-erbB2 scFv Intrabodies Enhance Tumor Cell Chemosensitivity

The basis whereby complete disease eradication was not achieved was next considered. One possible mechanism in this regard is the outgrowth of "resistant" tumor cells which no longer manifest sensitivity to anti-erbB-2 scFv-mediated cyto-

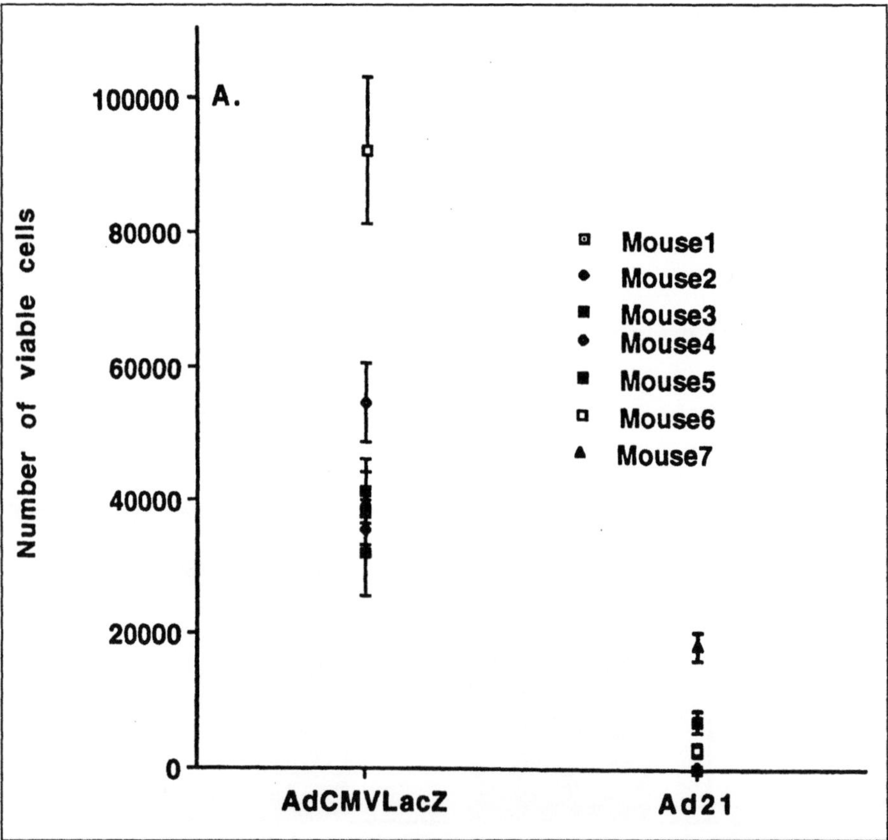

Fig. 6.13. Specific cytotoxicity of an anti-erbB-2 scFv encoding adenovirus against human ovarian cancer targets. In a murine model of malignant ascites, SCID mice were injected intraperitoneally with 10^7 SKOV3 cells. After 48 hours, animals were challenged intraperitoneally with 10^9 plaque forming units of Ad21 or AdCMVlacZ. After 96 hours, mobile cells were harvested by peritoneal lavage and analyzed for survival by the XTT assay.

toxicity. To address this possibility, tumor cells were harvested by lavage from Ad21-treated animals at late treatment times. These cells exhibited the same magnitude of sensitivity to scFv-induced cytoxicity as virgin cell line counterparts (data not shown). Therefore, it does not appear that resistance to erbB-2 scFv-mediated cytotoxicity, per se, is an operational factor in this context. Thus, two additional considerations appear relevant. Firstly, the net gene transfer efficacy may be limiting effective cell kill. Thus, strategies to augment the efficiency of in situ gene transfer to tumor cells appears warranted. Additionally, all genetically modified tumor cells may not be effectively eradicated. Data suggestive of this phenomenon were indeed noted in the context of the scFv-expressing stable clone derived in Table 6.1 and in the experiment in which transient expression induced cytotoxicity (Fig. 6.5). In both of these studies, a subset of tumor cells could be shown to survive despite expression of the anti-erbB-2 scFv gene. In this regard, maneuvers to increase gene transfer efficiency, per se, would not be predicted to be completely efficacious. Thus, strategies are required to address the tumor cell subset which can survive, despite expression of the anti-erbB-2 scFv.

In this regard, in addition to its direct role in neoplastic conversion, erbB-2 overexpression is associated with tumor cell resistance to chemotherapeutic agents. It has been noted that heterologous overexpression of human erbB-2 accomplished by genetic transduction can increase the chemoresistance of murine fibroblasts[63] and human lung carcinoma cells[64] to a variety of chemotherapeutic agents. These findings are corroborated by the clinical observation that erbB-2 overexpressing tumors possess a higher intrinsic chemoresistance and thus are associated with a shorter relapse-free interval.[65] Another line of evidence supporting the role of erbB-2 in modulating tumor cell chemoresistance is the observed therapeutic synergy between cisplatin (CDDP) and anti-erbB-2 monoclonal antibodies.[66-69] These studies have documented that anti-erbB-2 antibodies capable of downregulating the erbB-2 oncoprotein achieve enhanced tumor cell sensitivity to this chemotherapeutic agent. Several groups have demonstrated this phenomenon and studies have elucidated the mechanistic basis for this effect.[67,69] Thus, these lines of evidence support the concept that the erbB-2 oncoprotein plays a key role in determining tumor cell chemoresistance.

These findings suggested that scFv-based strategies to downregulate the erbB-2 oncoprotein might also have utility for enhancing tumor cell chemosensitivity. Based on this concept, we hypothesized the erbB-2 downregulation accomplished by the anti-erbB-2 scFv might enhance chemosensitivity of erbB-2 overexpressing tumors. This strategy was specifically conceptualized as an approach to the erbB-2 overexpressing tumor cells which were not directly eradicated by intracellular expression of the anti-erbB-2 scFv. We thus explored whether the anti-erbB-2 could directly effect tumor cell sensitivity to chemotherapeutic agents. As an initial study, the erbB-2 overexpressing tumor cells, SKOV3, were treated with either the anti-erbB-2 scFv (via transient transfection), the chemotherapeutic agent cisplatin (CDDP), or a combination of these agents. In this analysis (Fig. 6.14), it was noted that intracellular expression of the anti-erbB-2 scFv or CDDP could induce cytotoxicity, but a synergistic effect was noted when the two agents were employed in combination.

It thus appeared that the anti-erbB-2 scFv was capable of enhancing tumor cell sensitivity to a chemotherapeutic agent. To further explore this phenomenon, we developed an experimental model which would allow more direct analysis of the

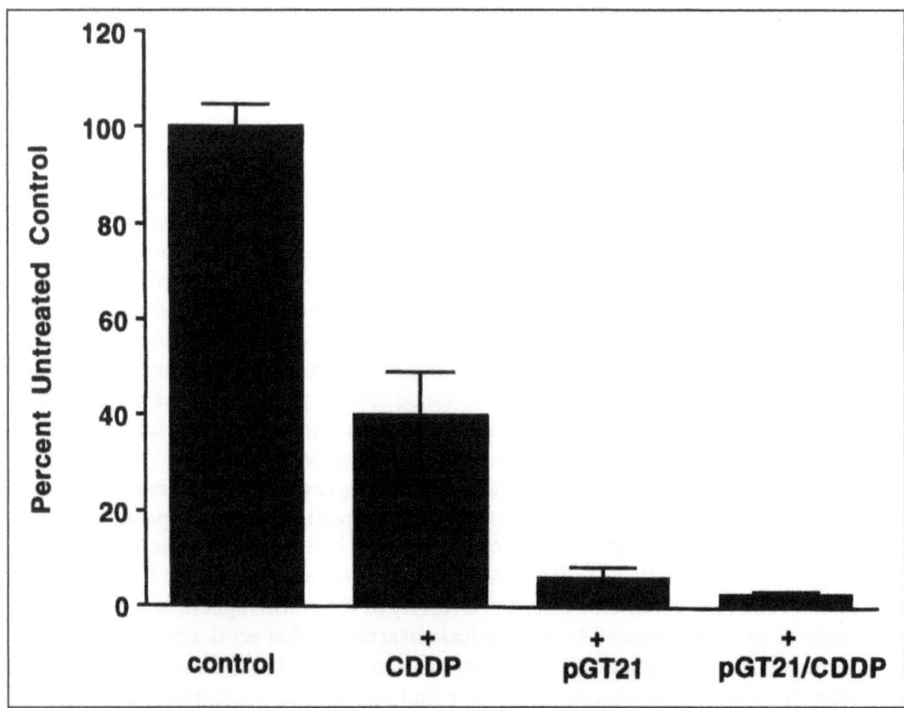

Fig. 6.14. Cytotoxic effect of an anti-erbB-2 scFv in combination with CDDP. The erbB-2 overexpressing ovarian carcinoma cell line SKOV3 was transfected with a plasmid construct encoding an ER form of an anti-erbB-2 scFv (pGT21) or a control plasmid construct (pCDNA3) and treated with CDDP (2 µg/ml). The cells were incubated for 72 hours and the number of viable cells determined by an MTS assay. Experiments were performed in triplicate and the results represent the mean +/- SEM.

anti-erbB-2 scFv mediated chemosensitization. For this analysis, we expanded and characterized the scFv-expressing SKOV3 clones derived in Table 6.1. Clonal cell populations were thus characterized for confirmation of expression of the anti-erbB-2 scFv. In addition, clonal populations of the parent cell line (SKOV3) the cytosolic anti-erbB-2 scFv (SKOV3/GT20), or the ER anti-erbB-2 scFv (SKOV3/GT21) which exhibited comparable growth kinetics were examined (data not shown). It was hypothesized that the parental clone and the cytosolic scFv clone would have comparable levels of cellular erbB-2. In addition, it would be predicted that the ER scFv clone would have reduced cellular erbB-2, based upon a level of scFv-mediated erbB-2 downregulation. These clones were thus evaluated for cellular erbB-2 by direct ELISA analysis (Fig. 6.15). In this study, it could be seen that the ER anti-erbB-2 scFv clone, SKOV3/GT21, was uniquely characterized by reduced erbB-2 levels, as predicted.

These clonal cell populations were further evaluated for their sensitivity to the chemotherapeutic agent CDDP. In this analysis, it could be seen that the cytosolic scFv expressing clone, SKOV3/GT20, did not differ in CDDP sensitivity when compared to the parental clone SKOV3. Thus, intracellular expression of the anti-erbB-2

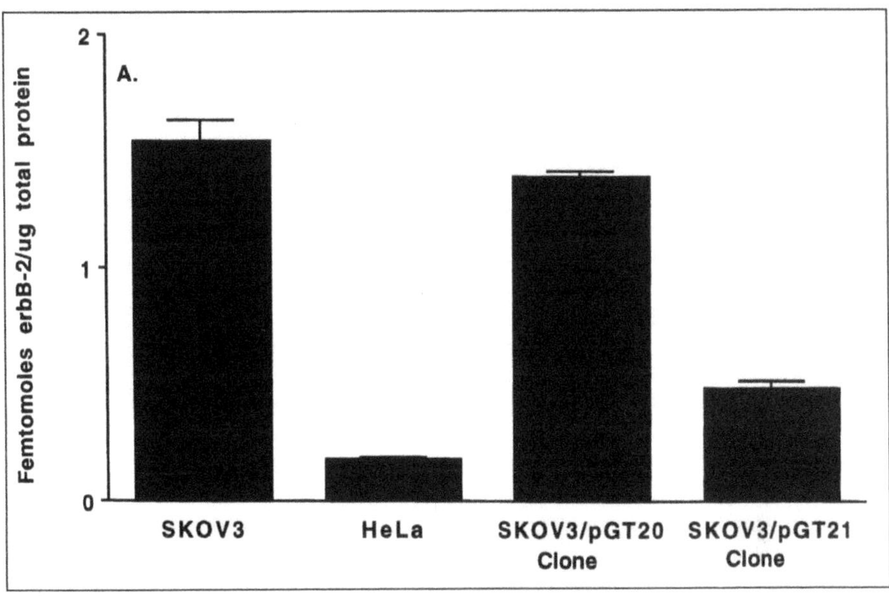

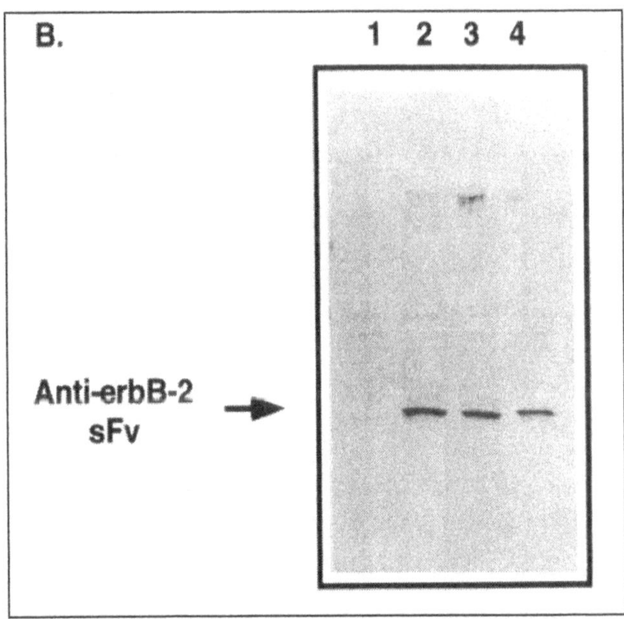

Fig. 6.15. Characterization of anti-erbB-2 scFv expressing SKOV3.ip1 clones. A) Determination of cell surface erbB-2 protein expression in stable clones as determined by an ELISA assay. Relative erbB-2 levels were calculated from a standard curve. SKOV3.ip1 cells are a positive control while HeLa, an erbB-2 negative human cervical cell line, served as a negative control. Results are expressed as mean +/- SEM. B) Determination of the presence of anti-erbB-2 scFv in stable clones by Western blot. Cells were solubilized and the samples electrophoresed with probe analysis using a polyclonal rabbit anti-erbB-2 scFv antibody. Lane 1: untransfected SKOV3 cells. Lane 2: SKOV3 cells transfected with pGT21. Lane 3: SKOV3/pGT20 clone. Lane 4: SKOV3/pGT21 clone.

scFv in the cellular cytosol has no effect on either erbB-2 levels (Fig. 6.15) or sensitivity to CDDP (Fig. 6.16). In marked contrast, the clonal population expressing the ER form of the anti-erbB-2 scFv exhibited significantly greater sensitivity to CDDP treatment than the parental clone. In this instance, the ER-scFv-mediated erbB-2 downregulation was associated with enhanced chemosensitivity.

Studies were also undertaken to determine whether this strategy could be of utility for erbB-2 positive tumors which were refractory to scFv-mediated cytotoxicity. In this regard, a subset of erbB-2 overexpressing tumor cell lines have been identified which were resistant to the effects of the Ad21 virus. In this instance, we hypothesized that if erbB-2 downregulation was achieved, cells might nonetheless be sensitized to a second apoptotic insult. The human carcinoma cell line PANC-1 was known from previous experiments to overexpress erbB-2 and be > 95% transducible with an adenoviral vector (data not shown). However, this human carcinoma cell line was resistant to the cytotoxic effects of the anti-erbB-2 scFv (Fig. 6.17, panel A). We therefore hypothesized that the erbB-2 downmodulation mediated by the anti-erbB-2 scFv in this cell line would sensitize the cells to a second insult with CDDP. As expected, the anti-erbB-2 scFv mediated inhibition of proliferation to a moderate degree (approximately 75% of untransfected control) in this carcinoma cell line. However, the addition of cisplatin to these cells, after

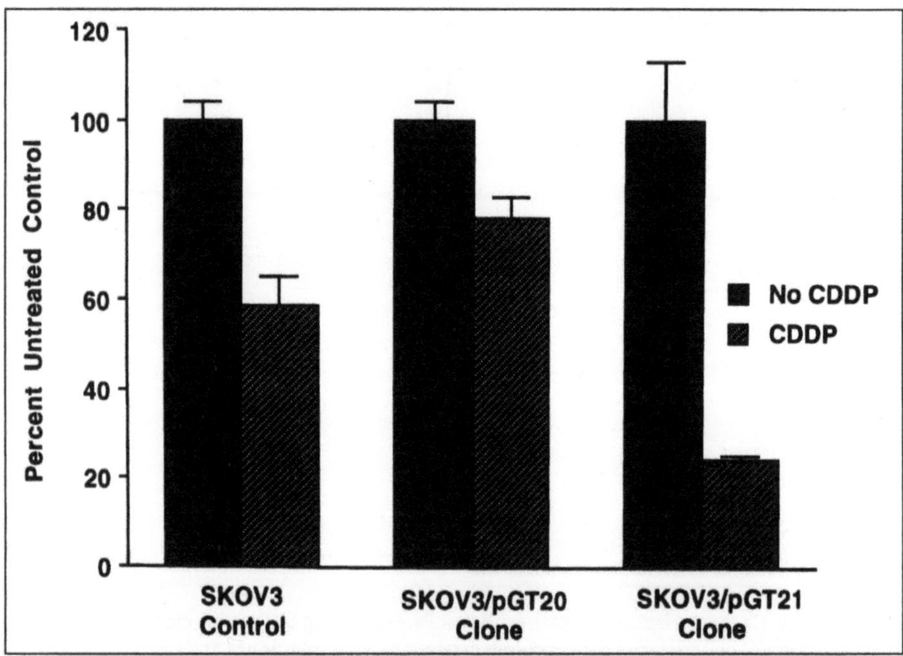

Fig. 6.16. Sensitivity of anti-erbB-2 scFv expressing SKOV3 clones to CDDP. SKOV3/ pGT21 clones expressing the ER form of the anti-erbB-2 scFv demonstrate enhanced chemosensitivity to CDDP. SKOV3/pGT21 clones were treated with CDDP (2 µg/ml) and incubated for 72 hours. Cell viability was then measured using an MTS assay. SKOV3 and SKOV3/pGT20 clones served as controls. Experiments were performed in triplicate and the results are reported as means +/- SEM.

transfection with Ad21, markedly enhanced the inhibition of proliferation of these cells over the range of cisplatin doses tested (Fig. 6.17, panel B). The augmented inhibition of proliferation was not observed when the control adenovirus (AdLacZ) was used in combination with CDDP (Fig. 6.17, panel C). These results suggest that in a "treatment-resistant" population of human carcinoma cells that overexpress erbB-2, an increase in sensitivity to conventional chemotherapeutic agents can be observed with the use of an anti-erbB-2 scFv. This finding thus suggests that this strategy may be a useful adjunct to erbB-2 overexpressing cells initially refractory to scFv-induced cytotoxicity.

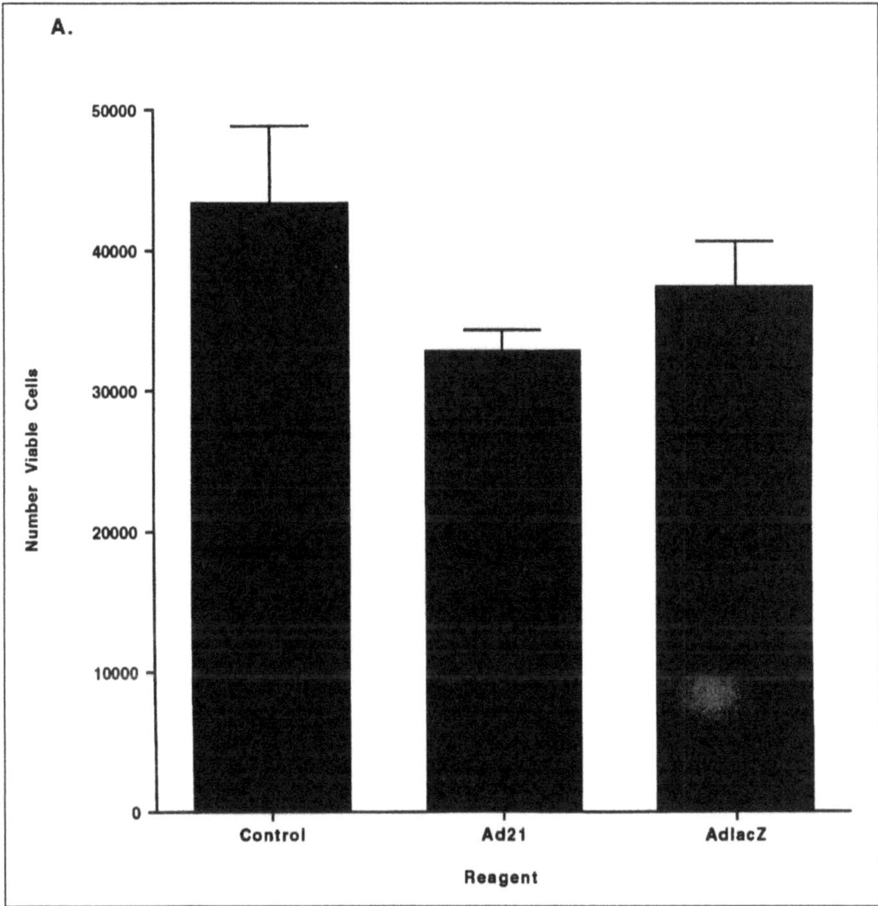

Fig. 6.17. Sensitivity of the pancreatic carcinoma cell line PANC-1 to anti-erbB-2 scFv-mediated chemosensitization. A) Determination of direct cytotoxicity of anti-erbB-2 scFv on human pancreatic cell line PANC-1. B) (next page) Determination of CDDP chemosensitization effects of anti-erbB-2 scFv encoding adenovirus Ad21. C) (next page) Determination of CDDP chemosensitization effects of control adenovirus AdLacZ. Cells were plated in 96 wells at 500C cells/well. After 24 hours, cells were infected with adenoviral vectors. Six hours postadenoviral infection, cisplatin (CDDP) was added at the indicated concentration. Cells were incubated 96 hours and cell viability then determined.

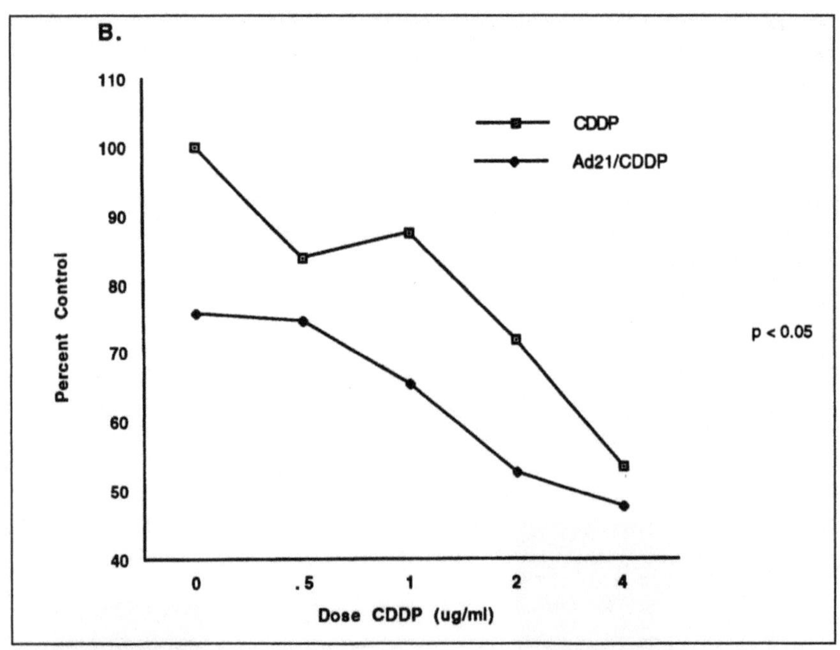
B.
Percent Control
110
100
90
80
70
60
50
40
CDDP
Ad21/CDDP
p < 0.05
0
.5
1
2
4
Dose CDDP (ug/ml)

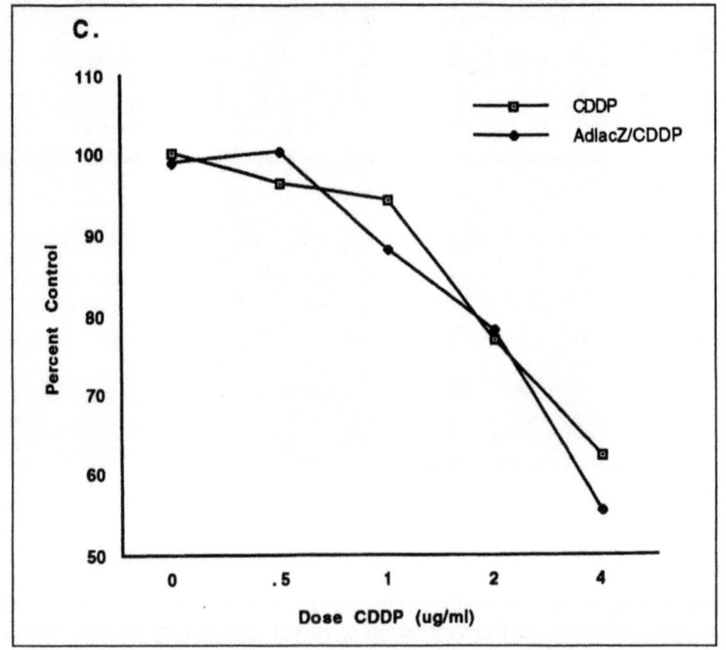
C.
Percent Control
110
100
90
80
70
60
50
CDDP
AdlacZ/CDDP
0
.5
1
2
4
Dose CDDP (ug/ml)

To establish in vivo efficacy of the anti-erbB-2 scFv (Ad21) in combination with cisplatin (CDDP), a murine model of human ovarian carcinoma was employed. In this regard, we had previously shown that the Ad21 vector administered by the IP route could prolong survival in this murine model of human ovarian carcinoma. In the present context, we hypothesized that a further survival advantage would accrue if the synergistic effects of the anti-erbB-2 scFv and CDDP allowed enhanced tumor cell kill in vivo. As shown in Figure 6.18, those animals that received Ad21 in combination with CDDP demonstrated a statistically significant enhanced survival over those animals receiving cells alone. This significant prolongation of

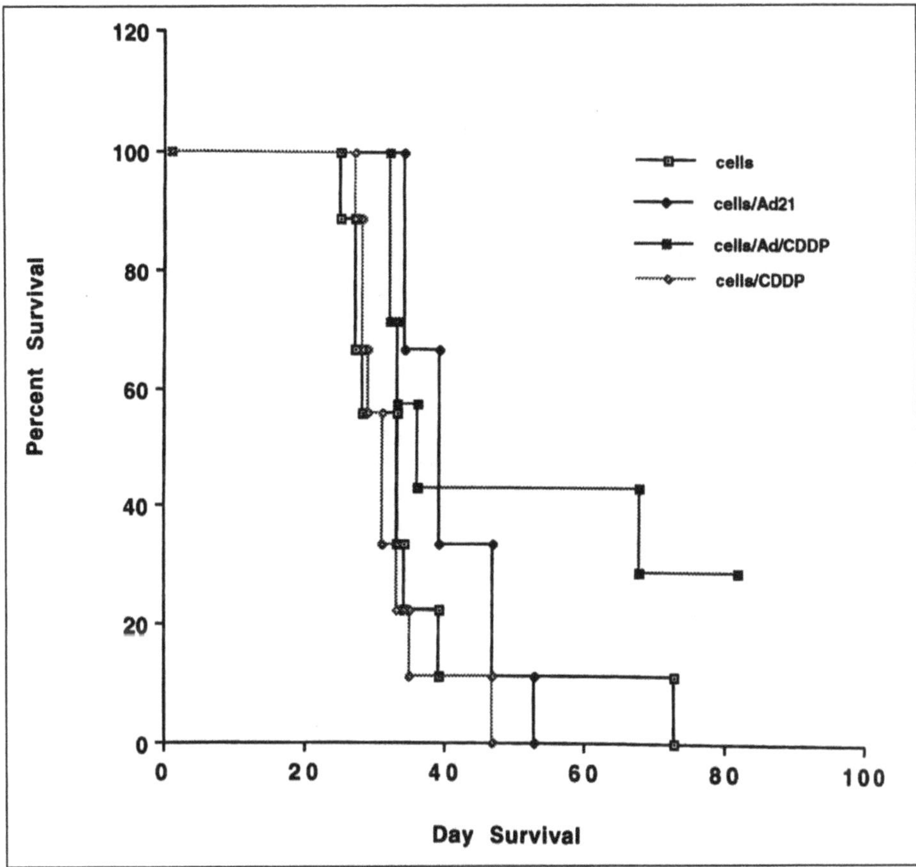

Fig. 6.18. In vivo efficacy of recombinant adenovirus encoding the ER form of the anti-erbB-2 scFv in combination with CDDP. On day 0, SCID mice were injected intraperitoneally with 1 x 10^7 SKOV3.ip1 cells (9 mice per group). The tumors were allowed to engraft and on day 5, Ad21 was administered at 100 pfu per cell. Following administration of Ad21, appropriate groups of animals received intraperitoneally CDDP (2 mg/kg) as a single dose on day 6. Animals were then followed for survival analysis. The following groups were represented; cells alone (9 mice), cells/ Ad21 (9 mice), cells/Ad21/CDDP (9 mice), cells/CDDP (9 mice), Ad21 along (4 mice), and CDDP alone (4 mice), media alone (4 mice).

survival was not achieved when animals received only CDDP or Ad21. These data further support the hypothesis that cells not directly killed by the anti-erbB-2 scFv are rendered susceptible to a second insult from CDDP. This finding may be of clinical utility in ultimately improving the anti-erbB-2 scFv gene therapy strategy.

Summary

We have developed a novel method of gene product ablation which has been employed to "knockout" a selected oncogene. The results from our preliminary data have demonstrated consequences from the standpoint of the development of a novel therapeutic modality. To achieve selective gene product ablation, an intracellular single-chain immunoglobulin was expressed in target cells. The intracellularly expressed scFv prevented the normal maturation of the transmembrane tyrosine kinase receptor erbB-2, which likely occurred on the basis of ER entrapment of the nascent protein during biosynthesis. The intracellular expression of an scFv has recently been described by Marasco et al.[34,35] In their study, the cellular expression of an anti-HIV scFv resulted in abrogation of the HIV infectious cycle with a functional reduction of viral production. We have also accomplished intracellular single-chain immunoglobulin-mediated knockout utilizing gene transfer of an scFv construct. In this instance, however, the consequence was ablation of transforming oncogene, as opposed to a xenogenic viral gene product. A consequence of this novel methodology is the definition of a potential therapeutic modality for the achievement of selective killing of specific target cells. In the context of anti-cancer therapeutics, it has been proposed that specific abrogation of specific target gene products can revert them from the malignant phenotype. In this regard, the overexpression of dominant oncogenes has been shown to be a critical determinant of neoplastic transformation and progression. Targeted disruption of selected oncogenes may accomplish reversion from the malignant phenotype or initiation of cell death. The utilization of intracellular single-chain immunoglobulins represents another strategy for these purposes. This methodology offers certain potential advantages over previous genetic techniques for therapeutic gene ablation. In this regard, many mAbs have been developed against a variety of cancer-related gene products. It would thus be logical to convert these reagents to scFvs which would possess the potential for therapeutic utility. In this schema, the ablation construct might be designed with a priori knowledge of its specific recognition of its cellular target. This is not the case with antisense methodologies, where no uniform roles predicting effective targeting of downstream processing exist. As in the example reported here, expression of intracellular immunoglobulins may possess the potential to achieve a highly selective effect on specific gene products with the end result being cell killing. In addition, the employment of DNA-based methods for delivery is implicit in the scFv strategy. This allows the reagents to be delivered employing a variety of high efficiency vehicles. In contrast, effective delivery of antisense constructs has represented a major limitation to their practical employment to date. Thus, the intracellular antibody strategy may offer significant practical advantages over antisense methods as a means to achieve selective tumor cytotoxicity based upon targeted knockout of oncogene or oncogene products.

Acknowledgments

We wish to acknowledge the expert editorial assistance of Christi Stewart. In addition, we are indebted to Gene P. Siegal, M.D., Ronald D. Alvarez, M.D., Jessy Deshane, Jon Grim, Mack Barnes, M.D., and Marci Wright. This work was supported by the following grants: National Institutes of Health-CA 68245 and National Institutes of Health-CA 69343-01.

References

1. Sikora, K. Gene therapy for cancer. Tibtech 1993; 11:197-201.

2. Dorudi S, Northover JMA, Vile RG. Gene transfer in cancer. Br J Surg 1993; 80:566-572.

3. Freeman SM, Zwiebel JA. Gene therapy of cancer. Cancer Investigation 1993; 11:676-688.

4. Vanchieri C. Opportunities "opening up" for gene therapy. JNCI 1993; 85:90-91.

5. Lemoine N, Sikora K. Interventional genetics and cancer treatment. BMJ 1993; 54:653-665.

6. Gutierrez AA, Lemoine NR, Sikora K. Gene therapy for cancer. The Lancet 1992; 339:715-721.

7. Karp JE, Broder S. New directions in molecular medicine. Cancer Research 1994; 54:653-665.

8. Rosenberg SA, Anderson WF, Blaese M et al. The development of gene therapy for the treatment of cancer. Annals of Surgery 1993; 218:455-464.

9. Baker SJ, Markowitz S, Fearon ER et al. Suppression of human colorectal carcinoma cells growth by wild-type p53. Science 1990; 249:912-915.

10. Huang H-JS, Yee J-K, Shew J-Y et al. Suppression of neoplastic phenotype by replacement of the RB gene in human cancer cells. Reports 1988; 242:1563-1566.

11. Bookstein R, Shew J-Y, Chen P-L et al. Suppression tumorigenicity of human prostate carcinoma cells by replacing a mutated RB gene. Science 1990; 247:712-716.

12. Fujiwara T, Grimm EA, Mukhopadhyay T et al. A retroviral wild-type p53 expression vector penetrates human lung cancer spheroids and inhibits growth by inducing apoptosis. Cancer Research 1993; 53:4129-4133.

13. Zhan W-W, Fang X, Mazur W et al. High-efficiency gene transfer and high-level expression of wild-type p53 in human lung cancer cells mediated by recombinant adenovirus. Cancer Gene Therapy 1994; 1:5-13.

14. Krol AV, Mol JNM, Stuitje AR. Modulation of eukaryotic gene expression by complement RNA or DNA sequences. Biotechniques 1988; 6:958-976.

15. Helene C, Toulme J-J. Specific regulation of gene expression by antisense, sense, and antigene nucleic acids. Biochemica et Biophysica 1990; 1049:99-125.

16. Zon G. Oligonucleotide as potential chemotherapeutic agents. Pharmaceutical Research 1988; 5:539-549.

17. Stein CA, Cheng Y-C. Antisense oligonucleotides as therapeutic agents—Is the bullet really magic? Science 1993; 261:1004-1012.

18. Wickstrom EL, Bacon TA, Gonzalez A et al. Human promyelocytic leukemia HL-60 cell proliferation and c-myc protein expression are inhibited by antisense pentadecadeoxynucleotide targeted against c-myc mRNA. Proc Natl Acad Sci USA 1988; 85:1028-1032.

19. Kashani-Sabet M, Funato T, Florenes VA et al. Suppression of a neoplastic phenotype in vivo by an anti-ras ribozyme. Cancer Research 1994; 54:900-902.

20. Ebbinghaus SW, Gee JE, Rodu B et al. Triplex formation inhibits HER-2/neu transcription in vivo. J Clin Invest 1993; 92:2433-2439.

21. Haseloff J, Gerlach WL. Simple RNA enzymes with new and highly specific endoribonuclease activities. Nature 1988; 334:585-591.

22. Kashani-Sabet M, Funato T, Tone T et al. Reversal of the malignant phenotype by an anti-ras ribosome. Antisense Research and Development 1992; 2:3-15.

23. Ratahczak MZ, Kant, JA, Luger SM et al. In vivo treatment of human leukemia in a scid mouse model with c-myb antisense oligonucleotides. Proc Natl Acad Sci USA 1992; 89:11823-11827.

24. Georges RN, Mukhopadhyay T, Zhang Y et al. Prevention of orthotopic human lung cancer growth by intratracheal instillation of a retroviral antisense k-ras construct. Cancer Research 1993; 53:1743-1746.

25. Iverson PL, Zhu S, Meyer A et al. Cellular uptake and subcellular distribution of phosphorothioate into cultured cells. Antisense Research and Development 1992; 2:211-222.

26. Citro G, Perrotti D, Cucco C et al. Inhibition of leukemia cell proliferation by receptor-mediated uptake of c-myb antisense oligodeoxynucleotides. Proc Natl Acad Sci USA 1992; 89:7031-7035.

27. Bergan R, Connell Y, Fahmy B, et al. Electroporation enhance c-myc antisense oligodeoxynucleotide efficacy. Nucleic Acids Research 1993; 21:3567-3573.

28. Theuer CP, Pastan I. Immunotoxins and recombinant toxins in the treatment of solid carcinomas. Amer J Surg 1993; 166:284-288.

29. Brinkmann U, Pai LH, FitzGerald DJ, et al. B3-(Fv)-PE38KDEL, a single chain immunotoxin that causes complete regression of a human carcinoma in mice. Proc Natl Acad Sci USA 1991; 88:8616-8620.

30. Colcher D, Bird R, Roselli M et al. In vivo tumor targeting of a recombinant single-chain antigen-binding protein. J Natl Cancer Institute 1990; 82:1191-1197.

31. Wawrzynczak EJ. Rational design of immunotoxins: current progress and future prospects. Anti-Cancer Drug Design 1992; 7:427-441.

32. Mykebust AT, Godal A, Fodstad O. Targeted therapy with immunotoxins in a nude rat model for leptomenineal growth of human small cell cancer. Cancer Research 1994; 54:2146-2150.

33. Friedman PN, Chance DF, Trail PA et al. Antitumor activity of the single-chain immunotoxin BR96 scFv-PE40 against established breast and lung tumor xenografts. J Immunol 1993; 150:3054-3061.

34. Marasco WA, Haseltine WA, Chen S-Y. Design, intracellular expression, and activity of human anti-human immunodeficiency virus type 1 gp120 single-chain antibody. Proc Natl Acad Sci USA 1993; 90:7889-7893.

35. Chen S-Y, Bagley J, Marasco WA. Intracellular antibodies as a new class of therapeutic molecules for gene therapy. Human Gene Therapy 1994; 5:595-601.

36. Yarden Y, Ullrich A. Growth factor receptor tyrosine kinases. Annu Rev Biochem 1988; 57:443-478.

37. Hudziak RM, Schlessinger J, Ulrich A. Increased expression of the putative growth factor receptor p185HER2 causes transformation and tumorigenesis of NIH 3T3 cells. Proc Natl Acad Sci USA 1987; 84:7159-7163.

38. Muller WJ, Sinn E, Pattengale PK et al. Single-step induction of mammary adenocarcinoma in transgenic mice bearing the activated c-neu oncogene. Cell 1988; 54:105-115.

39. Slamon DJ, Godolphin W, Jones LA et al. Studies of the HER-2/neu proto-oncogene in the human breast and ovarian cancer. Reports 1989;707-712.

40. Hynes NE. Amplification and overexpression of the erbB-2 gene in human tumors: its involvement in tumor development, significance as a prognostic factor, and potential as a target for cancer therapy. Cancer Biology 1993; 4:19-26.

41. Giovanella BC, Vardeman DM, Williams LJ et al. Heterotransplantation of human breast carcinomas in nude mice. Correlation between successful heterotransplants, poor prognosis and amplification of HER-2/neu oncogene. Int J Cancer 1991; 47:66-71.

42. Drebin JA, Link VC, Greene MI. Monoclonal antibodies specific for the neu oncogene product directly mediate anti-tumor effects in vivo. Oncogene 1988; 2:387-394.

43. Fendley BM, Winget M, Hudziak RM et al. Characterization of murine monoclonal antibodies reactive to either the human epidermal growth factor receptor or HER2/neu gene product. Cancer Research 1990; 50:1550-1558.

44. Carter P, Presta L, Gorman CM et al. Humanization of an anti-p185HER2 antibody for human cancer therapy. Proc Natl Acad Sci USA 1992; 89:4285-4289.

45. Hurwitz E, Stancovski I, Sela M et al. Suppression and promotion of tumor growth by monoclonal antibodies to ErbB-2 differentially correlate with cellular uptake. Proc Natl Acad Sci USA 1995; 92:5867-5871.

46. De Santas K, Slamon D, Anderson SK et al. Radiolabeled antibody targeting if the HER-2/neu oncoprotein. Cancer Research 1992; 52:1916-1923.

47. Batra JK, Kazpryzyk PG, Bird RE et al. Recombinant anti-erbB-2 immunotoxins containing pseudomonas exotoxin. Proc Natl Acad Sci USA 1992; 89:5867-5871.

48. Bertram J, Killian M, Brysch W et al. Reduction of erbB-2 gene product in mammary carcinoma cell lines by erbB-2 mRNA-specific and tyrosine kinase consensus phosphorothioate antisense oligonucleotides. Biochemical and Biophysical Research Communications 1994; 200:661-667.

49. Brysch W, Magal E, Louis J-C et al. Inhibition of p185c-erbB-2 proto-oncogene expression by antisense oligodeoxynucleotides downregulates p185-associated tyrosine-kinase activity and strongly inhibits mammary tumor-cell proliferation. Cancer Gene Therapy 1994; 1:99-105.

50. Curiel DT, Wagner E, Cotten M et al. High efficiency gene transfer mediated by adenovirus coupled to DNA-polylysine complexes via an antibody bridge. Human Gene Therapy 1992; 3:147-154.

51. Gerdes J, Lemke H, Baisch H et al. Cell cycle analysis of a cell proliferation-association human nuclear antigen defined by the monoclonal antibody Ki-67. J Immunol 1984;133:1710-1715.

52. Hinshaw VS, Olsen CW, Dybhahl-Sissoko N et al. Apoptosis: a mechanism of cell killing by influenza A and B viruses. Journal of Virology 1994; 68:3667-3673.

53. Duke RC, Cohen JJ. Morphological and Biochemical Assays of Apoptosis. Current Protocols Supplement 1992; 3.

54. Vaux DL. Toward an understanding of the molecular mechanisms of physiological cell death. Proc Natl Acad Sci USA 1993; 90:786-789.

55. Williams GT. Programmed cell death: apoptosis and oncogenesis. Cell 1991; 65:1097-1098.

56. Thompson CB. Apoptosis in the pathogenesis and treatment of disease. Science 1995; 267:1456-1462.

57. Steller H. Mechanisms and genes of cellular suicide. Science 1995; 267:1445-1449.

58. Carson WE, Haldar S, Baiocchi RA et al. The c-kit ligand suppresses apoptosis of human natural killer cells through the upregulation of bcl-2. Proc Natl Acad Sci USA 1994; 91:7553-7557.

59. Miyashita T, Reed JC. Bcl-2 gene transfer increases relative resistance of S49.1 and WEH17.2 lymphoid cells to cell death and DNA fragmentation induced by glucocorticoids and multiple chemotherapeutic drugs. Cancer Research 1992; 52:5407-5411.

60. Miyashita T, Reed JC. Bcl-2 oncoprotein blocks chemotherapy-induced apoptosis in a human leukemia cell line. Blood 1993; 52:5407-5411.

61. Hinshaw VS, Olsen CW, Dybdahl-Sissoko N et al. Apoptosis: a mechanism of cell killing by influenza A and B viruses. Journal of Virology 1994; 68:3667-3673.

62. Alvarez RD, Curiel DT. A phase I study of recombinant adenovirus vector-mediated delivery of an anti-erbB-2 single chain (sFv) antibody gene for previously treated ovarian and extraovarian cancer patients. Human Gene Therapy 1997; 8:229-242.

63. Yamada-Okabe T, Yamada-Okabe H, Kashima Y et al. Effects of oncogenes on the resistance to cis-diamminedichloroplatinum (II) and metallothionein gene expression. Toxicology and Applied Pharmacology 1995; 133:233-238.

64. Tsai C-M, Yu D, Chang K-T et al. Enhanced chemoresistance by elevation of p185neu levels in HER-2/neu-transfected human lung cancer cells. J Natl Cancer Institute 1995; 87:682-684.

65. Tsai C-M, Chang K-T, Perng R-P et al. Correlation of intrinsic chemoresistance of nonsmall-cell lung cancer cell lines with HER-2/neu gene expression but not with the ras gene mutations. J Natl Cancer Institute 1995; 85:897-901.

66. Hancock MC, Langton BC, Chan T et al. A monoclonal antibody against the c-erbB-2 protein enhances the cytotoxicity of cis-diamminedichloroplatinum against human breast and ovarian tumor cell lines. Cancer Research 1991; 51:4575-4580.

67. Arteaga CL, Winnier AR, Poirier MC et al. p185c-erbB-2 signaling enhances cisplatin-induced cytotoxicity in human breast carcinoma cells: association between an oncogenic receptor tyrosine kinase and drug-induced DNA repair. Cancer Research 1994; 54:3758-3765.

68. Langton-Webster BC, Xuan J-A, Brink JR et al. Development of resistance to cisplatin is associated with decreased expression of the gp185c-erbB-2 protein and alterations in growth properties and responses to therapy in an ovarian tumor cell line. Cell Growth and Differentiation 1994; 5:1367-1372.

Intracellular Targeting of Oncogenes: A Novel Approach for Cancer Therapy

Olivier Cochet, Isabelle Delumeau, Mireille Kenigsberg,
Nadège Gruel, Fabien Schweighoffer, Laurent Bracco,
Jean Luc Teillaud and Bruno Tocque

Introduction

It is now well established that cancer is a multistep disease. In the case of colon cancer, we know of at least five steps in the process of moving from a normal cell to a tumor cell.[1] Each step occurs at a very low frequency but can be accelerated or obviated by environmental or genetic factors. This process may be explained at a molecular level by the need for a tumor cell to sustain several mutations in its genome, each of which is required to alter a distinct target whose activation yields a subset of the phenotypic changes necessary for conversion to the fully tumorigenic state. Are there as many distinct mechanisms of transformation as there are oncogenes? In vitro and in vivo experiments showed that there are two central cellular growth regulatory pathways that must be disrupted in order for the cell to proceed to full oncogenic transformation. There is ample testimony to the fact that oncogenic retroviruses carrying single oncogenes are potently tumorigenic. The biological model of two genetic changes required for transformation is best exemplified by the cooperation between the ras and p53 genes[2] and provides a challenge to those studying the biochemistry of their encoded proteins. How can the biological effects of such mutated proteins be circumvented to induce tumor regression? Strategies for neutralizing the effects of these proteins by expressing synthetic genes in target tumors can now be envisioned. We will discuss in this chapter some of the approaches aimed at blocking the oncogenic functions of genes such as ras, and at neutralizing the oncogenic function of the p53 mutant forms using single chain antibodies.

Ras as a Target for Cancer Treatment

The discovery that ras genes are mutated in a large proportion of human tumors has generated considerable hope for cancer treatment (see ref. 3 for review). Indeed, mutated ras oncogenes have been associated with tumor progression,

Intrabodies: Basic Research and Clinical Gene Therapy Applications, edited by
Wayne A. Marasco. © 1998 Springer-Verlag and R.G. Landes Company.

metastasis and increased resistance to chemotherapy and radiotherapy.[4,5] One can expect some beneficial clinical input in the near future coming from any of the therapeutic approaches discussed below. Isoprenylation of Ras proteins is an absolute requirement for their membrane anchorage and transforming activity.[6] In a series of post-translational modifications, isoprenylation of cysteine[186] occurs first and is followed by the removal of the three last amino acids, and finally, by carboxymethylation of the isoprenylated and now C-terminal cysteine.[7] These findings opened up the possibility that specific inhibitors of Ras might be identified which have a selective toxicity for tumors constitutively expressing ras oncogenes.[8,9]

The most universally employed methodology to ablate expression of oncogenes is the utilization of "antisense" mlecules (DNA oligonucleotides or RNA antisense transcripts). A variety of experimental models have demonstrated the potential utility of the antisense approach as an anticancer therapy. Several studies have shown the ability to selectively block an oncogene with subsequent reversion of the malignant phenotype.[10-12] The in vivo demonstration of the effect could also be accomplished by direct, in vivo delivery of the antisense molecule.[13] However, this novel therapeutic approach is in practice associated with severe drawbacks which have hampered widespread use of this technology in human cancer gene therapy protocols. The tumor environment is deleterious to these molecules and it is often difficult to achieve effective intracellular concentrations. Different modifications of the antisense molecules as well as a number of vector approaches have been proposed to try to overcome these problems.[14]

As an alternative or a complement to the novel therapeutic approaches described above, vaccination with synthetic mutated Ras peptides may induce a beneficial immune response in patients with advanced tumors. It has been shown that HLA class II molecules can bind mutated Ras peptides and present them to human T cells, thereby initiating an immune response.[15,16] T cell subsets are able to specifically kill or inhibit the growth of colon cancer cell lines presenting the corresponding ras mutation.[17] Pilot clinical studies have been carried out, based on loading antigen-presenting cells from peripheral blood with a synthetic Ras peptide corresponding to the ras mutation found in the tumor from the same patient.[18] No systemic toxicity was reported and further clinical studies are in progress to demonstrate the efficacy of this approach.

To be envisioned as a anti-cancer agent, as depicted in Figure 7.1, the recombinant anti-Ras scFv should ideally abolish the tumorigenic properties, cause tumor shrinkage (possibly through an apoptotic mechanism), induce a bystander effect to tumor cells that do not express the transgene, and finally, synergize with common anti-cancer therapies. We have named such an scFv an Oncogene Neutralizing Binding Decoy (ONBD), and ideally, the toxicity of this ONBD should be selective for cancer cells.

Microinjection of the rat neutralizing anti-Ras monoclonal antibody (mAb) Y13-259[19] was shown to lead to phenotypic reversion of ras-transformed cells.[20] The epitope recognized by the antibody is a critical region that undergoes conformational changes upon GTP binding and is important for interaction of Ras with different effectors.[21] Thus we derived a single chain Fv fragment (scFv) from Y13-259 mAb and tested different modifications of this scFv (Fig. 7.2) for neutralization of oncogenic Ras pathways in different model systems. In addition, another scFv derived from Y13-238 mAb[19] which does not interfere with Ras oncogenic activity was used as a control.

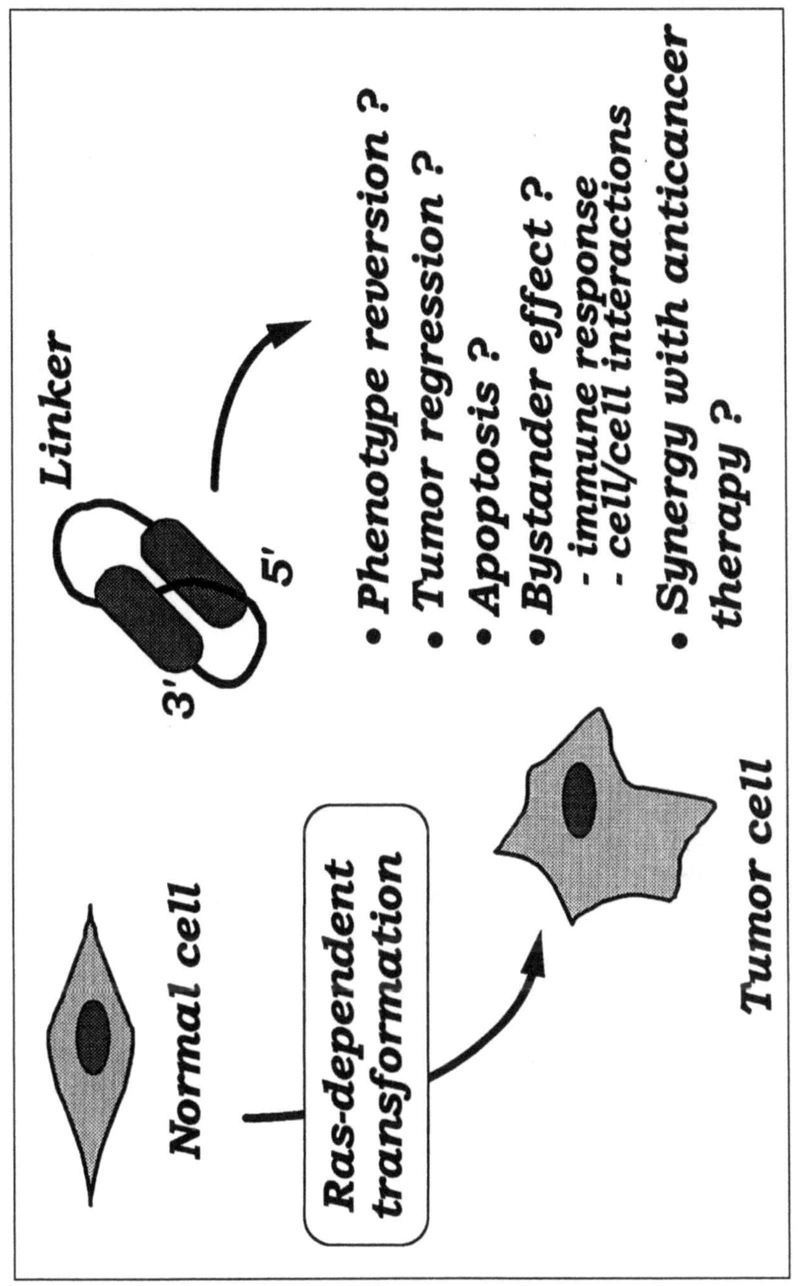

Fig. 7.1. Ras neutralizing scFv as an Oncogene Neutralizing Binding Decoy (see text for details).

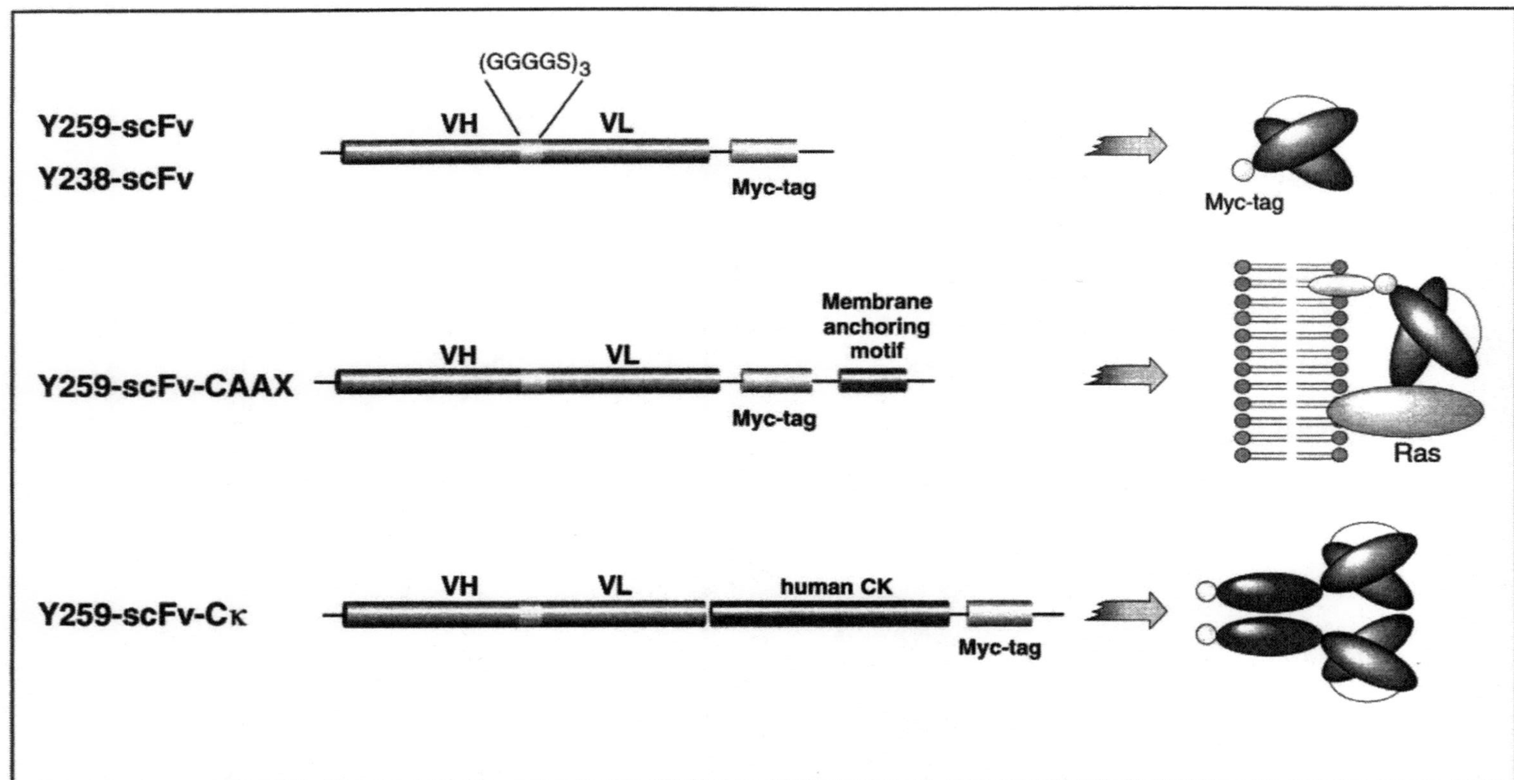

Fig. 7.2. Y259-scFv and derived molecules, Y259-scFv-CAAX and Y259-scFv-Cκ (see text for details).

Generation and In Vitro Characterization of scFvs Targeted to Ras

The generation of Y259-scFv was performed using a three-step PCR method based on the use of primers designed to maximize the mouse V_H and V_κ PCR product diversity and that cross-hybridize to rat immunoglobulin variable regions. Total RNA from rat Y13-259 hybridoma was used for cDNA synthesis. V_H and V_κ were separately amplified with mouse primers V_H1BACK,[22] V_H1FOR-2[23] and Vκ2BACK, Vκ4FOR,[24] respectively (Table 7.1). PCR conditions were 2 min at 94°C followed by 30 cycles (1 min 94°C, 1 min 60°C, 1 min 72°C) of amplification. Purified PCR products were then separately amplified with nested primers adding restriction sites to the 5' V_H and 3' V_κ ends and overlapping complementary sequences encoding the $(Gly_4S)3$ linker to the 3' V_H and 5' V_κ ends. The primers used were V_H1BACK-SfiI (containing a *Nco*I site), Vκ 4FOR-*Not*I[24] SCV$_H$FOR and SCVκBACK (Table 7.1). PCR conditions were 2 min at 94°C followed by 10 amplification cycles (1 min 94°C, 1 min 50°C, 1 min 72°C). The third PCR (assembly PCR) was conducted in two steps: purified V_H and Vκ were mixed in equimolar amounts and assembled by 10 PCR cycles. Amplification was then followed by 15 cycles (1 min 94°C, 2 min 74°C) after addition of V_H1BACK-SfiI and Vκ4FOR-*Not*I. The PCR product (about 780 bp) was gel-purified, digested with *Nco*I and *Not*I and ligated into the pUC119 derived vector pSW1[23] to generate the pY259-myc vector, where the Vκ cDNA sequence is fused to a c-myc tag-encoding sequence (refer to Figure 7.2). Y259-scFv was also subcloned into pSW1-His vector (containing a 6X histidine motif) for purification by Ion Metal Affinity Chromatography (IMAC). The insert was sequenced and shown to be identical to the sequence previously published.[25]

The cloning of Y238-scFv was more difficult to achieve as PCR amplification with primers used for generating Y259-scFv led to the assembly of the V_H of Y13-238 mAb with the Vκ produced by the secreting myeloma fusion partner Y3-Ag 1.2.3.[26] The sequence of the cloned V_H was confirmed to be a rat immunoglobulin V_H by alignment with sequences of rat variable regions using the Genebank database. The V_L region of Y13-238 was successfully cloned after designing primers based on the N-terminal peptide microsequence of the light chain of purified Y13-238 mAb (not shown). Alignment of the peptide sequence with sequences from Genebank database revealed that the V_L of Y13-238 belonged to a λ subgroup. De novo designed primers scRV$_L$2-Back and Not-RV$_L$-For (Table 7.1) were therefore used to amplify the Vλ region. Assembly with V_H and cloning into pSW1 were conducted as described for Y259-scFv.

We reasoned that intracellular targeting of the scFv to Ras could be significantly improved by adding a localization sequence that would target Y259-scFv in the vicinity of oncogenic Ras. As previously mentioned, Ras proteins are functional only when anchored to the inner face of the plasma membrane.[6] In the case of p21$^{K-Ras(2B)}$, following farnesylation of the C-terminal cysteine, a polybasic domain of six consecutive lysine residues further increases association with the plasma membrane through electrostatic interactions.[27] Thus, a DNA sequence encoding the 20 last amino acids of p21$^{K-Ras(2B)}$ (KMSKDGKKKKKKSKTK-CVIM) was removed by EcoRI digestion of the pA vector[28] (kindly provided by Pr. D. Louvard), and cloned into pSW1 at the 3' end of the Y259-scFv-c-myc sequence to give the Y259-scFv-CAAX construct (Fig. 7.2).

Efficient intracellular cytosolic expression of scFvs also faces another important challenge. Antibodies are normally assembled and folded in the secretory compartment, with the help of chaperone molecules such as BiP and GRP94.[29] In addi-

Table 7.1

V$_H$1FOR-2	5'-TGAGGAGACGGTGACCGTGGTCCCTTGGCCCC-3'
V$_H$1BACK	5'-AGGTS*MARCTGCAGSAGTCWGG-3'
Vκ4FOR (equimolar mix)	5'-CCGTTTGATTTCCAGCTTGGTGCC-3' 5'-CCGTTTTATTTCCAGCTTGGTCCC-3' 5'-CCGTTTTATTTCCAACTTTGTCCC-3' 5'-CCGTTTCAGCTCCAGCTTGGTCCC-3'
Vκ2BACK	5'-GACATTGAGCTCACCCAGTCTCCA-3'
SCV$_H$FOR	5'-AGAGCCACCTCCGCCTGAACCGCCTCCACCTGAGGAGACGGTGACCG-3'
V$_H$1BACK-SfiI	5'-TACTCGCGGCCCAACCGGCCATGGCCCAGGTSMARCTGCAGSAGTC-3'
Vκ4FOR-*NotI* (equimolar mix)	5'-GATATGAGATACTGCGGCCGCCCGTTTGATTTCCAGCTTGGTGCC-3' 5'-GATATGAGATACTGCGGCCGCCCGTTTTATTTCCAGCTTGGTCCC-3 5'-GATATGAGATACTGCGGCCGCCCGTTTTATTTCCAACTTTGTCCC-3' 5'-GATATGAGATACTGCGGCCGCCCGTTTCAGCTCCAGCTTGGTCCC-3'
SCVκBACK	5'-GGCGGAGGTGGCTCTGGCGGTGGCGGATCGGACATTGAGCTCACCCAG-3'
SCRV$_L$2BACK	5'GGCGGAGGTGGCTCTGGCGGTGGCGGATCGCAGKTTGTGCTTACTCAGCC AACC-3'
Not-RV$_L$-For	5'-GATATGAGATACTGCGGCCGCTAGGACAGWGAGCTTTGTTCCACC-3'
VκFCκ2	5'-AGATGGTGCAGCCACAGTCCGTTTGATTTCCAGCTT-3'
huCκBACK	5'-ACTGTGGCTGCACCATCT-3'
huCκF-*NotI*	5'-GATATGAGATACTGCGGCCGCCTCTCCCCTGTTGAAGCT-3'

* S = C or G, M = A or C, R = A or G, W = A or T, K = G or T

tion, the reducing environment of the cytosol is unfavorable to intrachain disulfide bonds formation;[30] yet recombinant whole antibodies or scFvs expressed in the cytosolic compartment proved to be functional.[25,31-34] It has also been reported that lower amounts of cytosol-targeted scFvs are detected in this compartment compared to the same scFv targeted to endoplasmic reticulum, suggesting a marked unstability of these antibody fragments in the cytosol.[30] Thus, we also investigated whether the neutralizing effect of Y259-scFv would be improved by fusing the human Cκ region to the V$_L$ C-terminal part of Y259-scFv. It had been previously reported that this region stabilizes recombinant fragment antibodies, possibly through dimerization.[34] The scFv-Cκ construct (Y259-scFv-Cκ) was generated in a two-step PCR: Y259-scFv cDNA inserted into pSW1 was amplified with V$_H$1BACK-SfiI and VκFCκ2 primers (Table 7.1). The human plasma leukemia cell line ARH-77 (obtained from the ATCC) was used to amplify the huCκ cDNA by RT-PCR using huCκBACK and huCkF-*NotI* primers (Table 7.1). Purified scFv and huCκ PCR products were then assembled by overlap PCR extension with primers V$_H$1BACK-SfiI and huCκF-*Not*I. The assembly PCR product was then cloned into the pSW1 vector as described above, giving rise to the scFv-Cκ-c-myc construct (Fig. 7.2). In order to express the anti-ras scFvs in eukaryotic cells, their DNAs were subcloned into pMT3[35] and pcDNA3. The pMT3 vector is specifically designed for expression in *Xenopus laevis* oocytes, while pcDNA3 allows high level expression in mammalian cells. The inserts were removed from pSW1 vector by *Nco*I/*Eco*RI digestion, blunt-ended and ligated into Sma*I*-digested pMT3 or EcoRV-digested pcDNA3.

In a first set of experiments, the scFv-encoding DNAs were expressed from PSW1 in the TG1 *Escherichia coli* strain and their binding to Ras was examined. The PelB sequence leader[23] allows the production of the scFvs into the periplasm. scFv fragments recovered from periplasmic lysates[36] were either purified (when the histidine-tag was present) or used as crude extracts. The presence of recombinant antibody fragments in periplasmic fractions was assessed by Western blot (Fig. 7.3A) with the 9E10 anti-c-myc tag mAb. IMAC purification yielded nearly pure scFvs (Fig. 7.3B), with production levels ranging from 0.5 to 1 mg/liter. The Ras binding ability of the Y13-259-derived scFvs was demonstrated by ELISA and Western blot (Fig. 7.3C and D).

As an alternative to the anti-c-myc 9E10 mAb, polyclonal rabbit anti-scFv antibodies were generated using a pool of scFvs as immunogens (Fig. 7.4). Four scFvs (anti-p21 Ras, anti-GD2, anti-FcγRII/CD32, and anti-CEA) containing an N-terminal histidine tail were purified from periplasmic fractions. Purified scFvs were pooled and the purity was checked by gel electrophoresis. Rabbits were immunized on days 1, 15, and 30 by subcutaneous injections of 50 mg of the scFv mix. A last boost was performed on day 60. Antibody response against the scFv mix was tested by ELISA. The anti-scFv recognition ability of the polyclonal antibodies was demonstrated using one scFv lacking the c-myc tag (Fig. 7.4). In addition, these antibodies bound purified mouse and human IgGs but not human Fcγ fragments, BSA or TG1 periplasmic fractions.

Functional Activity of the Anti-Ras scFvs in Xenopus Oocytes

Stage VI oocytes obtained from *Xenopus laevis* frogs can undergo meiotic maturation in response to insulin, which activates the Ras signaling pathway, or progesterone which does not requires Ras.[37] Germinal vesical breakdown (GVBD), a hallmark of maturation, is scored by the appearance of a white spot on the dark animal pole of the oocyte. Prior to GVBD, a cascade of kinase enzymes including maturation promoting factor or $p34^{cdc2}$ is activated.[38] We compared the three anti-Ras Y13-259 derived scFv constructs in these assays because microinjection of the parental antibody Y13-259 has been reported to block oocyte maturation induced by Ras as well as by insulin.[37] pMT3 plasmids encoding each construct were microinjected into the nucleus and expression allowed for 30 hours before addition of the progesterone or Ras stimulus. Both GVBD and $p34^{cdc2}$ were efficiently inhibited by all three scFvs (Fig. 7.5) in a dose-dependent manner (not shown). Targeting the scFv to the plasma membrane with a CAAX motif, or adding a Cκ sequence, which is expected to stabilize the molecule, did not result in improved activity in this model system. This lack of difference might be explained by the finding that the unmodified scFv is already associated with the membrane in oocytes.[32] In addition, all three scFvs were expressed at high levels and appeared to be quite stable (not shown).

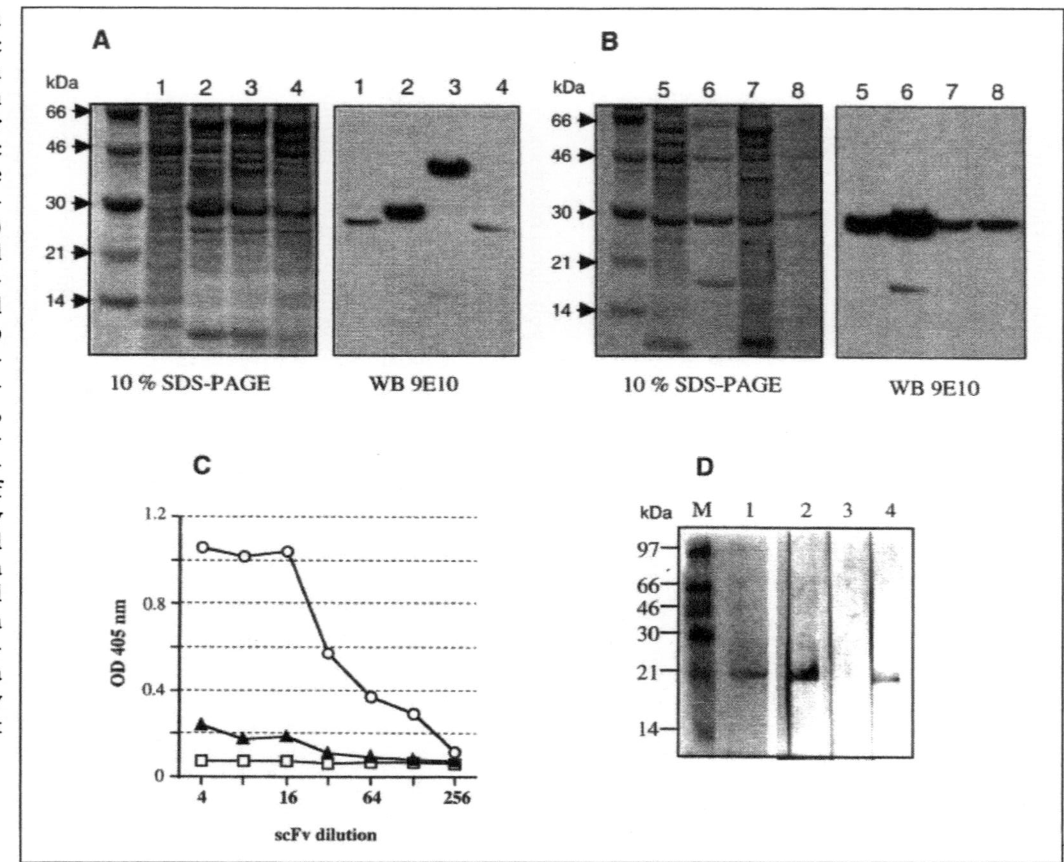

Fig. 7.3. Expression and characterization of Y259- and Y238-scFvs. (A) Periplasmic expression of the anti-Ras scFvs shown by Coomassie staining after separation in a 10% polyacrylamide gel (left panel) or detected by Western blot with anti-c-Myc 9E10 mAb (right panel). Y259-scFv (lane 1), Y259-scFv-CAAX (lane 2), Y259-scFv-C_κ (lane 3) and Y238-scFv (lane 4). (B) Purification of Y259- (lanes 5,6) and Y238-scFv (lanes 7,8) by IMAC. Periplasmic fractions (lanes 5,7) and purified scFvs (lanes 6,8) were detected with 9E10 mAb. (C) Binding of the scFvs to immobilized human p21Ha-ras by ELISA. Purified Y259-scFv (open circles), periplasmic fraction of Y259-scFv (triangles) and purified Y238-scFv were detected with 9E10 mAb. (D) Binding of Y259-scFv to p21Ha-Ras demonstrated by Western blot with 9E10 mAb. Purified p21Ha-ras stained after migration on a 10% polyacrylamide gel (lane 1), detected after transfer onto nitrocellulose with Y13-259 mAb (lane 2), an irrelevant control scFv (anti-CD3, lane 3) or a periplasmic extract containing Y259-scFv (1/5 dilution, lane 4). Molecular weight markers are indicated in kilodaltons.

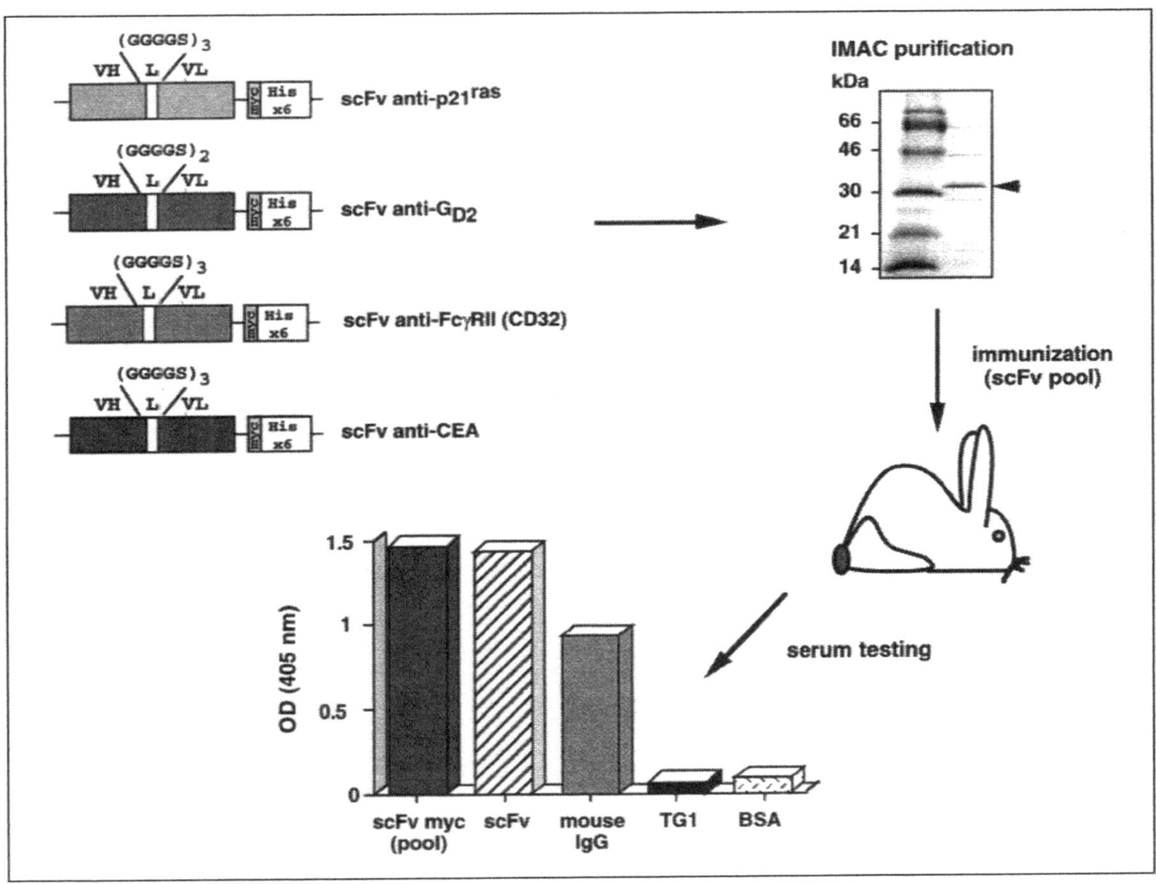

Fig. 7.4. Four IMAC-purified scFvs (anti-p21ras, anti-GD2, anti-FcgRII/CD32, and anti-CEA) containing an N-terminal histidine tail were injected into rabbits after evaluation of their purity by SDS-PAGE. Antibody response was assessed by ELISA against the scFv mix, a scFv lacking the c-Myc tag, purified mouse IgG, BSA or TG1 periplasmic fraction.

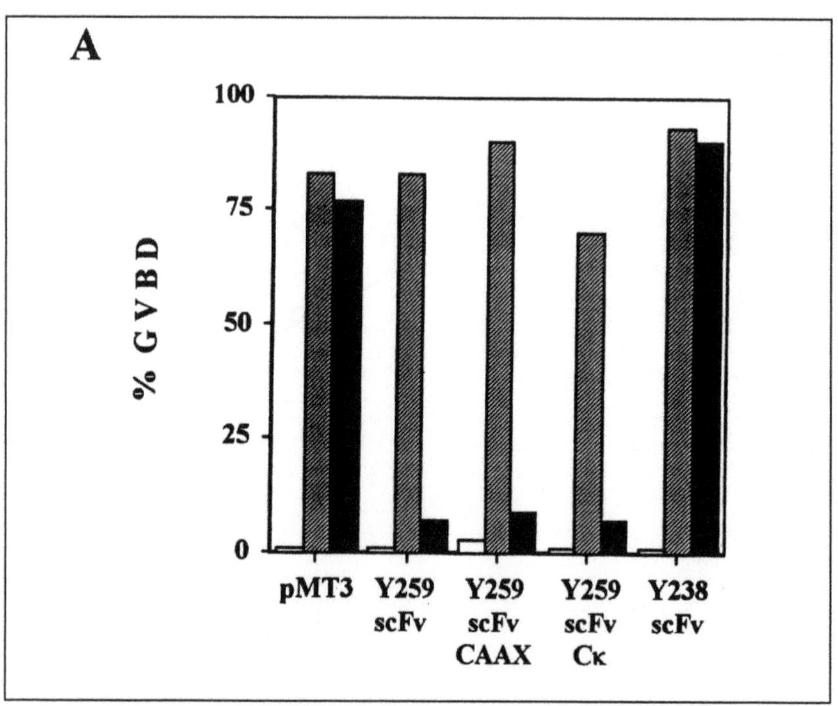

A
100
75
50
25
0
% GVBD
pMT3
Y259 scFv
Y259 scFv CAAX
Y259 scFv Cκ
Y238 scFv

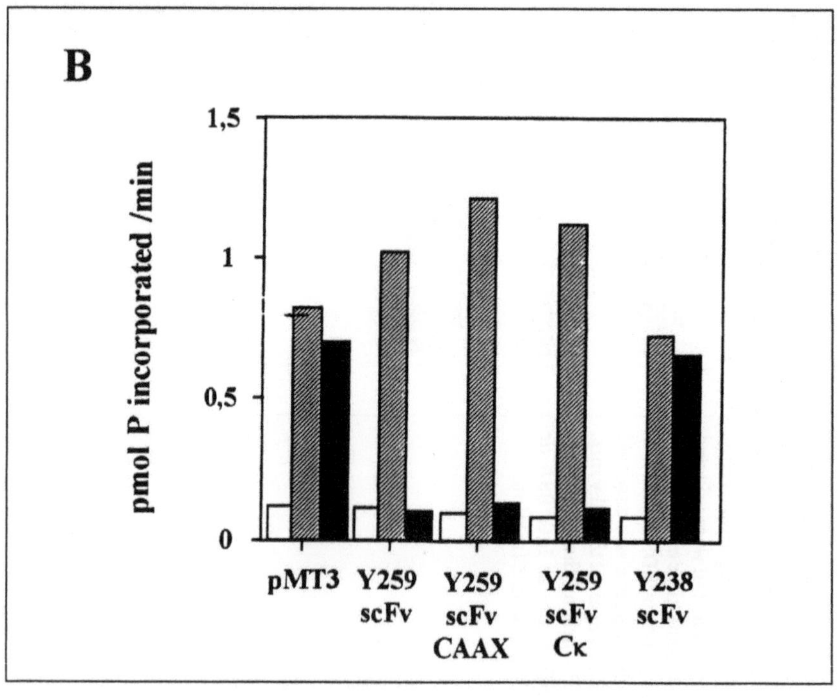

B
1,5
1
0,5
0
pmol P incorporated /min
pMT3
Y259 scFv
Y259 scFv CAAX
Y259 scFv Cκ
Y238 scFv

Functional Activity of the Anti-Ras scFvs in Mammalian Cells

Expression and activity of the scFvs were then tested in mammalian cells. The subcellular localization of the proteins expressed following transient expression in Swiss 3T3 cells was determined by immunofluorescence using the polyclonal anti-scFv antiserum. Addition of the CAAX box resulted in enhanced colocalization of Ras and the scFv (Fig. 7.6A) at the plasma membrane, although a large proportion of the scFv was still present in the cytosol. The scFv-Cκ (Fig. 7.6B) had the same cytosolic distribution as the nonmodified scFv (Fig. 7.6C), however, it could be detected for longer periods of time, suggesting that its overall stability was increased. Accordingly, Western blot analysis showed the presence of the full length Y259-scFv-Cκ at time points when the unmodified version was no longer detectable (not shown). Finally, as shown by electron microscopy, the scFvs were absolutely excluded from the nucleus (not shown).

Expression of oncogenic Ras has been shown to activate the transcriptional AP-1 complex.[39] To investigate the inhibitory potential of the scFvs on this pathway, NIH3T3 cells were cotransfected with a reporter plasmid encoding the enzyme Chloramphenicol Acetyl Transferase under the control of a Ras-responsive promoter (Py-CAT), along with pSV$_2$ expression vectors encoding activated Ha-Ras or oncogenic Raf, in the presence or absence of the scFv constructs in pcDNA3. As shown in Figure 7.7, the anti-Ras scFvs did not inhibit the basal activity of the Py promoter. As reported, however, the unmodified scFv derived from Y13-259, but not that derived from the control Y13-238 antibody, decreased transcription of the CAT gene induced by oncogenic Ras but not by v-Raf.[40] Interestingly, addition of the CAAX motif appeared to increase the ability of Y259-scFv to inhibit Ras-induced signaling, again, without affecting the Raf pathway. Addition of the Cκ moiety, on the other hand, did not significantly improve the neutralizing efficacy of the scFv (Fig. 7.7). To further assess the potential gain in unmodified scFv, the expression of the endogenous c-fos protein was monitored by immunofluorescence after serum stimulation of quiescent Swiss 3T3 cells. As shown in Figure 7.8, the scFv-CAAX was reproducibly more effective at preventing the expression of c-fos than the unmodified Y259-scFv.

Neutralization of Ras should result in phenotypic reversion of ras-transformed cells. Since Ras was shown to act as a suppressor of apoptosis in different cell systems,[45] we asked whether the expression of the anti-Ras scFv would result in the death of target tumor cells. Indeed, this turned out to be the case.[40] Table 7.2 scores the results obtained in terms of cell death 20 hours after injection of the plasmids encoding the different scFvs into human lung carcinoma H460 cells, which express

Fig. 7.5. (opposite) Y259-, but not Y238-scFv, interferes with Ras-dependent signaling in *Xenopus* oocytes. (A) Effect of Y259- and Y238-scFvs on Ras- and progesterone-induced GVBD. Oocytes were injected with the empty plasmid (pMT3) or with the plasmid expressing each scFv. 30 hours later, they were unstimulated (□), stimulated by progesterone (▨), or microinjected with p21 Ha-ras (■). GVBD was scored 18 hours later. (B) Effect of Y259- and Y238-scFvs on Ras- and progesterone-stimulated p34^{cdc2} kinase activity. Once GVBD was scored, oocytes were lysed in b glycerophosphate containing buffer,[41] and 10 µg of lysate was used for assaying p34^{cdc2} activity using Biotrak's kit as described by the manufacturer (Amersham). The results from one experiment representative of three are shown.

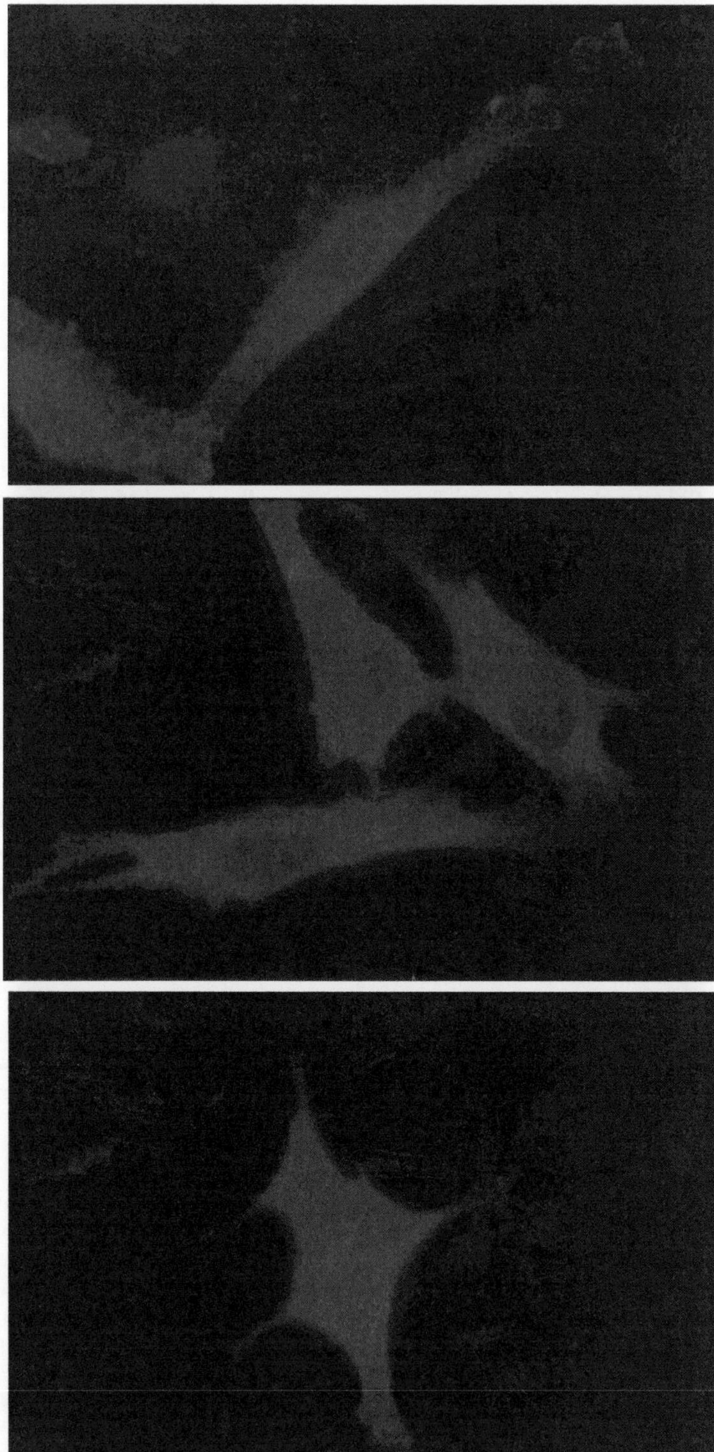

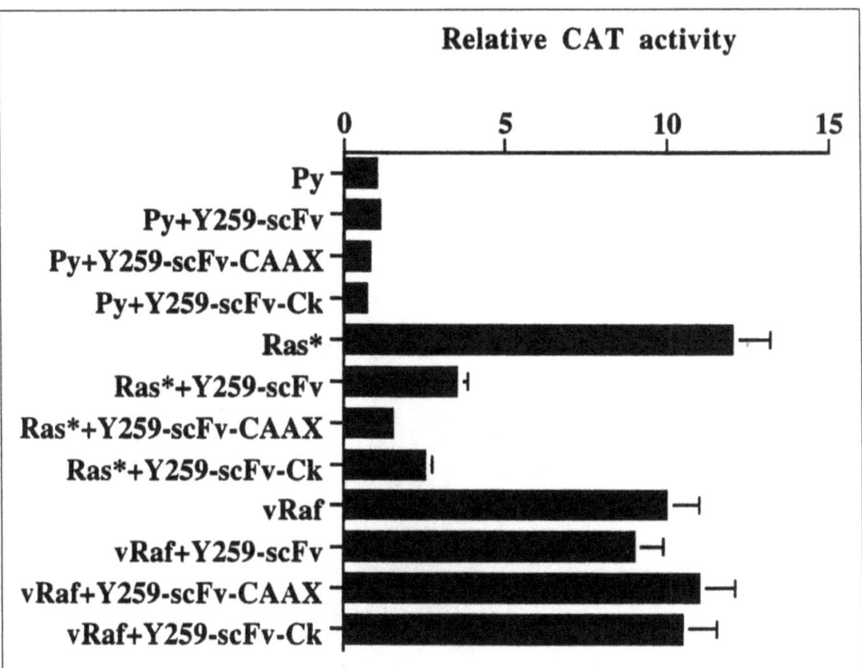

Fig. 7.7. Effect of the different anti-Ras scFvs on Ras-stimulated transactivation of a CAT reporter gene in NIH3T3 cells. CAT activity resulting from the expression of activated Ha-ras or v-Raf oncoproteins was measured 48 hours after transfection of cells with the scFv pcDNA3 expression vectors.[41] Results are shown as the mean ± sed of 3 transfections.

a mutant Ki-ras gene. Once again, the scFv-CAAX was the most effective at inducing apoptosis. To evaluate the in vivo efficacy of the anti-Ras scFv, we engineered an adenoviral vector to target mutant ras expressing tumor cells established in nude mice. We have shown that the anti-Ras scFv could be readily expressed in human lung and colon carcinoma cells. In addition, infection of tumor cells by direct intratumoral injection caused a pronounced tumor regression.[40] Although we have not yet tested the CAAX-modified scFv in this model, we can anticipate that this construct would be even more effective in vivo. How this ONBD works in vivo remains to be ascertained, but it likely triggers a potent bystander effect. Certainly, this scFv deserves more extensive analysis at the preclinical level to further evaluate its value as a candidate for cancer gene therapy.

Fig. 7.6. (opposite) Swiss 3T3 cells were microinjected with pcDNA3 plasmids (100 μg/ml) encoding Y259-scFv-CAAX (A), Y259-scFv-C_K (B), or Y259-scFv (C). Three hours later, cells were fixed in formaldehyde, permeabilized in Triton-X100 and stained with the polyclonal anti-scFv antiserum followed by incubation in anti-rabbit F(ab')$_2$ coupled to Texas Red.

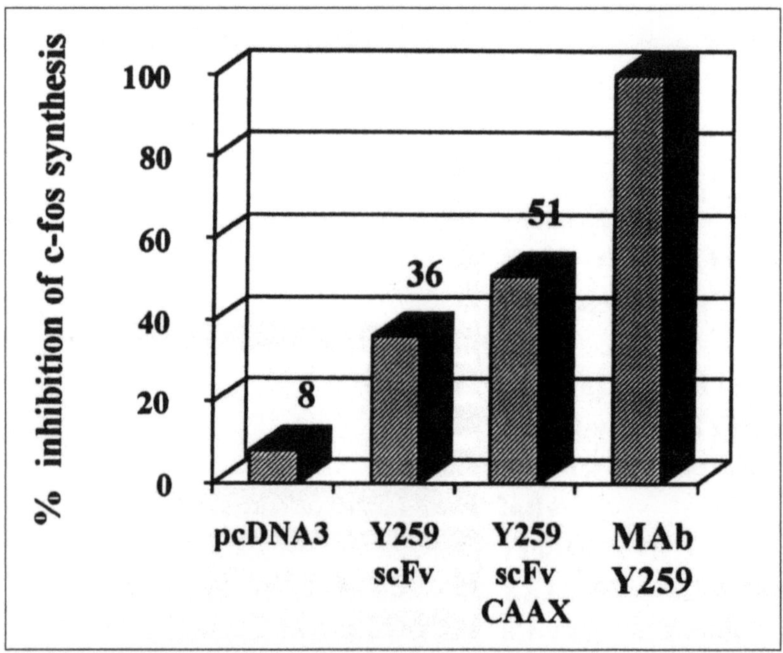

Fig. 7.8. Effect of Y259-scFv and Y259-scFv-CAAX on the expression of c-Fos stimulated by serum. Swiss 3T3 cells starved for 24 hours in 0.5% serum were microinjected with the indicated plasmids (1 mg/ml). A rat IgG was coinjected to facilitate the identification of injected cells. After a 3 hour-incubation, they were stimulated with 10% serum for an hour, then fixed and stained with a sheep anti-c-Fos antibody (Cambridge Research Biochemical) followed by an anti-sheep $F(ab')_2$ antibody coupled to fluorescein. Control cells were injected with Y13-259 mAb (10 mg/ml) prior to serum exposure. Detection of the cells injected with either the plasmids or Y13-259 mAb was carried out by staining with a Texas Red-labeled anti-rat antibody. This experiment was repeated twice with similar results.

Future Directions

As mentioned at the beginning of this chapter, p53 is clearly another target of choice for cancer therapy. However, even if p53-based gene therapy in the treatment of selected human cancers is becoming a reality,[41] we are still facing numerous challenges. In addition to the design of more efficient and specific vectors, the transgene to be expressed in the tumor has to function optimally. This may not always be the case for wild-type p53, which is known to be inactivated by several mechanisms. The most frequent target for p53-based gene therapy consists in tumors with a point mutation in one of the p53 alleles. The resulting p53 mutant is stabilized, adopting a conformation which, in some instances, has been shown to act as a neutralizing dominant negative molecule with respect to wild-type p53 through an hetero-oligomerization mechanism (Fig. 7.9). Upon binding to certain p53 mutants, some anti-p53 mAbs are able to restore wild-type p53-associated properties such as specific DNA binding and transcriptional activity. Selected anti-p53

Table 7.2. Y259-scFv and Y259-scFv-CAAX induce cell death by apoptosis in ras-expressing H460 tumor cells

Plasmid injected	Cells injected	TUNEL positive cells	% Apoptotic Cells
pcDNA3	129	6	4.6%
Y259-scFv	131	27	20.6%
Y259-scFv-CAAX	110	34	30.9%
Y238-scFv	143	8	5.6%

H460 cells plated on glass coverslips were microinjected with pcDNA3-Y238-, pcDNA3-Y259-scFv-CAAX or pcDNA3-Y259-scFv (0.25 mg/ml) along with a rat IgG. 20 hours later, cells were fixed in formaldehyde and permeabilized in Triton X-100. Coverslips were incubated with F(ab')₂ anti-rat-Texas Red (Jackson ImmunoResearch) to detect injected cells. Apoptotic cells were revealed by incorporation of FITC-conjugated dUTP into DNA breaks using the "in situ cell death detection kit" (TUNEL staining, Boehringer). The number of cells injected and the number of apoptotic cells were then determined.

scFvs may, therefore, serve to switch the functions of mutant p53 back to those of wild-type. Continuous intracellular expression of such scFvs may not only restore the transcriptional activity of p53 mutants, but also enable them to promote cell cycle arrest and/or apoptosis (Fig. 7.9).

ScFvs could also be used to obtain specific expression of killer genes in p53 mutated cells. Targeting transgene expression to the appropriate cells is critical for gene therapy. So far, most efforts have concentrated on the use of tissue-specific promoters isolated from genes which have a restricted expression pattern. A novel class of scFv-based chimeric molecules, the "trabodies", represented by the fusion between a scFv against a tumor-specific transcription factor and a DNA binding motif could be engineered. Since p53 mutants are found in most human cancers and possess a functional transactivation domain, we envision to combine these properties in order to specifically induce the expression of a deleterious gene in mutant p53-expressing tumor cells. These «trabodies» would activate transcription only in cells harboring p53 mutants, therefore limiting the possible toxicity of such therapeutic transgenes. We believe that the generation of synthetic ONBDs, such as the anti-ras scFv and the scFv-based p53 molecule, may not only increase the clinical efficacy and expand the range of cancers treatable by gene therapy, but will also highlight therapeutic points of intervention for conventional small molecular weight compounds. Moreover, these results should also shed light on strategies for identifying novel genes of therapeutic potential.

References

1. Fearon ER, Vogelstein B. A genetic model for colorectal tumorigenesis. Cell 1990; 61:759-767.
2. Zambetti GP, Olson D, Labow M et al. A mutant p53 is required for maintenance of the transformed phenotype in cells transformed with p53 plus ras cDNAs. Proc Natl Acad Sci USA 1992; 89:3952-3956.
3. Muleris M, Laurent-Puig P, Salmon RJ et al. Chromosome 12 alterations and c-Ki-ras mutations in colorectal tumors. Cancer Genet Cytogenet 1993; 69:161-216.

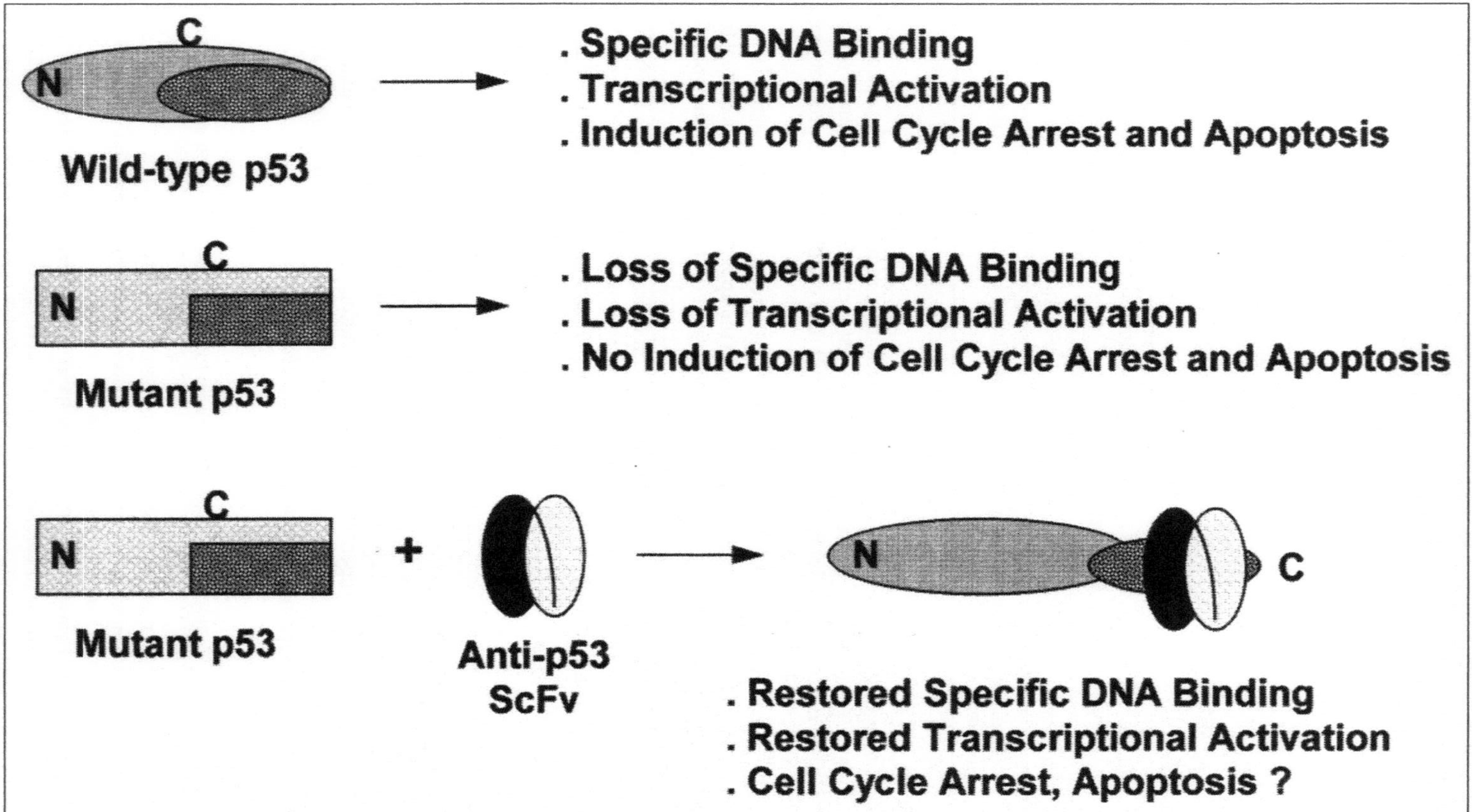

Fig. 7.9. Anti-p53 scFv: switching p53 mutants back to wild-type (see explanation in the text).

4. McKenna WG, Bernhard EJ, Markiewicz DA et al. Regulation of radiation-induced apoptosis in oncogene-transfected fibroblasts: influence of H-ras on the G2 delay. Oncogene 1996; 12:237-245.

5. Basu A, Cline JS. Oncogenic transformation alters cisplatin-induced apoptosis in rat embryo fibroblasts. Int J Cancer 1995; 63:597-603.

6. Kato K, Cox AD, Hisaka MM et al. Isoprenoid addition to Ras is the critical modification for its membrane association and transforming activity. Proc Natl Acad Sci USA 1992; 89:6403-6407.

7. Hancock JF, Cadwaller K, Marshall CJ. Methylation and proteolysis are essential for efficient membrane binding of prenylated p21$^{K-ras(B)}$. EMBO J 1991; 10:641-646.

8. Kohl NE, Omer CA, Conner MW et al. Inhibition of farnesyltransferase induces regression of mammary and salivary carcinomas in ras transgenic mice. Nature Med 1995; 1:792-797.

9. Kohl NE, Wilson FR, Mosser SD et al. Protein farnesyltransferase inhibitors block the growth of ras-dependent tumors in nude mice. Proc Natl Acad Sci USA 1994; 91:9141-9145.

10. Schwab G, Duroux I, Chavany C et al. An approach for new anticancer drugs: oncogene-targeted antisense DNA. Ann Oncol 1994; 5 (suppl.4):S55-S58.

11. Schwab G, Chavany C, Duroux L et al. Antisense oligonucleotides absorbed to polyalkylcyanoacrylate nanoparticles specifically inhibit mutated Ha-ras mediated cell proliferation and tumorigenicity in nude mice. Proc Natl Acad Sci USA 1994; 91:10460-10464.

12. Saison-Behmoaras T, Tocqué B, Rey I et al. Short antisense oligonucleotides directed against Ha-ras point mutation induce selected cleavage of the mRNA and inhibit T24 cells proliferation. EMBO J 1991; 10:1111-1118.

13. Wickstrom E. Strategies for administering targeted therapeutic oligodeoxy-nucleotides. TIB Tech 1992; 10:281-287.

14. Wagner, RW. The state of the art in antisense research. Nature Med 1995; 1:1116-1118.

15. Jung S, Schluesener HJ. Human T lymphocytes recognize a peptide of single point mutated, oncogenic ras proteins. J Exp Med 1991; 73:273-276.

16. Fossum B, Gedde-Dahl III T, Breivik J et al. p21 ras peptide-specific T cell responses in a patient with colorectal cancer. CD4$^+$ and CD8$^+$ T cells recognize a peptide corresponding to a common mutation (^{13}Gly to Asp). Int J Cancer 1994; 56:40-45.

17. Gedde-Dahl III T, Nilsen E, Thorsby E et al. Growth inhibition of a colonic adenocarcinoma cell line (HT29) by T cell specific for mutant p21 ras. Cancer Immunol Immunother 1994; 38:127-134.

18. Gjertsen MK, Bakka A, Breivik J et al. Ex vivo ras peptide vaccination in patients with advanced pancreatic cancer: results of a phase I/II study. Int J Cancer 1996; 65:450-3.

19. Furth ME, Davis LJ, Fleurdelys B et al. Monoclonal antibodies to p21 products of the transforming gene of Harvey murine sarcoma virus and the cellular ras gene family. J Virol 1982; 43:294-304.

20. Smith MR, DeGudicibus SJ, Stacey DW. Requirement for c-ras proteins during viral oncogene transformation. Nature 1986; 320:540-542.

21. Jurnack F, Heffron S, Bergmann E. Conformational changes involved in the activation of ras p21: implications for related proteins. Cell 1990; 60:525-528.

22. Orlandi R, Gussow DH, Jones PT et al. Cloning immunoglobulin variable domains for expression by the polymerase chain reaction. Proc Natl Acad Sci USA 1989; 86:3833-3837.

23. Ward ES, Güssow D, Griffiths AD et al. Binding activities of a repertoire of single immunoglobulin variable domains secreted from *Escherichia coli*. Nature 1989; 341:544-546.
24. Clackson T, Hoogenboom HR, Griffiths AD et al. Making antibody fragments using phage display libraries. Nature 1991; 352:624-628.
25. Werge TM, Biocca S, Cattaneo A. Intracellular immunization. Cloning and intracellular expression of a monoclonal antibody to the p21ras protein. FEBS Lett 1990; 274:193-198.
26. Crowe JS, Smith MA, Cooper HJ. Nucleotide sequence of Y3-Ag 1.2.3. rat myeloma immunoglobulin κ chain cDNA. Nucleic Acid Res 1989; 17:7992.
27. Hancock JF, Paterson H, Marshall J. A polybasic domain or palmitoylation is required in addition to the CAAX motif to localize p21ras to the plasma membrane. Cell 1990; 63:133-139.
28. Hancock JF, Magee AI, Childs JE et al. All ras proteins are polyisoprenylated but only some are palmitoylated. Cell 1989; 57:1167-1177.
29. Melnick J, Dul JL, Argon Y. Sequential interaction of the chaperones BiP and GRP94 with immunoglobulin chains in the endoplasmic reticulum. Nature 1994; 370:373-375.
30. Biocca S, Ruberti F, Tafani M et al. Redox state of single chain Fv fragments targeted to the endoplasmic reticulum, cytosol and mithochondria. Bio/Tech 1995; 13:1110-1115.
31. Biocca S, Neuberger MS, Cattaneo A. Expression and targeting of intracellular antibodies in mammalian cells. EMBO J 1990; 9:101-108.
32. Biocca S, Pierandrei-Amaldi P, Campioni N et al. Intracellular immunization with cytosolic recombinant antibodies. Bio/Tech 1994; 12:396-399.
33. Marasco WA, Haseltine WA, Chen S. Design, intracellular expression, and activity of a human anti-human immunodeficiency virus type 1 gp120 single-chain antibody. Proc Natl Acad Sci USA 1993; 90:7889-7893.
34. Mhashilkar AM, Bagley J, Chen SY et al. Inhibition of HIV-1 Tat-mediated LTR transactivation and HIV-1 infection by anti-Tat single chain intrabodies. EMBO J 1995; 14:1542-1551.
35. Tocqué B, Janicot M, Kenigsberg M. Oocyte microinjection assay for evaluation of Ras-induced signaling pathway. Methods Enzymol 1995; 255:426-435.
36. Skerra A, Plückthun A. Assembly of a functional immunoglobulin Fv fragments in *Escherichia coli*. Science 1988; 240:1038-1041.
37. Deshpande AK, Kung HF. Insulin induction of Xenopus laevis oocyte maturation is inhibited by monoclonal antibody against p21 ras proteins. Mol Cell Biol 1987; 7:1285-1288.
38. Pomerance M, Thang MN, Tocqué B et al. The Ras-GTPase activating protein SH3 domain is required for cdc2 activation and mos induction by oncogenic ras in Xenopus oocyte independently of mitogen-activated protein kinase activation. Mol Cell Biol 1996; 16:3179-3186.
39. Wasylyk C, Imler JL, Wasylyk, B. Transforming but not immortalizing oncogenes activate the transcription factor PEA1. EMBO J 1988; 7:2475-2483.
40. Cochet O, Kenigsberg M, Delumeau I et al. Intracellular expression of an antibody fragment neutralizing p21 Ras promotes tumour regression. submitted.
41. Roth J A, Nguyen D, Lawrence DD et al. Retrovirus-mediated wild-type p53 gene transfer to tumors of patients with lung cancer. Nature Med 1996; 2:985-991.

CHAPTER 8

Intrabodies Against the HIV-1 Regulatory Proteins: Tat and Rev as Targets for Gene Therapy

Abner M. Mhashilkar and Wayne A. Marasco

Introduction

Human Immunodeficiency Virus Type 1 (HIV-1) the causative agent of Acquired Immunodeficiency Syndrome (AIDS) is tightening its grip all around the globe. For example, in India, researchers estimate that by the year 2000 anywhere from 15-50 million people could be HIV positive. Half the prostitutes in Bombay are already infected and doctors report that the disease is spreading along major truck routes and rural areas, as migrant workers bring the virus home. In Central and Eastern Europe, countries that have largely escaped the AIDS epidemic are seeing an explosion in the number of cases, mainly among drug users and their heterosexual contacts.[1]

Since the onslaught of HIV-1, a number of potential strategies for anti-retroviral therapy of AIDS and its related diseases have emerged. The armamentarium of anti-HIV-1 is rapidly growing. Nucleoside and non-nucleoside inhibitors of HIV-1 reverse transcriptase (HIV-1 RT) have been identified to be active against HIV-1 in vitro and to a lesser extent, in vivo. Recently, HIV-1 protease inhibitors have illustrated notable therapeutic effects in clinical trials. However, drug-related toxicities, drug-resistant HIV-1 variants and drug-related side-effects are major hurdles which have not been surpassed for a successful long-term anti-HIV-1 therapy.[2]

In the last decade, genetic-based strategies for the treatment of HIV-1 have received increased attention due to lack of chemotherapeutic drugs or vaccines that show long-term efficacy in-vivo. Anti-HIV-1 gene-therapy approaches can be broadly categorized into three groups. The first category attempts to use different approaches to enhance body's immune response to HIV-1. The second is to engineer cells to secrete a factor that would either assist the body's resistance mechanisms against HIV-1 or inhibit HIV-1 replication. The third is to engineer HIV-1 infected cells, so as to either kill them directly or to inhibit the replication of HIV-1 within them.[3]

Intrabodies: Basic Research and Clinical Gene Therapy Applications, edited by Wayne A. Marasco. © 1998 Springer-Verlag and R.G. Landes Company.

The concept of "Intracellular Immunization (which aptly means intracellular resistance or intracellular inhibition)" was first introduced by David Baltimore[4] and has emerged as a novel strategy to counteract HIV-1. It utilizes molecular modulators such as anti-sense RNA, ribozymes, dominant negative mutants and intracellular antibodies, "intrabodies," for inhibiting viral gene expression within the cell.[5] The following paragraphs will summarize the efficacy of intrabodies directed against two important regulatory components of HIV-1, namely Tat and Rev.

HIV-1 Tat, an Important Regulatory Component of HIV-1

HIV-1 encodes a small (16 kDa) regulatory protein, Tat, a potent transcriptional activator of HIV-1 long terminal repeat (LTR). Tat activation of HIV-1 gene expression is dependent on a cis-acting element, Tat activation response element (TAR), which extends from -17 to +80 in the long terminal repeat (LTR).[6] TAR RNA forms a stable stem-loop structure and Tat binds to its bulge region with high efficiency.[7] This RNA-protein interaction is essential for Tat-mediated activation of HIV-1 LTR.[8] Other functions of Tat also include transactivation of the JC virus late promoter,[9] TAR-independent transactivation of the murine cytomegalovirus major immediate-early promoter,[10] upregulation of interleukin-2 secretion in activated T cells[11] and upregulation of interleukin-4 receptors on a human β-lymphoblastoid cell line.[12] Tat functions primarily to stimulate transcription initiation and increases transcriptional elongation. Recent evidence has also suggested that Tat has additional functions in the pathogenesis of AIDS, in part, because of its ability to be taken up by cells, enter the nucleus, and transactivate genes. Tat has also been shown to stimulate the growth of Kaposi's sarcoma cells; activate expression of tumor necrosis factor α and β, enhance gene expression and viral expression of human cytomegalovirus; and is also known to decrease the activity of MHC class 1 gene promoter, thereby providing a mechanism whereby HIV-1 infected cells might be able to avoid immune surveillance.[13] Various functions and characteristics of HIV-1 Tat protein are represented in Table 8.1.

Tat is likely to have both direct and indirect effects in the pathogenesis of AIDS through its multiple roles in the HIV-1 life cycle and on the immune system and thus represents an important target for genetic- based therapeutic intervention. Disruption of Tat protein interaction either with TAR RNA or the cellular factors that bind Tat protein may be an effective therapeutic approach to combat HIV-1 infection.

Downregulation of HIV-1 Tat Function Using Intrabodies

The effects of single-chain intrabodies (scFvs) directed against Tat protein has been examined.[15] As shown in Figure 8.1, Tat acts as a potent transactivator of HIV-1 LTR. To construct the scFvtat intrabodies, two different murine anti-HIV-1 Tat hybridomas were used as source of mRNA and cDNA. One monoclonal antibody, 1D9D5 (termed anti-tat1), recognized the N-terminal, proline-rich, exon 1 specific epitope EPVDPRLEWKHPGSQPKTA of HIV-1 Tat protein. This region is known to be an important epitope in transactivation of TAR containing mRNA. The other anti-tat mAb (termed anti-tat3) specifically recognized exon 2 region of Tat. PCR primers were designed such that the forward primers will anneal to the FR1 region of each of the murine heavy chain genes. Similarly, individual PCR primers were used to anneal to FR1 region of each of murine κ chain gene and a reverse C_κ with

---

Table 8.1. Characteristics and functions of HIV-1 Tat protein

---

• Potent transactivator of HIV-1 LTR.
•Concentrated in the nucleus and nucleolus of infected cells; increases frequency of RNA transcription, and enhances transcription elongation.
•Two-exon derived, 86-amino acid long protein; components include a proline-rich region (amino acids 3-18; transactivator domain), a cysteine-rich region (amino acids 22-37) and a basic region (amino acids 49-57, RNA binding, nuclear/nucleolar localization sequence).
•Additional role in AIDS pathogenesis with its ability to be taken up by cells, enter the nucleus, and transactivate genes.
•Activates expression of tumor necrosis factors α and β and interleukin-2, which individually activate HIV-1 replication and expression.
•Stimulates the growth of Kaposi's sarcoma cells.
•Several immunosuppressive effects of Tat implicated which include: increasing expression of TGF-β1; suppressing antigen-induced but not PHA-induced proliferation of T cells; decreasing activity of MHC-1 gene promoter, which could be one of the reasons that HIV-1 infected cells are able to avoid immune surveillance and recognition by specific T lymphocytes.
•Enhances gene expression and viral replication of human cytomegalovirus (HCMV); transactivates the late promoter of JC virus, the etiologic agent of progressive multifocal leukoencephalopathy (PML) in CNS glial cells.
•May play a direct role in cancer; additional roles implicated in downregulation of p53, and programmed cell death or apoptosis.

---

or without C-terminal SV40 nuclear localization motif TPPKKRKV was used.[16] A 93 basepair interchain linker (Gly$_4$Ser)$_3$ was used to link heavy and light PCR fragments via overlap extension[17] (see Fig. 8.2).

The coding sequences of all scFvtat intrabodies were modified to contain a strong Kozak sequence and start methionine immediately 5' to amino acid one of framework one of the heavy chain. A second series of scFvtat derivatives were constructed for cytoplasmic expression that had the entire human k chain constant region fused in frame with the scFv cassette (scFvtatC$_\kappa$). For nuclear targeting, as done with the scFvtat, the scFvtatC$_\kappa$ was also additionally modified to contain the carboxy-terminal SV40 nuclear localization signal (scFvtat-SV40 and scFvtatC$_\kappa$-SV40). All scFvtat constructs were cloned in pRc/CMV under the control of the cytomegalovirus Immediate-early promoter (CMVIE). Figure 8.3 illustrates schematically the different scFvtat intrabodies that were made for eukaryotic expression.

Stably expressed scFvtat1C$_\kappa$ intrabody directed against the N-terminal, exon 1 specific trans-activation domain of Tat effectively sequestered Tat in the cytoplasm, and blocked its transport to the nucleus of the cell (Fig. 8.4). This resulted in inhibition of Tat-mediated trans-activation of the HIV-1 LTR. The anti-viral activity of scFvtat1C$_\kappa$ intrabody scFv was significantly enhanced by addition of a human C$_\kappa$ domain at the C-terminus of the scFv molecule, which might have a role in stability and may also promote dimerization.[18] T lymphocytes (Sup T cells) stably transduced with the scFvtat1C$_\kappa$ intrabody demonstrated long-term resistance, as long

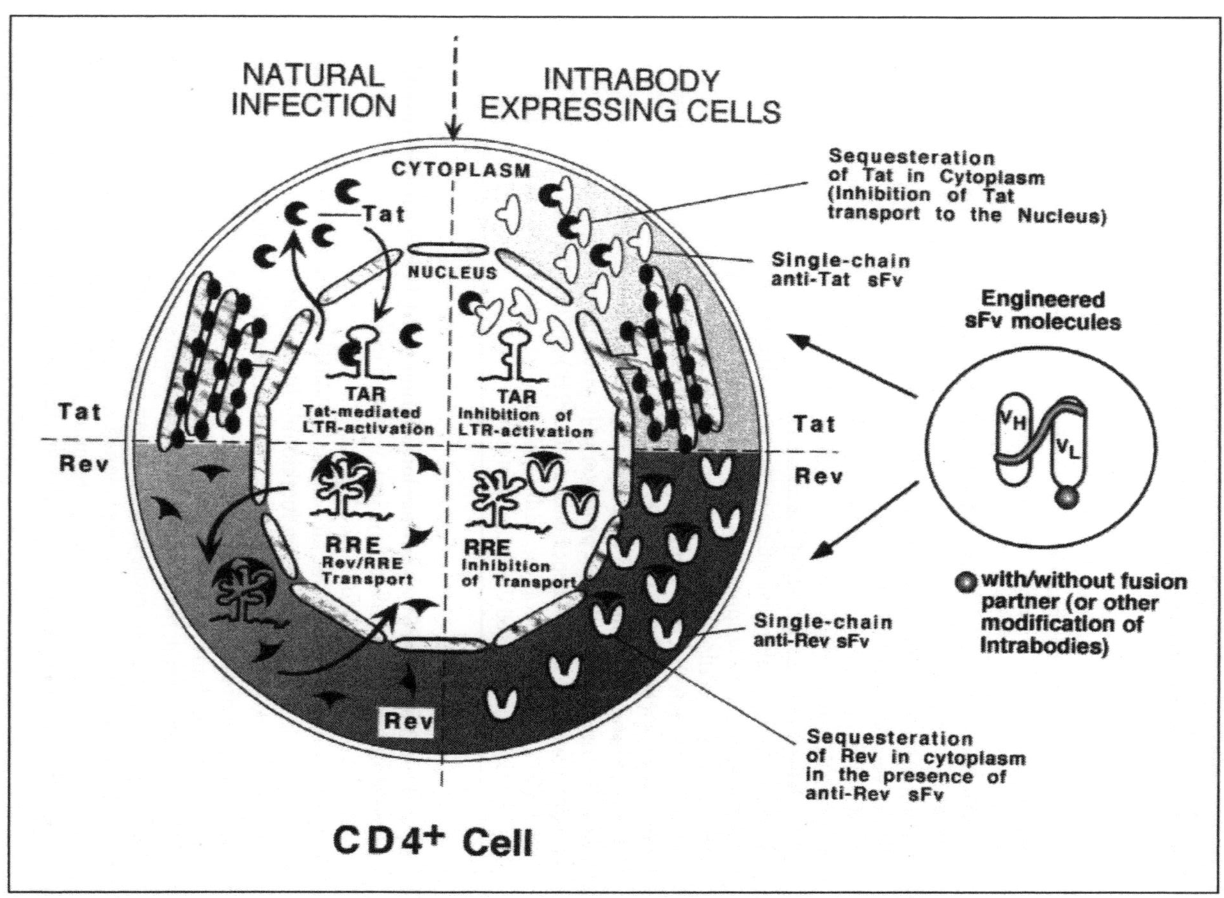

Fig. 8.1. (a) (top) Strategy for inhibition of Tat-mediated transactivation of viral gene expression by scFvtat1C$_\kappa$ intrabodies. Tat trans-activates HIV-1 LTR via its binding to transactivation response element (TAR) and other cellular factors, and initiates transcription. SFvtat1C$_\kappa$ scFvs targeted to the cytoplasm and/or the nucleus can thus inhibit Tat-mediated functions; (b) (bottom) Strategy for inhibition of Rev-mediated cytoplasmic transport of RRE containing viral mRNAs by anti-Rev single-chain antibodies. Rev protein functions in trans to rescue unspliced and singly spliced HIV-1 specific mRNA from nucleus of infected cells. Cytoplasmic and/or nuclear expression of anti-Rev scFvs can inhibit Rev-dependent mechanisms.

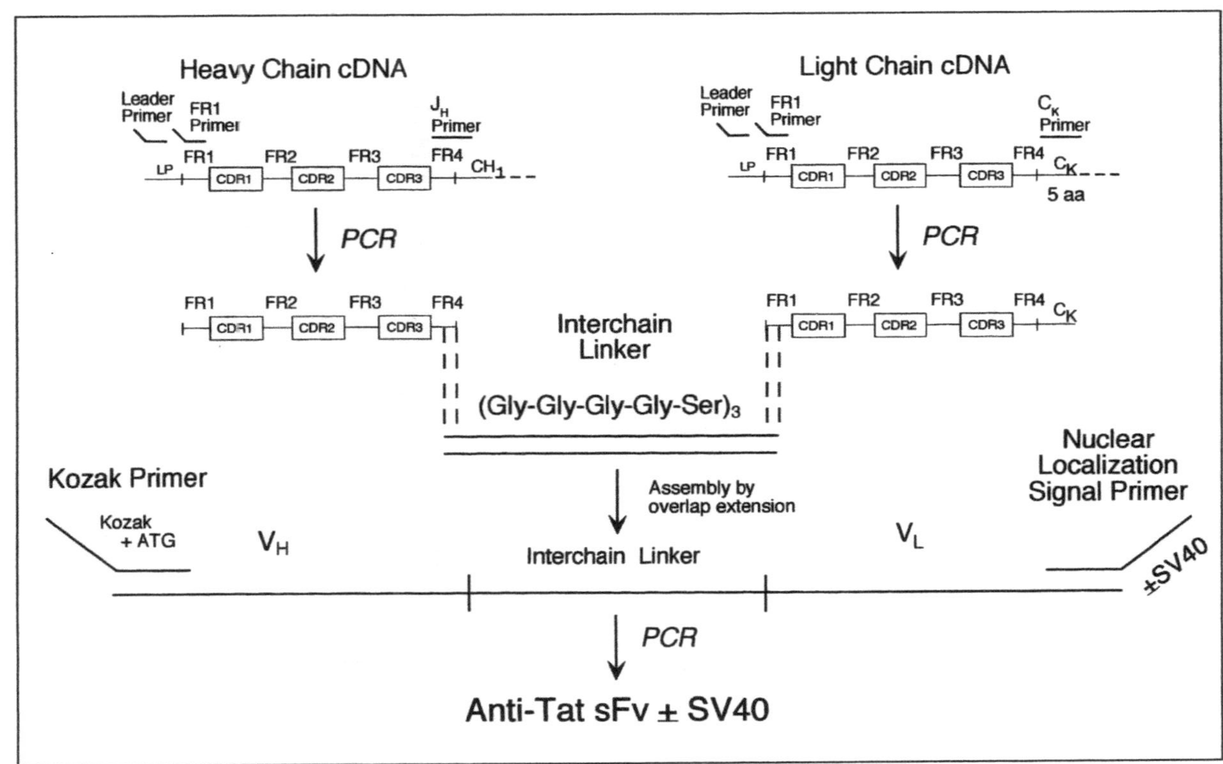

Fig. 8.2. Strategy for production of scFvtat1C_K for eukaryotic expression in the cytoplasm and nucleus. After PCR amplification of the individual heavy and light chains and assembly of the interchain linker by overlap extension, the scFvs are again PCR amplified with a new 5′ primer that contains the Kozak sequence and ATG start codon 5′ to amino acid 1 in V_H FR1 and in some cases with a new 3′ primer that anneals to the five most proximal C_K amino acids and which contain nucleotides that encode for the SV40 nuclear localization signal.

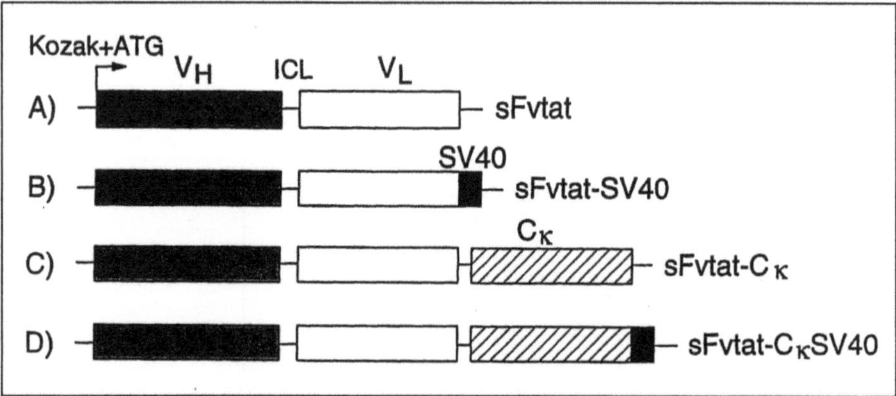

Fig. 8.3. sFvTat1 constructs. Schematic representation of different scFvtat1 that were constructed for eukaryotc expression. Single chain intrabody comprising of the V_H and V_L chains linked together with 15 amino acid long interchain linker (ICL)(A); with C-terminal SV40 nuclear localization signal (NLS) (B); with an entire C_κ domain fused in frame to the last amino acid in the rearranged V_κ gene (C); and C_κ fused domain with an SV40 NLS site (D).

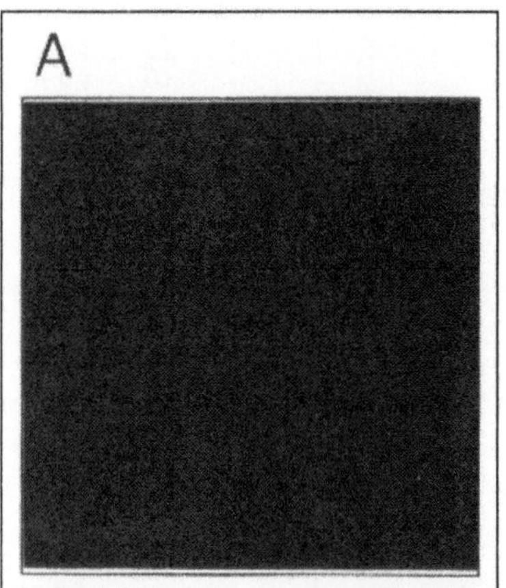

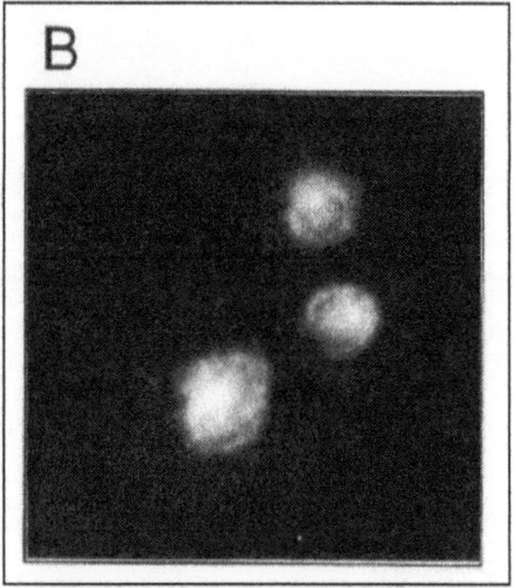

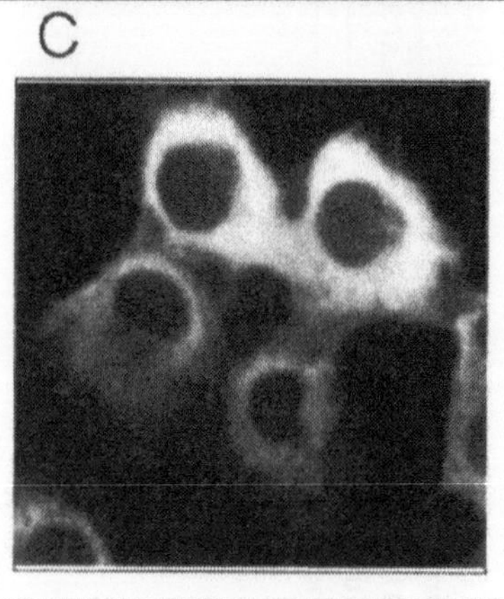

Fig. 8.4. Inhibition of nuclear transport of Tat in stable Cos-scFvtat1C_κ cells. Immuno-fluorescence (IF) staining showing inhibition of Tat transport from cytoplasm to nucleus. Stable Cos-Vector and Cos-scFvtat1C_κ cells were transfected with the Tat-expressor plasmid (pSVTat) by the DEAE Dextran method. After 48 hours the cells were fixed with ethanol/acetic acid and stained first with cell supernatant of hybridoma cell line J1D2C9 (exon 2 specific anti-Tat mAb). The cells were washed with PBS and stained with FITC-conjugated anti-mouse IgG and examined for IF. (A) Control Cos-cells untransfected, (B) Cos-Vector cells transfected with pSVTat and (C) Cos-scFvtat1C_κ cells transfected with pSVTat.

as 5-6 weeks, to challenge against laboratory isolates (Fig. 8.5),[15] and several primary isolates of HIV-1 (unpublished data). The multiplicity of Infective (MOI) units used in these experiments were in the range from 0.2-0.5. In parallel, another scFv intrabody, recognizing the IL-2α chain (scFvTac) was used as an irrelevant control in the challenge experiment.

An alternative scFvtat intrabody directed against the C-terminus of Tat was unable to suppress HIV-1 replication, indicating the critical importance of the specific targeted epitope. This is because one exon Tat is made early and two exon Tat is made late during the replication cycle of HIV-1.[19]

Ongoing studies are now determining the efficacy of scFvtat1C intrabody scFvs in transduced PBMCs challenged with primary isolates of HIV-1. Retrovirally transduced PBMCs expressing scFvtat1C$_k$ as both whole cell population and CD4$^+$ subsets (Fig. 8.6), are able to resist viral infection as long as 2-4 weeks, as observed by p24 assay (unpublished data).

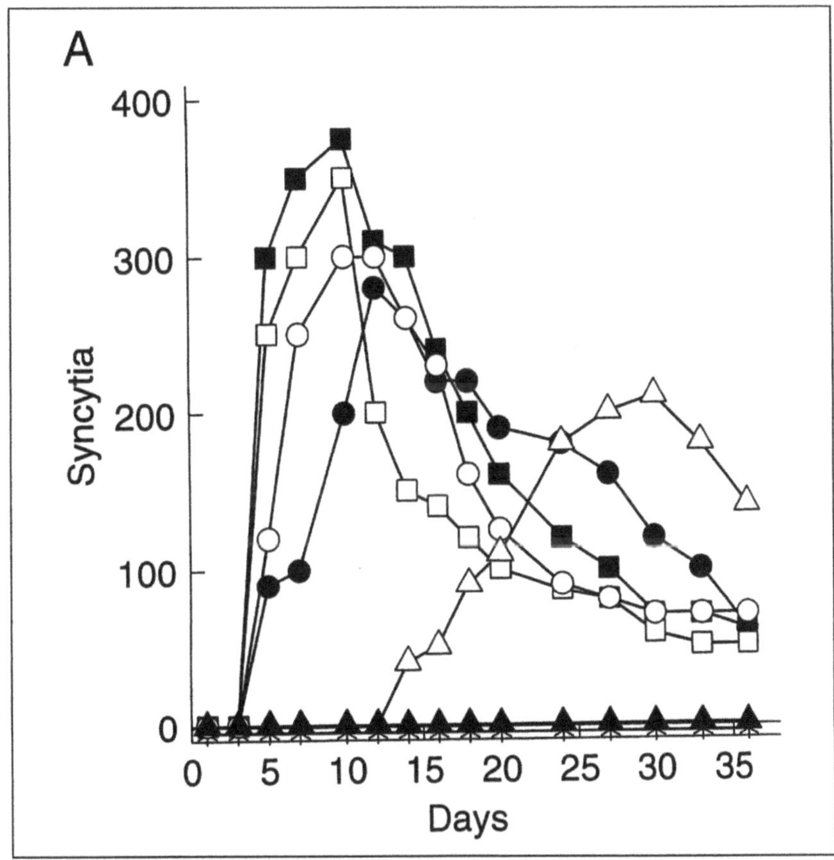

Fig. 8.5. HIV-1 infection in scFvtat expressing transformed CD4$^+$ SupT1 cells. Cells (1 x 10^6) stably transformed with vector only (□), exon 1 specific scFvtat1 (△), scFvtat1C$_k$ (▲), exon 2 specific scFvtat3 (○), scFvtat3C$_k$ (●), anti-IL-2α chain specific scFvTac (■) as irrelevant control, were infected with HIV-1 (HXIIIB) (0.5 MOI); p24 levels were determined in the cell-free supernatants that were collected on alternate days after challenge with the virus. Uninfected Sup T1 cells were also used in parallel as negative control (✳).

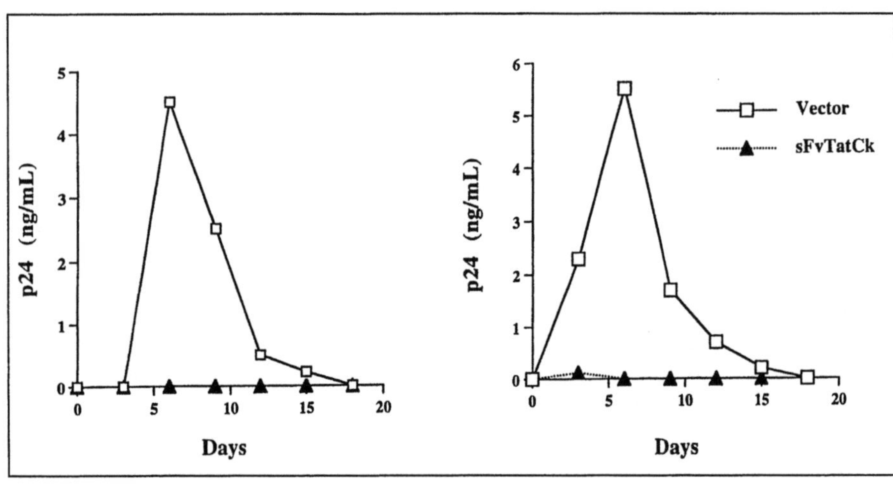

Fig. 8.6. HIV-1 infection in scFvtat1C$_\kappa$ transduced PBMCs and selected CD4$^+$ mononuclear cells. Transduced cells with scFvtat1C$_\kappa$ (▲) and CMV vector (□) were processed as described in the text of Figure 8.5.

In another study, CD4$^+$ T cells from HIV-positive patients and uninfected donors were transduced with MuLV based vectors encoding β-galactosidase and scFvtat1C$_\kappa$. The expression of marker gene (β-galactosidase) and the effects of scFvtat1C$_\kappa$ were monitored in vitro in unselected transduced CD4$^+$ T cells. Efficiency and stability of transduction varied during the course of HIV infection; CD4$^+$ T cells from asymptomatic individuals were transducible at higher efficiencies and stabilities than CD4$^+$ T cells from patients with AIDS. Expression of scFvtat1C$_\kappa$ effectively inhibited HIV-1 replication in transduced cells from HIV-1 infected individuals (work in press).

Downregulation of HIV-1 Replication Using a Novel Combination of scFvtat1C$_\kappa$ Intrabody and NFκB Antagonists

Recently, studies from our laboratory have demonstrated that inhibitors of NF-κB and Tat in combination result in a more profound and durable inhibition of HIV-1 replication than inhibitors of either NF-κB or Tat alone. CD4$^+$ T lymphocytes or PBMCs transduced with scFvtatlC$_\kappa$ were challenged with laboratory or primary isolates of HIV-1 in the presence or absence of two NF-κB antagonists. The combined treatment resulted in more durable inhibition of HIV-1 replication for as long as 50 days in T lymphocytes and more than 2 weeks in transduced PBMCs, than was seen with scFvtat1C$_\kappa$ alone (3-4 weeks in T lymphocytes and 7-10 days in PBMCs) and NF-κB inhibitors alone (7-14 days in T lymphocytes and 4-7 days in PBMCs) (unpublished data). Such combination of pharmacologic and genetic approaches may enhance the efficacy and/or reduce the toxic effect of individual therapies; they may also minimize or retard the emergence of drug-resistant or escape variants of HIV-1. In future clinical gene therapy trials that target the Tat/Rev or other HIV-1 proteins, the addition of a pharmacologic arm, for example directed to inhibition of NF-κB, may improve the survival of the transduced cells and prolong the clinical benefits to the patient.

Rev, the Regulator of HIV-1 Protein Expression

Rev protein controls the expression of the viral structural and late regulatory proteins.[20] Infectious virions are not made from a Rev-defective genome.[21] Rev protein functions in a pivotal position of the HIV-1 life cycle and acts in trans to rescue unspliced and singly spliced HIV-1 specific RNA from the nucleus of the infected cells.[22] Multimerization,[23] and effective functional threshold levels,[24] are required for Rev function. In the absence of functional Rev protein, only fully spliced HIV-1 mRNA is expressed.[25-27] Rev mutants of HIV-1 are incapable of inducing the synthesis of the viral-structural proteins and are therefore replication defective.[28,29] Other characteristics of the Rev protein are described in Table 8.2.

Downregulation of HIV-1 Rev Function Using Intrabodies

Equally promising results have been obtained by using scFvs directed against HIV-1 Rev. The Rev protein shuttles between the nucleus and cytoplasm of the infected cells (Fig. 8.1). Duan and others[30] were able to show potent inhibition of HIV-1 replication by an intracellular anti-Rev scFv (D8) directed against the C-terminus effector (or activation) domain of HIV-1 Rev, a region implied to be the most immunogenic portion of the Rev protein. This intrabody strongly antagonized Rev function by sequestering the Rev protein in the cytoplasm of the stably transduced HeLa-T4 cells, which led to a sustained resistance to challenge by multiple strains of HIV-1.[30,31] They epitope mapped D8 to the leucine-rich motif from amino acid 70-84 in the HIV-1 Rev protein sequence. Wu and others,[32] compared D8 with another scFv D10, which recognized the C-terminal, nonactivation region (amino acids 96-109) of Rev. Although binding affinity assays demonstrated that D8 had extracellular binding affinity significantly lower than D10, the scFvD8 demonstrated a very potent activity in inhibiting virus production in human T cell lines and peripheral blood mononuclear cells than D10.

Presently in our laboratory, further studies are underway to examine the efficacy of other scFvRev that were constructed against different antigenic epitopes of HIV-1 Rev. Recently, an scFvRev construct made from monoclonal antibody mapped against the leucine motif (amino acids 70-88 of HIV-1 Rev primary amino acid sequence) has shown efficient capability to inhibit Rev-responsive element (RRE)-mediated activation of chloramphenicol acetyl-transferase activity (Fig. 8.7). Experiments are now underway to study the effect of these anti- scFvRev, singularly and in combination with other scFvs, against HIV-1 infection in CD4+ lymphocytes and PBMCs. In addition, another scFvRev constructed against the C-terminal epitope of Rev (amino acids 91 to 110) was unable to effectively inhibit Rev function as compared to the scFv intrabody targeted to the leucine motif region of Rev.

Humanization and Use of Human scFvtat and scFvrev Intrabodies

Humanization of the murine scFvtat by CDR grafting are underway, and are being analyzed for their therapeutic potential in clinical trials. Based on the murine scFvtat1C$_K$ sequences, human frameworks have been identified that should accept the murine CDRs with minimal change in the secondary structure of the variable region. CDR grafting has been completed and these molecules are being analyzed for their binding and functional characteristics.

Table 8.2. Characteristics and functions of HIV-1 Rev protein

•Regulator of virion protein expression, which has a profound effect on the fate of primary RNA transcripts within the nucleus.

•Two-exon derived, 116 amino acid long protein. Two distinct domains essential for Rev function.(1) The N-terminal 50 amino acid long stretch comprising of Arginine-rich central core (amino acids 35-50; nuclear/nucleolar localization signal), surrounded by amino acids required for multimerization of Rev which is critical for its function, in vivo.(2) Another domain centered around amino acids 73-83 (Leucine motif) may interact directly with the nuclear RNA transport or splicing machinery.

•In the absence of Rev protein, only small, multiply-spliced RNAs accumulate in the cytoplasm; thus only regulatory proteins are produced. In the presence of Rev, unspliced and singly spliced viral mRNAs accumulate in the cytoplasm, which are translated to produce structural and envelope glycoproteins of the virus.

•Rev binds to a complex stem-loop structure, the Rev responsive region (RRE), within the envelope coding region of HIV-1 envelope glycoprotein gene. Its binding to RRE depends on correct presentation of a primary nucleotide sequence in the RNA stem-loop structure. Binding of the Rev protein to the RNA sequences alters the splicing fate of the primary transcripts.

•Viral RNA, which contains cis-acting repression sequences (CRS), cannot serve as templates for protein synthesis in the absence of Rev.

Pilkington and others[33] have been successful in making a human Fab phage display library from peripheral blood lymphocytes of an asymptomatic individual after 10 years with HIV-1. The library was panned against Rev and Tat protein and several clones (3 for Rev and 4 for Tat) with binding constants varying from 10^{-6} M to 10^{-8} M were isolated. The anti-Rev Fab were directed to sites adjacent to the Rev basic nuclear localization sequence (residues 52-64) and to Rev activation domain residues 75-88. The anti-Tat Fab were directed to the functional domain between amino acid 22-33 of Tat molecule. The authors concluded that these newly described human antibody fragments to HIV-1 regulatory proteins may be critical moieties for gene therapeutic protocols to control HIV-1 replication in human cells.

Other Protein and RNA-Based Therapeutic Strategies

In addition to intrabodies (see Table 8.3), several other protein and RNA-based inhibitors of HIV-1 have been reported,[2,3] which include mutant Tat and Rev proteins that interfere dominantly with the activity of wild type Tat or Rev protein,[34-36] respectively. RNA-based inhibitors include anti-sense RNA,[37] ribozymes[38,39] and RNA decoys.[40-42] Several of these strategies have been used to interfere with Tat/ TAR or Rev/RRE interactions. RNA and ribozyme technology may be beneficial in some regards while other aspects may be rate-limiting, such as insufficient expression of the RNA inhibitor in the cell, insufficient sub-cellular colocalization of target and inhibitor RNAs and inaccessibility of the target RNA sequence caused by the formation of secondary and tertiary structure or binding of proteins. Overall, based on the important roles of HIV-1 Tat and Rev in HIV-1 pathogenesis, effective application of anti-Tat and anti-Rev molecules, alone or in combination, may pose as good strategy to inhibit Tat and Rev-mediated functions. Future studies using

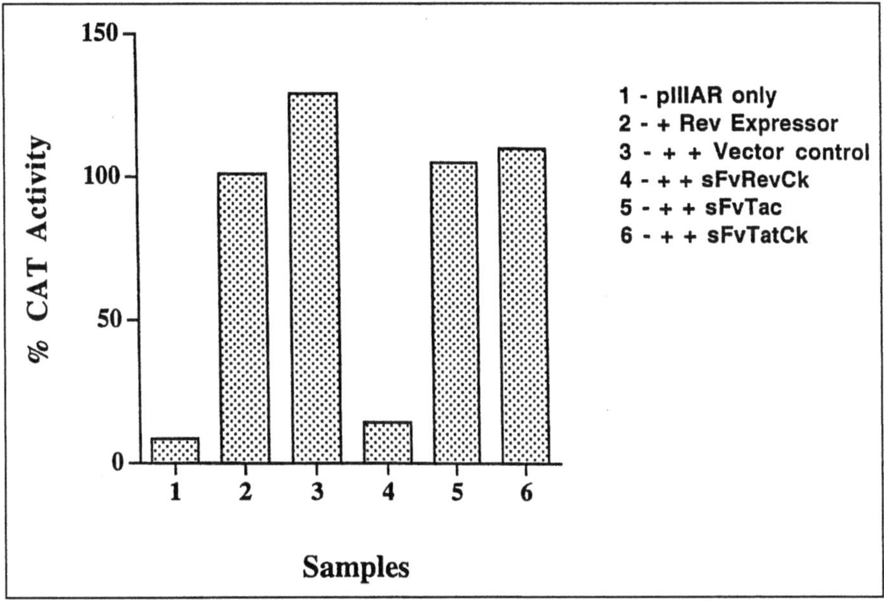

Fig. 8.7. Inhibition of Rev-dependent CAT expression by scFvrev intrabody. COS-1 cells were transiently transfected with pIIIAR plasmid (Rev-dependent RRE- mediated CAT expressor). In the absence (1) and presence of Rev expressor (2). In parallel, pIIIAR plasmid was also cotransfected with Rev expressor and CMV vector control (3), or with scFvrevC$_\kappa$ (4), or scFvTac or scFvtat1C$_\kappa$ (5 and 6 as irrelevant antibody controls). 48 hours post-transfection cells were analyzed for CAT activity.

combinations of these molecules and approaches are underway to determine if an additive or possibly synergistic effect will be obtained in providing protection to T lymphocytes and PBMCs against HIV-1 infection.

Future Perspectives

Recent clinical studies have suggested the possibility of using combination of anti-HIV-1 chemotherapy-based strategies to decrease viral loads and perhaps prolong survival of patients in the face of HIV-1 infection. A logical extension of these therapeutic approaches would be to use combinations of intrabodies that will be directed against different HIV-1 proteins (structural or regulatory), and will likely enhance the efficacy and may also minimize or retard the emergence of escape variants of HIV-1. It may also be also possible to construct bispecific intrabodies with sequences coding for two different scFv intrabodies that recognize different viral proteins (for example Tat and Rev, or p17 and integrase, or integrase and RT).

The immunoglobulin variable region gene repertoire has the capacity to recognize an enormous range of diverse antigenic moieties. Advanced antibody engineering technologies such as phage display libraries, repertoire cloning, affinity maturation and guided selection will be useful in the future to identify intrabodies with desirable functions in terms of high binding affinity and specificity. Further issues pertaining to effective use of intrabodies in clinical trials include the maintenance of sustained levels of expression of the intrabodies in specific cell types,

Table 8.3. Protein and RNA-based suppressors used against HIV-1

I. Protein-based Suppressors.
  a. Intracellular single-chain antibodies and other antibody fragments used for sequestration of viral proteins in appropriate compartments.
    • scFv-F105, anti-gp120.
    • scFvtat.
    • scFvrev.
    • anti-reverse transcriptase.
  b. Transdominant-Negative mutant proteins.
    • Mutant Rev protein.
    • Mutant Tat protein.
    • Double transdominant fusion (Tat+Rev) mutants.
    • Gag sequences.
II. RNA-based Suppressors.
  a. TAR decoys.
    • chimeric tRNAmet-TAR transcription unit.
    • TAR containing repeats.
    • chimeric TAR-antisense Tat construct.
  b. RRE decoys.
  c. Ribozymes.
    • Gag sequence specific.
    • HIV-1 LTR sequence specific.
    • Tat sequence specific.
  d. Antisense RNA.
    • Tat sequence specific.

together with nontoxic and nonimmunogenic characteristics. The design of more effective gene delivery systems (both viral and nonviral) for transduction of genes in lymphohematopoetic stem cells or primary lymphocytes may produce superior results in clinical trials. As an alternate to transduction of quiescent stem cells, repeated treatment of circulating lymphocytes may be feasible, especially if resistant cells have a growth advantage under the selective pressure of viral infection. Furthermore, the testing of potential gene therapies should require the use of improved primate animal models; for example, it will be worthwhile to demonstrate efficacy of anti-HIV-1 intrabodies against the SHIV virus (HIV-1 env, Tat, and Rev in SIV backbone) in a rhesus monkey infection model.

In conclusion, anti-HIV-1 intrabodies can inhibit HIV-1 infection by disrupting both early or late functional events of HIV-1 life cycle. In the future, the successful application of intrabody-based gene therapy against HIV-1 and other infectious agents will depend on effective basic research coupled with strict adherence to the principles of controlled clinical trials.

Acknowledgments

We would like to thank Drs. Adriaan de Bruïne and Xuguang Li for their critical review of this chapter and helpful comments, and Christine Naugle for her assistance with the artwork.

References

1. Purvis A. The global epidemic. Time Magazine, Dec 30 1996/Jan 6 1997; Vol 148:1-2.
2. Mhashilkar AM, Marasco, WA. Intrabody-based gene-therapy approaches against HIV-1. In: Advancing a New Agenda for AIDS Research. Poznansky MC, Coker RJ, Gilles P (eds.) Groundworks Publishing 1996.
3. Mhashilkar AM, Agarwal S, Levin R et al. Genetic-based strategies for control of HIV-1: Tat and Rev as potential targets. Drug News & Perspectives 1996; 9(4):220-227.
4. Baltimore D. Intracellular immunization. Nature 1988; 335:395-396.
5. Marasco WA. Intracellular antibodies "Intrabodies" as research reagents and therapeutic molecules for Gene Therapy. Immunotechnology 1995; 1:1-19.
6. Sodroski J, Rosen C, Wong-Staal et al. Trans-acting transcriptional regulation of the human T cell leukemia type-III long terminal repeat. Science 1985; 227:171-173.
7. Roy S, Parkin NT, Rosen C et al. Structural requirements for trans-activation of human immunodeficiency virus type 1 long terminal repeat-directed gene expression by Tat: importance of base pairing, loop sequences and bulges in the Tat-responsive sequences. J Virol 1990; 64:1402-1406.
8. Berkhout B and Jeang K-T. Transactivation of human immunodeficiency virus type-1 is sequence specific for both the single-stranded bulge and loop of the trans acting responsive hairpin: a quantitative analysis. J Virol 1989; 63:5501-5504.
9. Tada H, Rappaport J, Lashgari M et al. Transactivation of the JC virus late promoter by the tat protein of type 1 immunodeficiency virus in glial cells. Proc Natl Acad Sci USA 1990; 87:3479-3483.
10. Kim YS and Risser R. TAR-independent transactivation of the murine cytomegalovirus major immediate-early promoter by tat protein. J Virol 1993; 67:239-248.
11. Westendorp MO, Li-Weber M, Frank RW et al. Human immunodeficiency virus type 1 Tat up-regulates Interleukin-2 secretion in activated T cells. J Virol 1994; 68:4177-4185.
12. Puri RK and Aggarwall BB. Human immunodeficiency virus type 1 tat gene up-regulates Interleukin 4 receptor on human β-lymphoblastoid cell line. Cancer Research 1992; 52:3787-3790.
13. Howcroft TK, Strebel K, Martin MA et al. Repression of MHC-1 gene promoter activity by two-exon Tat of HIV. Science 1993; 260:1320-1322.
14. Goldstein G. HIV-1 Tat protein as a potential AIDS vaccine. Nat Med 1996; 1(9):960-964.
15. Mhashilkar AM, Bagley J, Chen S-Y et al. Inhibition of HIV-1 Tat mediated LTR transactivation and HIV-1 infection by anti-Tat single-chain antibodies. EMBO J 1995; 14:1542-1551.
16. Yoneda Y, Semba T, Kaneda Y et al. A long synthetic peptide containing a nuclear localization signal and its flanking sequences of SV40 T-antigen directs the transport of IgM into the nucleus efficiently. Exper Cell Res 1992; 201:313-320.
17. Clackson T, Hoogenboom HR, Griffiths AD and Winter G. Making antibody fragments using Phage Display libraries. Nature 1991; 352:624-628.
18. McGregor DP, Molly PE, Cunningham C et al. Spontaneous assembly of bivalent single chain antibody fragments in *Escherichia coli*. Mol Immunol 1994; 31(3):219-226.
19. Cullen BR. Regulation of HIV gene expression. AIDS 1995; 9(Suppl. A):519-532.
20. Haseltine WA. Molecular biology of human immunodeficiency virus type-1. FASEB J 1991; 5:2349-2360.
21. Sodroski J, Goh WC, Rosen C et al. A second posttranscriptional trans-activator gene required for HTLV-III replication. Nature 1986; 321:412-417.

22. Cullen BR. Mechanism of action of regulatory proteins encoded by complex retroviruses. Microbio Rev 1992; 56(3):375-394.

23. Olsen HS, Cochrane AW, Dillon PJ et al. Interaction of human immunodeficiency virus type 1 Rev protein with a structured region in env mRNA is dependent on multimer formation mediated through a basic stretch of amino acids. Genes Dev 1990; 4:1357-1364.

24. Malim MH, Cullen BR. HIV-1 structural gene expression requires the binding of multiple Rev monomers to the viral RRE: implications for HIV-1 latency. Cell 1991; 65:241-248.

25. Zapp ML, Hope TJ, Parslow TG et al. Oligomerization and RNA binding domains of the type 1 human immunodeficiency virus Rev protein: a dual function of an arginine-rich binding motif. Proc Natl Acad Sci USA 1991; 88:7734-7738.

26. Emerman M, Vazeux R, Peden K. The rev gene product of the human immuno-deficiency virus affects envelope specific RNA localization. Cell 1989; 57:1155-1165.

27. Felber BK, Hadzopoulou-Cladaras M, Cladaras C et al. Rev protein of human immunodeficiency virus type 1 affects the stability of and transport of the viral mRNA. Proc Natl Acad Sci USA 1989; 86:1495-1499.

28. Sadaie MR, Rappaport J, Bentner T et al. Missense mutations in an infectious human immunodeficiency viral genome: functional mapping of Tat and identification of the Rev splice acceptor. Proc Natl Acad Sci USA 1988; 85:9224-9228.

29. Terwilliger E, Burghoff R, Sia R et al. The art gene product of human immuno-deficiency virus is required for replication. J Virol 1988; 63:655-658.

30. Duan L, Bagasra O, Laughlin MA et al. Potent inhibition of human immuno-deficiency virus type 1 replication by an intracellular anti-Rev single chain antibody. Proc Natl Acad Sci USA 1994; 91:5075-5079.

31. Duan L, Zhang H, Oakes JW et al. Molecular and virological effects of intracellular anti-Rev single-chain variable fragments on the expression of various human immunodeficiency virus-1 strains. Hum Gen Ther 1994; 5:1315-1324.

32. Wu Y, Duan L, Zhu M et al. Binding of intracellular anti-rev single chain variable fragments to different epitopes of human immunodeficiency virus Type 1 rev: variation in viral inhibition. J Virol 1996; 70:3290-3297.

33. Pilkington GR, Duan L, Zhu M et al. Recombinant human Fab antibody fragments to HIV-1 Rev and Tat regulatory proteins: Direct selection from a combinatorial phage display library. Mol Immunol 1996; 33:439-450.

34. Bridges SH, Sarver N. Gene Therapy and immune restoration for HIV disease. The Lancet 1995; 345:427-432.

35. Malim MH, Bohnlein S, Hauber J et al. Functional dissection of the HIV-1 Rev transactivator-derivation of a trans-dominant repressor of Rev function. Cell 1989; 58:205-211.

36. Nabel GJ, Fox BA, Post L et al. A molecular genetic intervention for AIDS-effects of a transdominant negative forms of Rev. Hum Gen Ther 1994; 5:79-92.

37. Lo KMS, Biasolo MA, Dehni G et al. Inhibition of replication of HIV-1 by retroviral vectors expressing tat-antisense and anti-Tat ribozyme RNA. Virol 1992; 190:176-183.

38. Sarver N, Cantin PS, Chang PS et al. Ribozymes as potential HIV-1 therapeutic agents. Science 1990; 247:1222-1225.

39. Yu M, Ojwang J, Yamada O et al. A hairpin ribozyme inhibits expression of diverse strains of human immunodeficiency virus type 1. Proc Natl Acad Sci USA 1993; 90:6340-6344.

40. Lisziewicz J, Sun D, Smythe J et al. Inhibition of human immunodeficiency virus type 1 replication by regulated expression of a polymeric Tat activation response RNA decoy as a strategy for Gene Therapy in AIDS. Med Sci 1991; 90:8000-8004.

41. Sullenger B, Gallardo HF, Ungers HE et al. Overexpression of TAR sequences renders cells resistant to human immunodeficiency virus replication. Cell 1990; 63:601-608.
42. Lisziewicz J, Sun D, Lisziewicz A et al. Anti-tat gene therapy: a candidate for late-stage AIDS patients. Gene Ther 1995; 2:218-222.

CHAPTER 9

Gene Therapy for HIV-1 Using Intracellular Antibodies Against HIV-1 Gag Proteins

Isaac J. Rondon and Wayne A. Marasco

Introduction

Human immunodeficiency virus (HIV-1), a member of the lentivirus subfamily (retrovirus group), is considered to be the etiologic agent of Acquired Immunology Syndrome (AIDS).[1] AIDS develops after a long incubation period of the virus, resulting in immunologic disorders and neurological disease.[2] The inability to control the progress of HIV-1 infection with traditional therapies has compelled the medical community to search for alternative therapies. Gene therapy is a new form of molecular medicine that utilizes DNA as a therapeutic agent. The goal of gene therapy is to deliver and to express the therapeutic gene in cells for the treatment of the disease. When considering HIV-1 infection, genetic alteration of the host cell could potentially confer permanent suppression of viral replication after infection, or protection against viral infection.

Genetic-Based Strategies Used to Inhibit HIV-1 Replication

Different innovative strategies have been investigated to control the progress of HIV-1 infection.[3-7] Among these strategies are immune reconstitution, nucleic acid-based therapeutic vaccines and intracellular immunization. Immune reconstitution or adoptive cell therapy for HIV-1 infection involves ex vivo expansion of selected, and sometimes genetically modified T cell populations [e.g. CD8 cytotoxic T lymphocytes (CTL) and CD4 lymphocytes], followed by their reinfusion into the HIV-1-infected patient.[8] Nucleic acid-based therapeutic vaccines involve direct delivery of HIV-1 genes (e.g. env) to mimic viral infection; the expression of viral proteins encoded by these nucleic acids elicits both cellular and humoral responses.[9] Lastly, intracellular immunization is the transfer of a therapeutic gene into target cells to render them resistant to viral replication. The resistant cells will then limit the spread of the virus in the patient.[10] Different modalities of intracellular immunization have been developed and they can be broadly divided into two categories: RNA-based and protein-based suppressors. The RNA-based suppressors include antisense RNA, ribozymes[11] and RNA decoys.[12] Antisense RNA and ribozymes hybridize to target viral transcripts to inactivate them. RNA decoys interact and sequester regulatory proteins (e.g. Tat, Rev) essential for virus replication. The RNA suppressors are limited to the cytoplasmic compartment. The protein-based suppressors include trans-dominant mutant proteins,[13] suicide molecules,[14] and intracellular antibodies.[15] Trans-dominant mutant proteins are altered viral proteins that compete with the native viral protein. Suicide molecules, unlike other approaches, do not protect the cells but rather destroy the infected cell. Intracellular antibodies or intrabodies are synthesized by the cell and directed to a particular cellular compartment to inactivate a target molecule in a highly specific manner. This chapter will address the use of intrabodies against HIV-1 Gag proteins as a novel strategy to control the progress of HIV-1 infection.

Intracellular Antibodies Against HIV-1

The HIV-1 genome, like that of other retroviruses, contains *gag* (group antigen), *pol* (polymerase) and *env* (envelope) structural genes. HIV-1 also encodes three regulatory proteins, Rev (regulator of virion protein), Tat (transactivator) and Nef (negative regulator factor), and three proteins believed to be involved in virus maturation and release, Vif (virion infectivity factor), Vpu (viral protein U) and Vpr (viral protein R).[16,17] Different anti-HIV-1 intrabodies have been investigated as a treatment strategy. These anti-HIV-1 intrabodies have targeted structural as well as regulatory proteins essential in the life cycle of the virus. Anti-Env intrabodies expressed in HIV-1 infected cells cause the production of less infectious virions.[15] Anti-Tat and anti-Rev intrabodies inhibit HIV-1 replication.[18,19] Anti-Tat intrabodies exert their action sequestering Tat in the cytoplasm, thus, inhibiting Tat-mediated Long Terminal Repeat (LTR) transactivation of the HIV-1 genome. Anti-Rev intrabodies sequester Rev in the cytoplasm and inhibit the rescue of unspliced and singly spliced HIV-1 specific RNA from the nucleus of infected cells. Lastly, intrabodies against some of the proteins present in the preintegration complex [reverse transcriptase (RT), integrase (IN) matrix (MA)] inhibit viral integration[20-22] and virion infectivity.[22] Due to the diverse functions of the targeted proteins, intrabodies can be used to control replication at different stages of the viral life cycle. Targeting proteins involved in early as well as late stages of the virus life cycle offer great advantage when considering intracellular immunization strategies. The transduction of CD4$^+$ peripheral blood T lymphocytes or CD34$^+$ stem cells with an intrabody that inhibits both viral integration and virion infectivity could result in a marked reduction of virus spread in an infected patient. Two Gag proteins (p17 and p7) play important roles during early and late stages of the virus life cycle. Therefore, these two Gag proteins are good targets for intracellular immunization strategies used to control HIV-1 replication.

Matrix (p17)

The Gag polyprotein forms the core of an HIV-1 virion, it is composed of 4 nucleocapsid proteins: p24 (capsid), p17 (matrix), p7 (nucleocapsid) and p6. Each of which is proteolytically cleaved from a 55 kDa Gag polyprotein precursor by the HIV-1 protease.[23,24] The HIV-1 matrix (MA) comes from the amino-terminus of p55 and is myristylated at its N-terminus. MA is believed to be involved in two critical stages of the viral life cycle, it is required for both nuclear import of the viral preintegration complex and for particle assembly. MA contains a nuclear localization signal (NLS) which appears to play a role in the nuclear transport of the viral preintegration complex, especially in nondividing cells.[25,26] MA also plays a critical role in particle formation, a myristylated residue and charged N-terminal amino acids of MA directs the Gag polyprotein precursor to the plasma membrane. This targeting is essential for the proper assembly of viral particles[27,28] and for their release into the extracellular space.[29,30] The anti-HIV-1 activity of anti-MA Fab intrabodies has been investigated.[22] The binding site for the 3H7 monoclonal antibody, used to clone the Fab, had been previously epitope mapped to the amino acid sequence KKAQQAAADT (residues 113-122) near the carboxy terminus of MA.[31] This epitope is associated with the Clade B HIV-1 genotype which appears to be distributed globally.[32]

Construction of Anti-MA Fab Expression Vectors

A bicistronic expression vector pCMV-Fab-IRES under the control of the cytomegalovirus immediate early (CMVIE) promotor that allows near stoichiometric coexpression of the Fd and light chain of a Fab was constructed (Fig. 9.1). An internal ribosomal entry site (IRES) corresponding to the 5' untranslated leader region (UTR) of the encephalomyocarditis virus (EMCV) was placed between the two chains to obtain cap-independent ribosomal binding and high level translation of the light chain.[33] Two versions of the anti-MA Fab3H7 constructs with and without native immunoglobulin leader sequences were designed, amplified, and cloned. The sequence of the primers used to PCR the immunoglobulin chains were antibody specific primers.[34] For secretion of anti-MA Fab, the construct with the native immunoglobulin leader sequences was used. While for cytoplasmic expression of the anti-MA Fab, the native immunoglobulin leader sequences of the heavy and κ chains were removed, and the individual chains were modified to contain a Kozak consensus sequence and an ATG initiation codon immediately preceding amino acid one of framework one of the heavy chain and amino acid minus four of the leader sequence of the κ chain.

To determine if the Fd and κ chains of the Fab3H7 fragments could be expressed in the bicistronic vector, in vitro transcription and translation were performed using a rabbit reticulocyte system. The Fd and κ chains were expressed in near stoichiometric amounts, regardless of whether the native immunoglobulin leader sequences were present or absent (data not shown).

Expression of Anti-MA Fab 3H7 Fragments

COS-1 cells as well as Jurkat cells were transfected by DEAE-Dextran with pCMV-Fab3H7-IRES plasmids, which encode Fd and k chains with or without immunoglobulin leader sequences. Stably transfected clones were generated with G418 selection and limiting dilution. Mouse Ig was detected using radioimmunoprecipitation of ^{35}S-cysteine-labeled proteins with rabbit anti-mouse IgG. Two protein bands (approximately 30 and 28 kDa) that corresponded to Fd and κ chains of Fab3H7 were detected in the lysate, while corresponding bands were not detected in parental cells or vector transfected cells (data not shown).

The culture medium of individual stably transfected Jurkat-Fab3H7 cell subclones was used to determine if the secreted anti-MA Fab3H7 fragments were able to bind to recombinant MA coated onto ELISA plates. As shown in Figure 9.2, medium from a representative Jurkat-Fab3H7 subclone demonstrated binding to MA at 10-fold higher levels compared with the medium used from Jurkat-vector cells. These experiments demonstrate that the bicistronic expression vector pCMV-Fab-IRES allows the simultaneous expression of Fab3H7 Fd and κ chains and that these Fab3H7 fragments are able to bind HIV-1 MA.

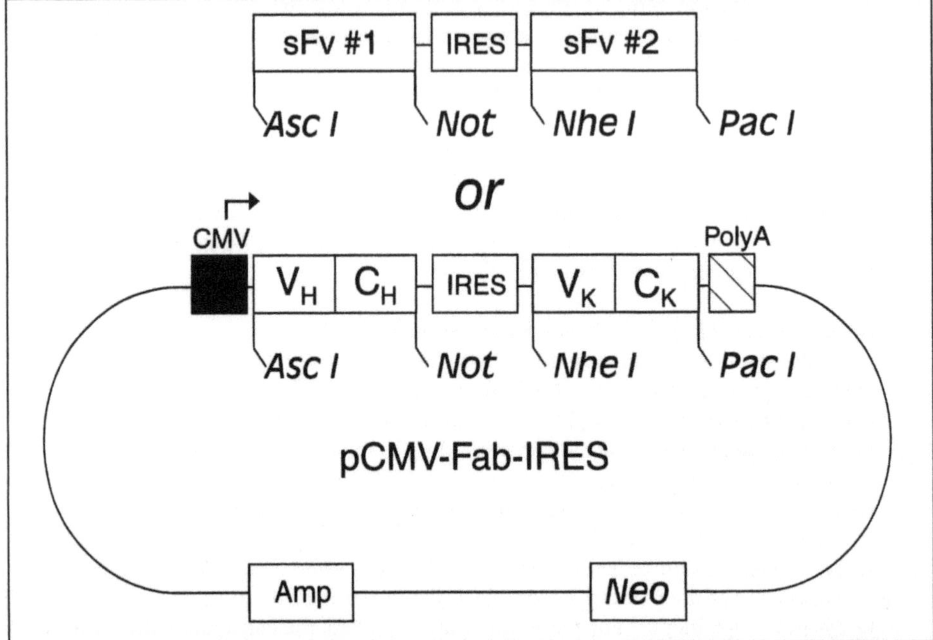

Fig. 9.1. Bicistronic expression vector pCMV-Fab-IRES. The 5' UTR of the encephalomyocarditis virus (EMCV), corresponding to an internal ribosomal entry site (IRES), has been cloned into the vector. Unique restriction enzyme sites allow the cloning of either Fd and light chains of Fab intrabodies or two scFv intrabodies.

Early Stages in the Virus Life Cycle

Inhibition of HIV-1 Infection in CD4[+] Jurkat T Cells Stably Expressing Cytoplasmic Anti-MA Intrabodies

To examine the role of the anti-MA intrabodies in the preintegration phase of the viral life cycle, the ability of cytoplasmic Fab3H7 intrabodies to inhibit HIV-1 gene expression was investigated. An HIV-1 CAT virus that has a chloramphenicol acetyl-transferase (CAT) gene replacing the *nef* gene and a deletion in the *env* gene was used. Wild type HIV-1 envelope was supplied in trans, so that the virions had the genetic capacity for only a single round of infectivity.[35] Following infection and integration of viral DNA, newly synthesized Tat protein transactivates the expression of the CAT gene. If integration is blocked, Tat-mediated gene expression can still occur presumably from unintegrated viral DNA, however, it is relatively inefficient[36-40] and in general, there is an insufficient level of HIV-1 gene expression to support a spreading viral infection.[37-41] When stably transfected Jurkat-Fab3H7 cells were challenged with the CAT virus, marked inhibition of CAT activity was seen (data not shown). This inhibition could occur as a result of inhibition of any critical step in the afferent arm of the virus life cycle that leads to integration of the HIV-1 provirus.

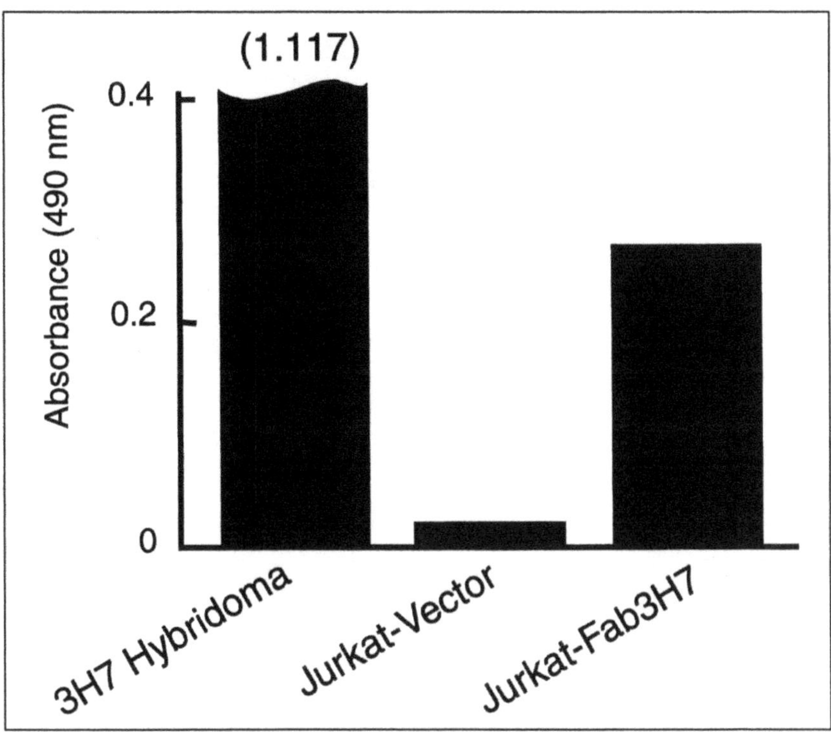

Fig. 9.2. Secretion of functional anti-MA Fab3H7 fragments from stable CD4$^+$ Jurkat T cells. Fresh medium from individual stable Jurkat-Fab3H7 cell clones were harvested after 24 hr incubation, concentrated 10X in an Amicon concentrator (Beverly, MA), and added to ELISA wells in triplicate that were coated with recombinant HIV-1 MA (Agmed) at 100 ng/well. Binding of Fab3H7 fragments was detected by HRP-labeled rabbit anti-mouse IgG as the secondary antibody and the standard deviations were <5%. Supernatants from the following sources were tested: Lane 1: 3H7 hybridoma cells; Lane 2: Jurkat-pRC/CMV-vector cells; Lane 3: stable Jurkat-Fab3H7 subclone 15/72-6.

To determine if the stable Jurkat-Fab3H7 cells were resistant to HIV-1 infection, stable Jurkat-Fab3H7 cells were challenged with both laboratory strains of HIV-1 and European, syncytium-inducing (SI) primary isolates that had been passaged on activated PBMCs and screened for their ability to directly infect Jurkat cell lines. The results of several experiments demonstrated that there was a marked delay of virus infection (as measured by p24 release) over the 20-25 day experiments in Jurkat-Fab3H7 cells compared to parental Jurkat cells or stable Jurkat-vector cells (Fig. 9.3). Finally, results in Figure 9.3E demonstrate that the inhibition is specific for HIV-1 since very little inhibitory effect is seen where these stable cells are infected with HIV-1. HIV-1 and HIV-2 are divergent in amino acid sequences in this region of MA.[42] Thus, these experiments demonstrated that actively dividing CD4$^+$ Jurkat T cells stably expressing anti-MA Fab intrabodies are

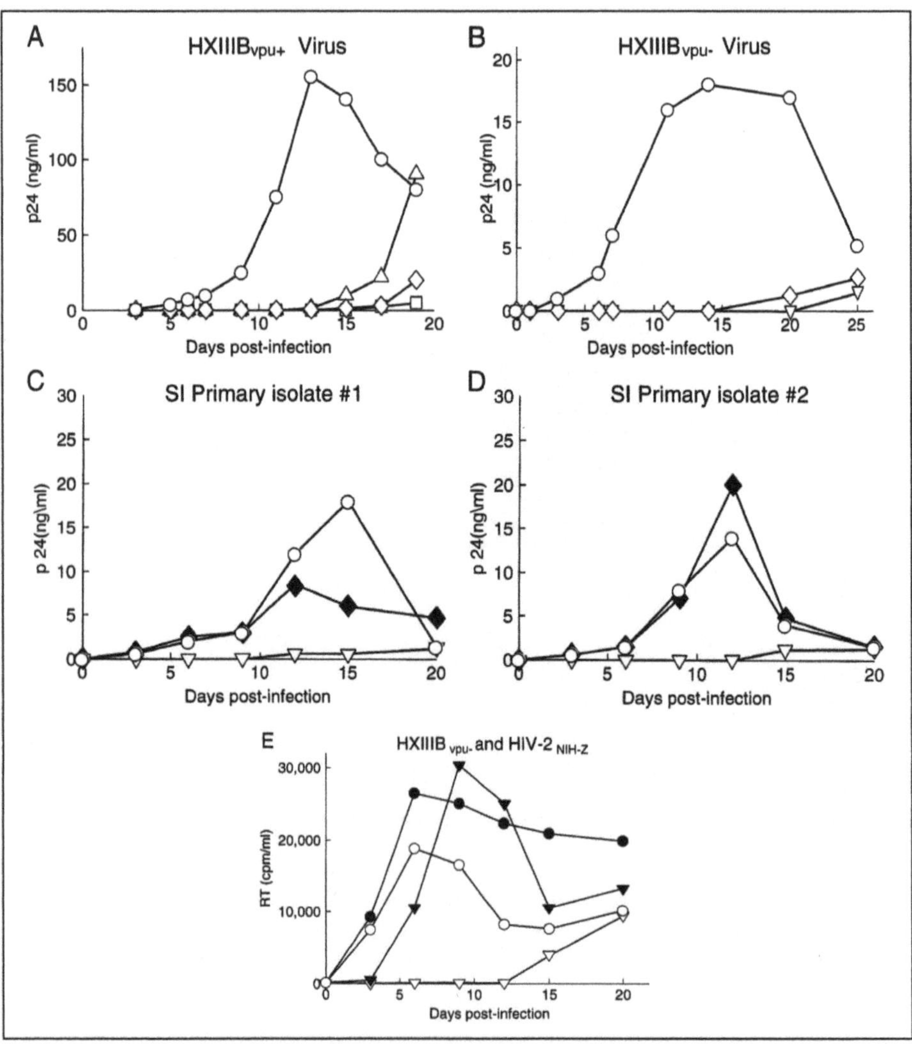

Fig. 9.3. Inhibition of HIV-2 infection in stably transfected CD4$^+$ Jurkat T cells expressing cytoplasmic anti-MA Fab3H7 intrabodies. Subclones of CD4$^+$ Jurkat-Fab3H7 cells (1 x 10^6) were infected with either of two isogeneic laboratory strains of HIV-1 that differed in the presence or absence of the accessory gene, *vpu* or two European SI primary isolates. Cell-free p24 levels were recorded on the indicated days. (A) Infection with HXIIIB$_{vpu+}$ at 2,000 cpm/ml RT activity (MOI Ç0.5) of Jurkat-pRC/CMV-vector cells (○), Jurkat-Fab3H7 (clone 621-1) (▲), Jurkat-Fab3H7 (clone 621-2) (□), and Jurkat-Fab3H7 cells (clone 621-3) (◇). (B) Infection with HXIIIB$_{vpuå}$ at 15,000 cpm/ml RT activity (MOI Ç6.5) of Jurkat-pRc/CMV-vector cells (○), Jurkat-Fab3H7n (clone 621-3) (◇), and Jurkat-Fab3H6 cells (clone 725-3) (▽). (C) and (D) Parental Jurkat cells (◆), Jurkat-vector cells (○), and Jurkat-Fab3H7 cells (clone 725-3) (▽) (1 x 10^6) were infected with 25 ng p24 of two different European, syncytia-inducing (SI) primary isolates that were previously screened for their ability to infect parental CD4$^+$ Jurkat T cells. (C) Infection with primary isolate #1. (D) Infection with primary isolate #2. (E) Infection with HIV-2$_{NIH-Z}$ or HXBIIIB$_{vpuå}$ (MOI Ç0.5). Open symbols represent infection with HXIIIB$_{vpuå}$; solid symbols, infection with HIV-2$_{NIH-Z}$; circles, Jurkat-vector cells; and triangles, Jurkat-Fab3H7 cells (clone 723-5).

resistant to HIV-1 infection when challenged with both laboratory strains and primary viral isolates. However, this resistance is relative and is not absolute with the stable cell lines and HIV-1 strains that were examined.

Inhibition of HIV-1 Integration in CD4$^+$ Jurkat T Cells Stably Expressing Cytoplasmic Anti-MA Intrabodies

The efficiency of HIV-1 proviral (HXIIIB$_{vpu'}$) integration into the cellular DNA was assessed using a PCR-based procedure in which viral-cellular DNA junction fragments of define size (1-3.5 kb) were isolated from a 1% agarose gel after restriction enzyme digestion of cellular DNA. After religation of the 5' and 3' LTR-cellular DNA fragments and PCR amplification of the ligated and reconstructed LTR fragments, a 561 bp fragment representing the integrated viral DNA is detected.[43] Southern blot analysis used a ^{32}P-labeled LTR probe to confirm the identify of the PCR product (data not shown). As seen in Figure 9.4, DNA from HIV-1-infected Jurkat cells (lane 3) show the 561 bp integration product 4 days postinfection, whereas, DNA from stable Jurkat-Fab3H7 subclones expressing the cytoplasmic anti-MA Fab3H7 intrabodies (lanes 4 and 5) do not. At 10 days postinfection, a strong integration band is seen only in the control cells (lane 6), and it is not seen in the anti-MA intrabody expressing subclones (lanes 7 and 8). The integration band is seen on day 13 in the anti-MA Fab intrabody expressing subclones (lanes 10 and 11). Thus in these experiments, where multiple rounds of infection are occurring, integration was delayed but not prevented in the anti-MA Fab intrabody-expressing cells. Similar analysis of DNA obtained from uninfected Jurkat cells spiked with viral DNA fragments obtained by restriction enzyme digestion outside of the 1-3.5 kb size range yielded no specific PCR products, indicating that the separation of DNA fragments of the LMP agarose gel and the extraction of these DNA fragments in the 1-3.5 kb region effectively removed viral DNA fragments outside the 1-3.5 kb range of DNA (lane 2). In addition, no specific signal was obtained if the 1-3.5 kb fragments were not ligated prior to PCR amplification, thus demonstrating that the 561 bp fragments were not an artifact resulting from contaminating linear or circular viral DNA fragments (data not shown).

These experiments demonstrate that the anti-MA Fab3H7 interferes with the afferent arm of the HIV-1 life cycle (Fig. 9.5). Although the mechanism of this interference was not determined in these studies, several processes may be involved. First, since most of the MA has been localized to the periphery of mature HIV-1 particles, where it is bound to the inner leaflet of the virus lipid bilayer[44] and in association with the envelope glycoprotein, it is possible that the anti-MA Fab intrabodies are interfering with viral uncoating, a step in the viral life cycle that is poorly understood. Second, the anti-MA Fab intrabodies may be binding to the carboxyl-terminus of MA while it is a part of a high-molecular-weight preintegration complex and they may inhibit either reverse transcription indirectly or the transport of the preintegration complex through the nucleopore to the nucleus directly.[45] In the latter case, however, because the HIV-1-infected Jurkat-Fab3H7 cells are actively dividing in culture, the breakdown of the nuclear envelope at mitosis should allow the preintegration complex to come in contact with cell chromosomes.[46,47] Therefore, for this direct mechanism of inhibition to occur, binding of the anti-MA Fab intrabodies to the MA-associated preintegration complex would have to render the complexes unstable. Decreased stability of viral RNA has been reported following HIV-1 entry in quiescent primary lymphocytes.[48]

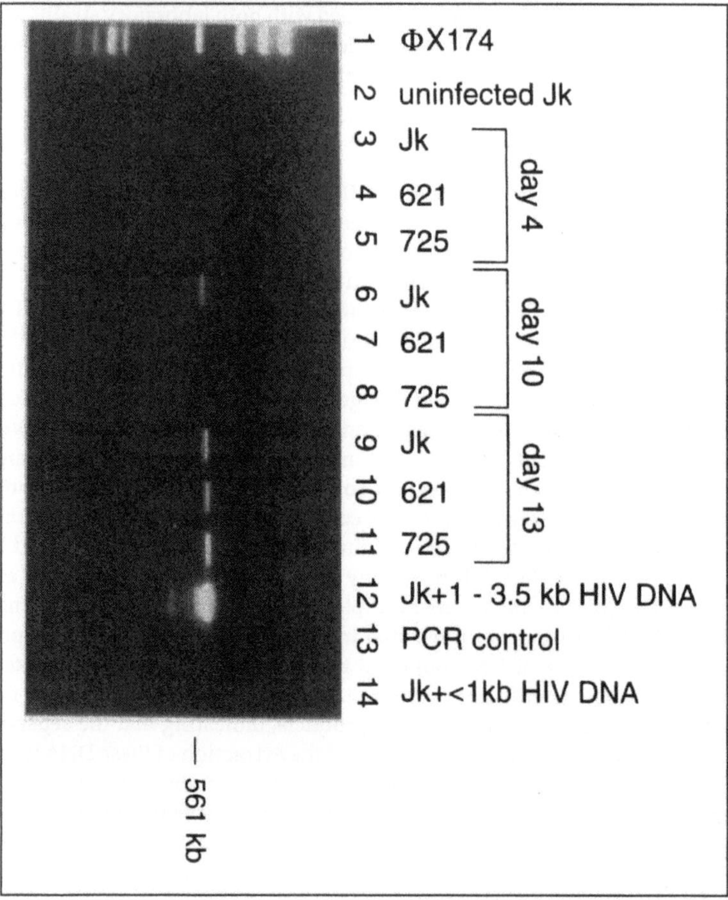

Fig. 9.4. HIV-1 DNA integration in Jurkat control and Jurkat-Fab3H7 cells from an experiment similar to that in Figure 9.3B. Lane 1: *Hae*III digested fX174 DNA molecular weight standards; Lane 2: Uninfected Jurkat cells: Lanes 3, 6, 9: Infected Jurkat cells; Lanes 4, 7, 10: Infected Jurkat-Fab3H7 cells (clone 621-3); Lane 5, 8, 11: Infected Jurkat-Fab3H7 cells (clone 725-3). Lanes 2-5: Day 4; Lanes 6-8: Day 10; Lanes 9-11: Day 13; Lane 12, Uninfected Jurkat cells spiked with 1-3.5 kb fragments from *SacI/XbaI* digested HXBc2 plasmid that contains the SVC$_{vpu+}$ HIV-1 provirus; Lane 13: PCR primers alone, no cellular DNA; Lane 14: Same as Lane 12 but spiked with proviral DNA fragments <1 kb.

Finally, it is possible that the anti-MA Fab intrabodies can prevent integration of the viral double-stranded DNA intermediate through steric interferences with other components of the preintegration complex (e.g. integrase). Indeed, a direct physical interaction between MA and integrase has been recently reported.[49]

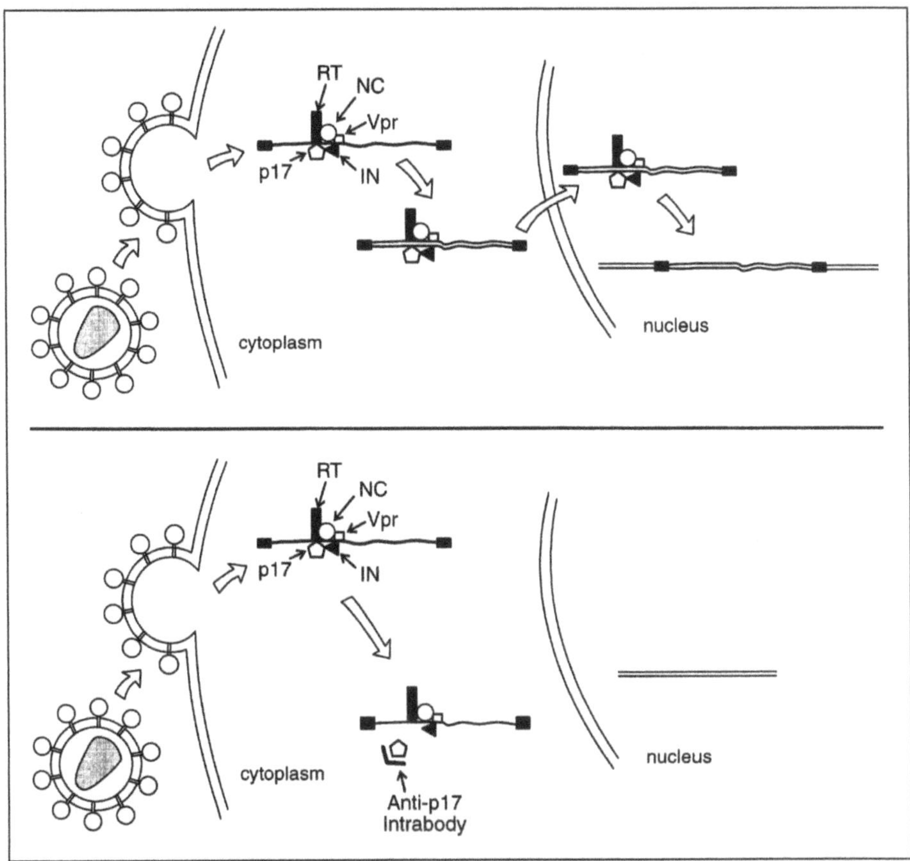

Fig. 9.5. Top panel, representation of normal integration of HIV-1 provirus. Bottom panel, representation of the same process in the presence of an anti-MA (anti-p17) intrabody. There is no viral integration.

Late Stages in the Virus Life Cycle

Inhibition of Infectious Virus Formation in CD4$^+$ Jurkat T Cells Stably Expressing Cytoplasmic Anti-MA Fab3H7 Intrabodies

The infectivity of HIV-1 virions produced from the infected anti-MA-expressing Jurkat-Fab3H7 cells was investigated by comparing the TCID$_{50}$ of the cell-free culture supernatant normalized for Gag p24. On the first day after infection of cells when p24 was detectable in the supernatants, equal amounts of p24 were used to initiate infection of fresh CD4$^+$ H9 cells to determine the infectivity (TCID$_{50}$ units) of the virions.[50] Supernatants from two to four subclones of HIV-1-challenged, Fab3H7 expressing Jurkat cells were examined. As shown in Table 9.1, marked inhibition of virion infectivity was seen with virions released from the Fab3H7 intrabody expressing cells compared to control cells in all three viruses

Table 9.1. Infectivity of virus particles released from stably transfected CD4$^+$ Jurkat T cell lines expressing anti-HIV-1 MA Fab3H7 intrabodies*

Virus	Cell Line	Day of p24 Harvest	TCID$_{50}$/ml	% Inhibition
HXIIIB$_{vpu+}$	Vector	9	1.85×10^6	–
	Fab3H7 clone 621-1	17	4.09×10^4	97.8
	Fab3H7 clone 621-3	19	2.90×10^4	98.5
HXBIII$_{vpu}$	Vector	15	2.60×10^6	–
	Fab3H7 clone 621-1	25	4.60×10^5	82.3
	Fab3H7 clone 621-3	20	6.30×10^6	76.0
	Fab3H7 clone 725-1	20	1.60×10^5	93.8
	Fab3H7 clone 725-3	20	1.16×10^5	96.0
SI Primary Isolate #1	Vector	15	1.80×10^6	–
	Fab3H7 clone 621-1	30	1.15×10^5	93.6
	Fab3H7 clone 621-3	30	2.80×10^4	98.5
	Fab3H7 clone 725-1	20	1.15×10^5	93.6
	Fab3H7 clone 725-3	20	4.00×10^4	97.7

* -0.5, 0.5, and 0.2 M.O.I. of HXIIIB$_{vpu+}$, HXIIIB$_{vpu}$ ', and SI Primary Isolate #1 were used to infect cells, respectively.

examined. This inhibition of virion infectivity ranged from 76% to 98.5%, and most subclones showed > 93% inhibition compared to the virions released from control cells.

Thus, these experiments demonstrate a significant decrease in the infectivity of the virions released from these HIV-1-infected, Fab3H7 intrabody-expressing cells (Fig. 9.6). Although the mechanism(s) of this inhibition were not uncovered in these studies, preliminary radioimmunoprecipitation studies of viral proteins from HIV-1-infected Jurkat-Fab3H7 cells showed no clear differences in the levels of p55 *gag* precursor or its cleavage product when cell lysates, supernatant, and cell-free pelleted virions were examined (data not shown). Thus, inhibition of a process other than p55 *gag* maturation could be occurring. It is possible that the inhibition may result from other specific events, such as: envelope association with MA,[51] MA dimerization,[52] viral genomic RNA association with the p55 *gag* precursor, or tyrosine phosphorylation of MA by a virion-associated cellular protein kinase reported to lead to preferential targeting of phosphorylated MA to the nucleus of target cells.[49,53,54]

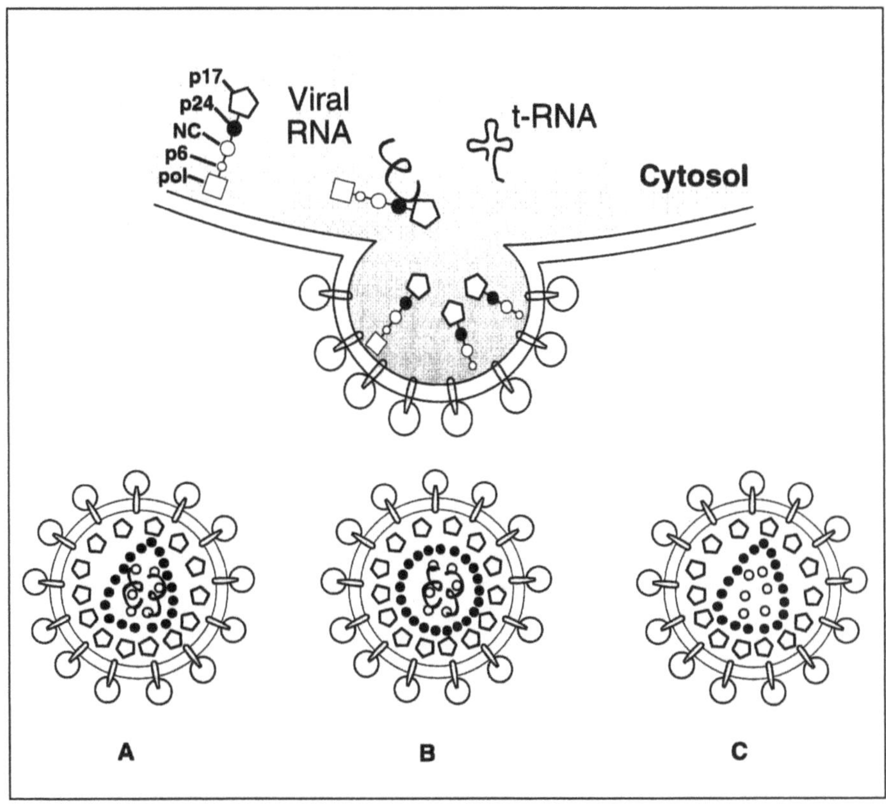

Fig. 9.6. Representation of possible assembled virions. (A) Normal conditions. (B) Immature virions. (C) Lack of viral RNA.

Nucleocapsid (p7)

The inner core of most retroviruses is formed by a shell of capsid protein molecules (p24) surrounding the dimeric RNA genome in close association with approximately 2,000 molecules of nucleocapsid protein (NC), 20 to 50 molecules of
reverse transcriptase and integrase,[55] as well as tRNAs, 5S RNA, 7S RNA and ribosomal RNAs.[56] The NC protein contains two highly conserved zinc finger motifs.[57]
In vitro studies have shown that the NC is required for genomic RNA dimerization, correct encapsidation, annealing of the tRNA primer to its complementary
binding site (PBS) on the genomic RNA, and for the minus-strand DNA transfer
during reverse transcription.[58,59] These observations indicate that the NC plays an
essential role in the afferent as well as the efferent arms of the HIV-1 life cycle.
Preliminary results have been reported showing the anti-HIV-1 activity of anti-NC
scFv intrabodies.[60]

Epitope Mapping of Murine Anti-NC Monoclonal Antibodies

A panel of monoclonal antibodies (mAbs) was produced through conventional hybridoma technology from the spleen of Balb/c mice immunized with either HIV-1 lysate or recombinant p24-15 antigen.[61] These mAbs were originally generated to identify immunogenic amino acid *gag* sequences. Complete epitope mapping of the panel of mAbs was carried out using a series of overlapping 15-mer peptides. Two mAbs (EC6 and M12) were shown to be directed to the nucleocapsid (NC) protein. They recognize a conserved epitope present in the second zinc finger of NC. This epitope was mapped to the amino acid sequence PRKKGCWKCG (residues 408-417) and is associated with the Clade B HIV-1 genotype.[32]

Analysis of NC mutants with altered zinc finger amino acid sequences has shown that retroviruses require intact zinc fingers for specific packaging of viral RNA otherwise, the resulting virus is noninfectious and deficient in packaged viral RNA.[62-66] These observations have been further supported by the evacuation of electrophilic compounds that displace the zinc and modify the zinc fingers of retroviral NC proteins.[67,68] A dithiane compound (NSC 624151) was shown to inactivate HIV-1 virions and blocked production of infectious virus from cells harboring integrated proviral DNA.[68] However, studies have also shown that the two zinc fingers present in HIV-1 NC are not functionally equivalent. The data suggest that the first zinc finger of HIV-1 NC plays a more prominent role in RNA selection and packaging. The data also indicate that both zinc fingers in the mature NC protein play roles in viral assembly or early stages of the viral infection process.[69]

Construction of Anti-NC Intrabodies

Using standard PCR conditions, the Fd and light chains of the mAb M12 were amplified from the hybridoma cDNA. The sequence of the primers used to PCR the immunoglobulin chains were antibody specific primers.[34] Subsequently, nested PCR was performed to eliminate the leader and constant sequences, and to introduce a Kozak consensus sequence with an ATG initiation codon immediately preceding amino acid one of their framework one. These variable domains were then assembled into the scFv format using the interchain linker sequence as overlap extension in a PCR reaction. In addition to the scFv M12 constructed, an scFv with a constant κ domain (C_κ) was also built. The C_κ domain has previously shown to increase the activity of scFvs expressed in the cytoplasm of mammalian cells.[18]

Half-Life of Cytoplasmic Expressed Intrabodies

Half life determination was performed to compare the stability of scFv NC with that of scFvCk NC in the cytoplasm of mammalian cells. The sequence encoding for the hemagglutinin (HA) epitope (YPYDVPDYA) was fused to the carboxy-terminus of scFv NC and scFvCk NC in order to facilitate detection. These constructs (scFv NC-HA and scFvCk NC-HA) were used to transect COS-7 cells. Forty-eight hours post-transfection, cells were incubated in methionine-free media for 60 minutes. ^{35}S-methionine was then added at 100 Ci/ml for 60 minutes, followed by replacement of complete medium supplemented with excess methionine. Samples were collected at 0, 10, 30 and 120 minutes. The intrabodies were immunoprecipitated using a monoclonal antibody specific for HA TAG (BabCO), and electrophoresed in 12% SDS-Polyacrylamide gels. As seen in Figure 9.7, the scFv NC as well as the scFvCk NC appear to have similar half-lives, approximately 45 minutes. These data suggest that the Ck domain does not increase the half life of the scFv, at least of the scFv NC. It is of interest to note that the anti-Tat scFvCk shows an increase in

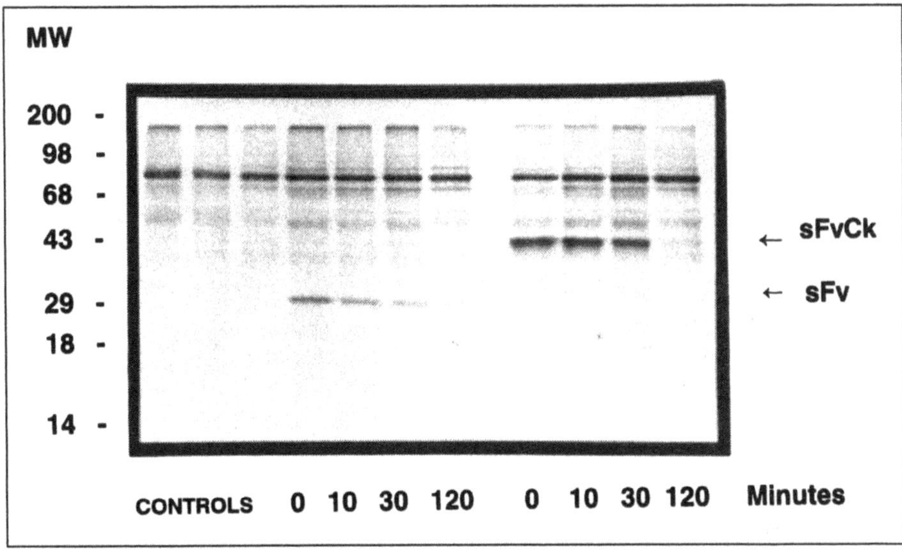

Fig. 9.7. Half life determination of scFvNC and scFvCk NC in COS-7 cells.

the activity over anti-Tat scFv when expressed in the cytoplasm.[18] This increase in activity seen with the anti-Tat scFvCk could be attributed to formation of multimeric forms[70] which translate to an avidity gain. Alternatively, the Ck domain might favor proper folding of the scFv in the cytoplasm without affecting its half life; however, this situation might be different for each individual scFv.

Early Stages in the Virus Life Cycle

Inhibition of HIV-1 Infection in CD4$^+$ Jurkat T Cells Stably Expressing Cytoplasmic Anti-NC Intrabodies

To examine the role of the anti-NC intrabodies in the preintegration phase of the virus life cycle, stably transfected Jurkat-scFv NC cells were challenged with single round replication CAT virus (similar to the way described for Jurkat-Fab3H7 cells). A marked inhibition of CAT activity was seen in the majority of stably transfected Jurkat-scFv NC clones compared to stably transfected Jurkat-vector clones. This inhibition could occur as a result of inhibition at any critical step in the afferent arm of the virus life cycle that leads to integration of the HIV-1 provirus (see Fig. 9.8). Further studies are on the way to determine the mechanism of inhibition. It is hypothesized that the anti-NC intrabody interferes with the function of NC during reverse transcription. In vitro studies suggest that the NC facilitates tRNA annealing to the viral genomic RNA and that it cooperates in HIV-1 RT catalyzed strand-transfer reactions.[58,59,71]

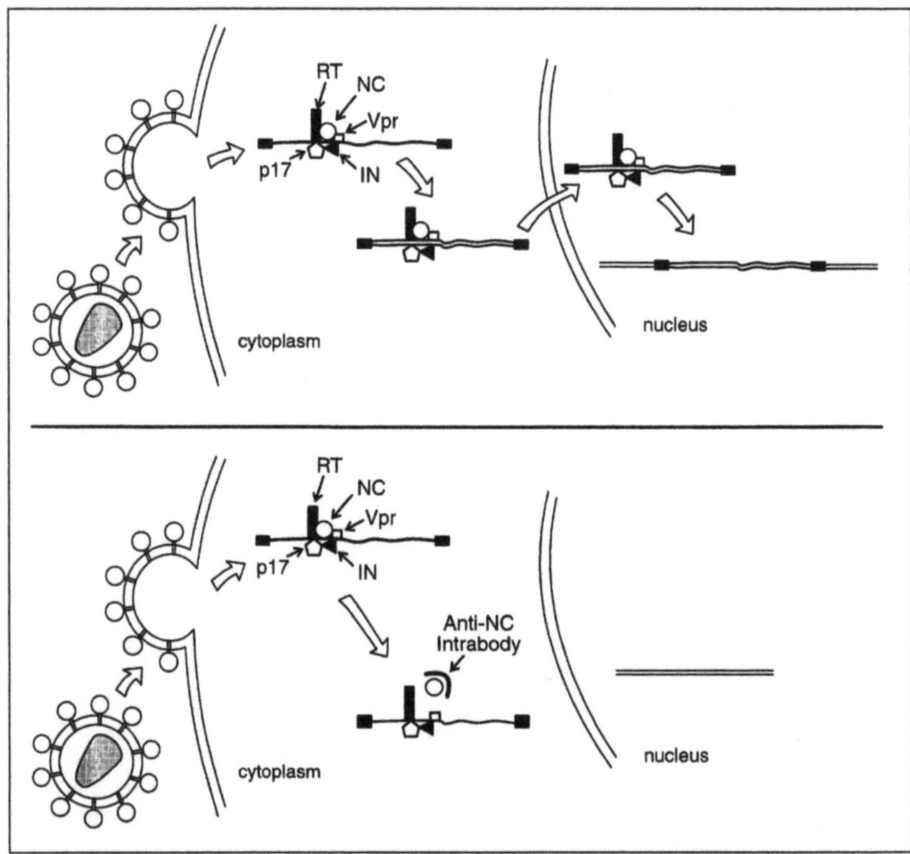

Fig. 9.8. Top panel, representation of normal integration of HIV-1 provirus into the host chromosome. Bottom panel, representation of the same process in the presence of an anti-NC intrabody. There is not complete reverse transcription; thus, there is no viral integration.

Late Stages in the Virus Life Cycle

Inhibition of Virus Reproduction in Cells Expressing Cytoplasmic Anti-NC Intrabodies

Either HeLa or COS-7 cells transfected with plasmids, needed to produce single round CAT virus, and an intrabody expressor vector, showed a decrease in production of viral particles. This decrease in particle formation could be due to incorrect processing of p55 *gag* precursor, or interference at steps that involve viral RNA packaging (see Fig. 9.6). Further experiments will delineate the mechanism of inhibition.

Conclusion

The data discussed support the choice of MA and NC as targets to evaluate intracellular immunization against HIV-1. Intrabodies against MA and NC affect early and late stages of HIV-1 life cycle. The effectiveness of these intrabodies might depend on their interacting epitopes. A particular epitope might be more important for activity during the early stage than for the late stage of the virus life cycle, thus, causing more of an inhibitory effect in one stage. Ultimately, however, only a comparison of all the different anti-HIV-1 intrabodies, tested under the same conditions, will determine which one is more efficacious reducing the spread of the virus.

References

1. Popovic M, Sarngadharan MG, Read E et al. Detection, isolation, and continuous production of cytopathic retroviruses (HTLV-III) from patients with AIDS and preAIDS. Science 1984; 224:497-500.
2. Levy JA. HIV and the pathogenesis of AIDS. JAMA 1989; 261:2997-3006.
3. Bridges SH, Sarver N. Gene therapy and immune restoration for HIV disease. The Lancet 1995; 345:427-432.
4. Crystal RG. Transfer of genes to humans: early lessons and obstacles to success. Science 1995; 270:404-410
5. Gilboa E, Smith C. Gene therapy for infectious diseases: the AIDS model. Trends Genet 1994; 10:139-144.
6. Dropuli´c B, Jeang K-T. Gene therapy for human immunodeficiency virus infection: genetic antiviral strategies and targets for intervention. Human Gene Therapy 1994; 5:927-939.
7. Rondon IJ, Marasco WA. Intracellular antibodies (intrabodies) for gene therapy of infectious diseases. Ann Rev Microbiol 1997; 51:257-283.
8. Roberts MR, Qin L, Zhang D et al. Targeting of human immunodeficiency virus-infected cells by CD8+ T lymphocytes armed with universal T-cell receptors. Blood 1994; 84:2878-2889.
9. Wang B, Ugen KE, Srikantan V et al. Gene inoculation generates immune responses against human immunodeficiency virus type 1. Proc Natl Acad Sci USA 1993; 90:4156-4160.
10. Baltimore D. Intracellular immunization. Nature 1988; 335:395-396.
11. Sarver N, Cantin EM, Chang PS et al. Ribozymes as potential anti-HIV-1 therapeutic agents. Science 1990; 247: 1222-1225.
12. Sullenger BA, Gallardo HF, Ungers GE et al. Analysis of trans-acting response decoy RNA- mediated inhibition of human immunodeficiency virus type 1 transactivation. J Virol 1991; 65 (12): 6811-6816.
13. Malim MH, Bohnlein S, Hauber J et al. Functional dissection of the HIV-1 Rev transactivator-derivation of a trans-dominant repressor of Rev function. Cell 1989; 58:205.
14. Brady HJM, Miles CG, Pennington DJ et al. Specific ablation of human immunodeficiency virus Tat-expressing cells by conditionally toxic retroviruses. Proc Natl Acad Sci USA 1994; 91:365-369.
15. Marasco WA, Haseltine WA, Chen SY. Design, intracellular expression, and activity of a human anti-human immunodeficiency virus type 1 gp120 single-chain antibody. Proc Natl Acad Sci USA 1993; 90:7889-7893.
16. Varmus H. Retroviruses. Science 1988; 240:1427-1435.
17. Weiss R, Teich N, Varmus H et al (eds). RNA Tumor Virus. Cold Spring Harbor, New York: Cold Spring Harbor Laboratory, 1985.
18. Mhashilkar A, Bagley J, Chen SY et al. Inhibition of HIV-1 Tat-mediated LTR transactivation and HIV-1 infection by anti-Tat single chain intrabodies. EMBO J 1995; 14:1542-1551.

19. Duan L, Bagasra O, Mark A et al. Potent inhibition of human immunodeficiency virus type 1 replication by an intracellular anti-Rev single-chain antibody. Proc Natl Acad Sci USA 1994; 91:5075-5079.

20. Shaheen F, Duan L, Zhu M et al. Targeting human immunodeficiency virus Type 1 reverse transcriptase by intracellular expression of single-chain variable fragments to inhibit early stages of the viral life cycle. J Virol 1996; 70: 3392-3400.

21. Levy-Mintz P, Duan L, Zhang H et al. Intracellular expression of single-chain variable fragments to inhibit early stages of the viral life cycle by targeting human immunodeficiency virus Type 1 integrase. J Virol 1996; 70:8821-8832.

22. Levin R, Mhashilkar AM, Dorfman T et al. Inhibition of early and late events of the HIV-1 replication cycle by cytoplasmic Fab intrabodies against the matrix protein, p17. In press

23. Kaplan AH, Swanstrom R. Gag proteins are processed in two cellular compartments. Proc Natl Acad Sci USA 1991; 88:4528-4532.

24. Sheng N, Erickson-Viitranen S. Cleveage of p15 in vitro by HIV-1 protease is RNA dependent. J Virol 1994; 68:6207-6214.

25. Bukrinsky M, Sharova N, Stevenson M. Human immunodeficiency virus Type 1 2-LTR circles reside in a nucleoprotein complex which is different from the preintegration complex. J Virol 1993; 67: 6863-6865.

26. von Schwedler U, Kornbluth RS, Trono D. The nuclear localization signal of the matrix protein of human immunodeficiency virus type 1 allows the establishment of infection in macrophages and quiescent T lymphocytes. Proc Natl Acad Sci USA 1994; 91:6992-6996.

27. Spearman P, Wang J-J, Vander Heyden N et al. Identification of human immunodeficiency virus type 1 Gag protein domains essential to membrane binding and particle assembly. J Virol 1994; 68:3232-3242.

28. Zhou W, Parent LJ, Wills JW et al. Identification of a membrane-binding domain within the amino terminal region of human immunodeficiency virus type 1 Gag protein which interacts with acidic phospholipids. J Virol 1994; 68:2556-2569.

29. Bryant M, Ratner L. Myristylation-dependent replication and assembly of human immunodeficiency virus 1. Proc Natl Acad Sci USA 1990; 87:523-527.

30. Gottlinger HG, Sodroski JG, Haseltine WA. Role of capsid precursor processing and myristylation in morphogenesis and infectivity of HIV-1. Proc Natl Acad Sci USA 1989; 86:5781-5785.

31. Niedrig M, Hinkula J, Weigelt W et al. Epitope mapping of monoclonal antibodies against human immunodeficiency virus type 1 structural proteins by using peptides. J Virol 1989; 63:3525-3528.

32. Louwagie J, McCutchan FE, Peeters M et al. Phylogenetic analysis of gag genes from 70 international HIV-1 isolates provides evidence for multiple genotypes. AIDS 1993; 7:769-780.

33. Elroy-Stein O, Fuerst TR, Moss B. Cap-independent translation of mRNA conferred by encephalomyocarditis virus 5' sequence improves the performance of the vaccinia virus/bacteriophage T7 hybrid expression system. Proc Natl Acad Sci USA 1989; 86:6126-6130.

34. Kabat EA, Wu TT, Perry HM et al. Tabulation and analysis of amino acid and nucleic acid sequences of precursors, V-regions, C-regions, J-chain, T-cell receptors for antigen, T-cell surface antigens, Thy-1, complement, c-reactive protein, thymopoietin, integrins, postgamma globulin, 2-macroglobulins and other related proteins. In: Sequences of Proteins of Immunological Interest. U.S. Department of Health and Human Services, NIH Publication No. 91-3242, 1991, Washington, D.C.

35. Helseth E, Kowalski M, Gabuzda D et al. Rapid complementation assays measuring replicative potential of human immunodeficiency virus type 1 envelope glycoprotein mutants. J Virol 1990; 64:2416-2420.

36. Lewis P, Hensel M, Emerman M. Human immunodeficiency virus infection of cell arrested in the cell cycle. EMBO J 1992; 11:3053-3058.

37. Sakai H, Kawamura M, Sakuragi J-I et al. Integration is essential for efficient gene expression of human immunodeficiency virus type 1. J Virol 1993; 67:1169-1174.

38. Wiskerchen M, Muesing MA. Human immunodeficiency virus type 1 integrase: Effects of mutations on viral ability to integrate, direct viral gene expression from unintegrated viral DNA templates, and sustain viral propagation in primary cells. J Virol 1995; 69:376-386.

39. Engelman A, Englund G, Orenstein JM et al. Multiple effects of mutations in human immunodeficiency virus type 1 integrase on viral replication. J Virol 1995; 69:2729-2736.

40. Ansari-Lari MA, Donehower LA, Gibbs RA. Analysis of human immunodeficiency virus type 1 integrase mutants. Virol 1995; 211:332-335.

41. Stevenson M, Haggerty S, Lamonica CA et al. Integration is not necessary for expression of human immunodeficiency virus type 1 protein products. J Virol 1990; 64:2421-2425.

42. Blomberg J, Medstrand P. A sequence of the carboxyl terminus of the HIV-1 matrix protein is highly similar to sequences in membrane-associated proteins of other RNA viruses: Possible functional implications. New Biologist 1990; 2:1044-1046.

43. Li G, Simm M, Potash MJ et al. Human immunodeficiency virus type 1 DNA synthesis, integration, and efficient viral replication in growth-arrested T cells. J Virol 1993; 67:3969-3977.

44. Gelderblom HR, Hausman EHS, Özel M et al. Fine structure of human immunodeficiency virus (HIV) and immunolocalization of structural proteins. Virol 1987; 156:171-176.

45. Bukrinsky MI, Sharova N, Dempsey MP et al. Active nuclear import of human immunodeficiency virus type 1 preintegration complexes. Proc Natl Acad Sci USA 1992; 89:6580-6584.

46. Lewis PF, Emerman M. Passage through mitosis is required for oncoretroviruses but not for human immunodeficiency virus. J Virol 1994; 68:510-516.

47. Roe TY, Reynolds TC, Yu G et al. Integration of murine leukemia virus DNA depends on mitosis. EMBO J 1993; 12:2089-2108.

48. Zack JA, Arrigo SJ, Weitsman SR et al. HIV-1 entry into quiescent primary lymphocytes: Molecular analysis reveals a labile, latent viral structure. Cell 1990; 61:213-222.

49. Gallay P, Swingler S, Song J et al. HIV nuclear import is governed by the phosphotyrosine-mediated binding of matrix to the core domain of integrase. Cell 1995; 83:569-576.

50. Johnson VA, Byington RE. Quantitative assays for virus infectivity. In: Aldovini A, Walker BD, eds. Techniques in HIV Research Stockholm, NY:Stockton Press, 1990 71-96.

51. Lodge R, Göttlinger H, Gabuzda D et al. The intracytoplasmic domain of gp41 mediates polarized budding of human immunodeficiency virus type 1 in MDCK cells. J Virol 1994; 68:4857-4861.

52. Morikawa Y, Kishi T, Zhang WH et al. A molecular determinant of human immunodeficiency virus particle assembly located in matrix antigen p17. J Virol 1995; 69:4519-4523.

53. Bukrinskaya AG, Ghorpade A, Heinzinger NK et al. Phosphorylation-dependent human immunodeficiency virus type 1 infection and nuclear targeting of viral DNA. Proc Natl Acad Sci USA 1996; 93:367-371.

54. Gallay P, Swingler S, Aiken C et al. HIV-1 infection of nondividing cells: C-terminal tyrosine phosphorylation of the viral matrix protein is a key regulator. Cell 1995; 80:379-388.

55. Chen M, Garon C, Papas T. Native ribonucleoprotein is an efficient transcriptional complex of avia myeloblastosis virus. Proc Natl Acad Sci USA 1980; 77:1296-1300.

56. Dickson C, Eisenman R, Fan H et al. Protein biosynthesis and assembly. In: Weiss R, Teich N, Varmus H, Coffin J, eds. RNA Tumor Viruses. Cold Spring Harbor, NY:Cold Spring Harbor Laboratory Press, 1985:513-648.

57. Bess Jr JW, Powell PJ, Issaq HJ et al. Tightly bound zinc in human immunodeficiency virus Type 1. Human T-cell leukemia virus Type 1, and other retroviruses. J Virol 1992; 66:840-847.

58. Peliska JA, Balasubramanian S, Giedroc DP et al. Recombinant HIV-1 nucleocapsid protein accelerates HIV-1 reverse transcriptase catalyzed strand transfer reactions and modulates RNase H activity. Biochemistry 1994; 33:13817-13823.

59. Prats AC, Sarih L, Gabus C et al. Small finger protein of avian and murine retroviruses has nucleic acid annealing activity and positions the replication primer tRNA onto genomic RNA. EMBO J 1988; 7:1777-1783.

60. Rondon IJ, Marasco WA. Inhibition of HIV-1 replication by a scFv intrabody against the nucleocapsid, NCp7. IBC's Seventh Annual International Conference on Antibody Engineering, December 2-4, 1996. Coronado, CA (Abstract).

61. Hinkula J, Rosen J, Sundqvist V-A et al. Epitope mapping of the HIV-1 Gag region with monoclonal antibodies. Mol Immunol 1990; 27:395-403.

62. Aronoff R, Hajjar AM, Linial ML. Avian retroviral RNA encapsidation: reexamination of functional 5' RNA sequences and the role of nucleocapsid Cys-His motifs. J Virol 1993; 67:178-188.

63. Dupraz P, Oertle S, Méric, P et al. Point mutations in the proximal Cys-His box of Rous sarcoma virus nucleocapsid protein. J Virol 1990; 64:4978-4987.

64. Gorelick RJ, Henderson LE, Hanser JP et al. Point mutants of Moloney murine leukemia virus that fail to package viral RNA: evidence for specific RNA recognition by a "zinc finger-like" protein sequence. Proc Natl Acad Sci USA 1988; 85:8420-8424.

65. Gorelick RJ, Nigida, Jr. SM, Bess JW et al. Noninfectious human immunodeficiency virus type 1 mutants deficient in genomic RNA. J Virol 1990; 46:3207-3211.

66. Méric C, Gouilloud E, Spahr P.-F. Mutations in Rous sarcoma virus nucleocapsid protein p12 (NC): deletions of Cys-His boxes. J Virol 1988; 62:3328-3333.

67. Rice WG, Supko JG, Malspeis L et al. Inhibitors of HIV nucleocapsid protein zinc fingers as candidates for the treatment of AIDS. Science 1995; 270: 1194-1199.

68. Rice WG, Baker DC, Schaeffer CA et al. Inhibition of multiple phases of human immunodeficiency virus type 1 replication by a dithiane compound that attacks the conserved zinc fingers of retroviral nucleocapsid proteins. Antimicro Agents and Chemo 1997; 41:419-426.

69. Gorelick RJ, Chabot DJ, Rein A et al. The two zinc fingers in the human immunodeficiency virus type 1 nucleocapsid protein are not functionally equivalent. J Virol 1993; 67:4027-4036.

70. McGregor DP, Molloy PE, Cunningham C et al. Spontaneous assembly of bivalent single chain antibody fragments in *Escherichia coli*. Molec Immunol 1994; 31:219-216.

71. Li X, Quan Y, Arts EJ et al. Human immunodeficiency virus Type 1 nucleocapsid protein (NCp7) directs specific initiation of minus-strand DNA synthesis primed by human tRNA,³ᵐ in vitro: Studies of viral RNA molecules mutated in regions that flank the primer binding site. J Virol 1996; 70:4996-5004.

CHAPTER 10

Single Chain Variable Fragment-Based Strategies for Anti-HIV-1 Gene Therapy: Targeting the Viral Preintegration Complex and Combination Molecular Approaches

Lingxun Duan and Roger J. Pomerantz

Introduction

The magnitude of the worldwide acquired immune deficiency syndrome (AIDS) pandemic and the difficulty of developing and testing an effective vaccine against human immunodeficiency virus type I (HIV-1) have led to proposals for alternative therapies which might prove applicable for infected patients. The HIV-1 virus specifically targets and destroys human $CD4^+$ T lymphocytes, which are critical for the proper functioning of the immune response against viral infection. With a highly mutable viral envelope protein, gp120, revere-transcriptase (RT) and proteinase, it is quite difficult to develop an effective vaccine which is solely based on the viral envelope. Most chemotherapeutic agents quickly lose their potency because of viral resistance developed via mutations in the RT or proteinase gene, during a short period of treatment. During HIV-1 infection, as the virus infects human $CD4^+$ cells, the proviral DNA can be integrated into the host cell genome. As such, HIV-1 infection can actually be defined as an acquired genetic disease.

Gene therapy is an approach for the treatment and prevention of inherited and acquired genetic diseases, in which genetic elements are introduced into cells to produce specific proteins or RNA, required to selectively correct or modulate disease conditions. Gene therapy may allow the replacement of defective genes, or add functional genes into specific cells of the body. In the case of AIDS, because the $CD4^+$ cells, which are a major targets for HIV-1 infection, are thought to be derived from common precursor cells in the bone marrow, it may be possible to

Intrabodies: Basic Research and Clinical Gene Therapy Applications, edited by Wayne A. Marasco. © 1998 Springer-Verlag and R.G. Landes Company.

introduce protective genes into most or all of the T lymphocytes and monocyte/ macrophages in an individual and greatly reduce HIV-1 replication. The idea that cells could be made resistant to intracellular pathogens by genetic modification was derived from previous studies which demonstrated that overexpression of mutant viral proteins, in the host cells, can interfere with wild-type viral protein functions in the same cells. Such strategies have now become widely known by the term "intracellular immunization".[33]

HIV-1 Replication

HIV-1 is an enveloped particle containing two copies of a single-stranded RNA genome. After binding to a high affinity receptor on a cell's surface, the CD4 molecule, as well as chemokine receptors which act as viral coreceptors, the virion penetrates and is uncoated. The virally-encoded RT then catalyzes the synthesis of a double-stranded DNA intermediate (provirus) in a preintegration complex. After transport to the nucleus, the provirus is integrated, via the viral integrase enzyme, into the host cell's genome. Viral transcription leads to the expression of HIV-1 specific genomic RNA and mRNA. After translation of viral mRNA, the virion particles assemble and bud from the cell surface. Of note, a plethora of viral regulatory proteins control the complex expression patterns of HIV-1.[7]

The genomic size of HIV-1 is approximately 9.8 kb, with open reading frames encoding for several viral proteins (Fig. 10.1). The primary transcript of HIV-1 is a full-length viral mRNA, which is translated into the Gag and Pol proteins. The Gag precursor p55 gives rise, by proteolytic cleavage, to several smaller proteins, p24, p17, p9 and p6. The Pol precursor protein is cleaved into products consisting of the RT, protease, and integrase proteins. The viral protease processes the Gag and Pol polyproteins, and the viral integrase is involved in viral DNA integration into host cell genome DNA. Viral splicing of RNA products many subgenomic mRNAs, which are important for the synthesis of other viral proteins.[43] The relative quantities of unspliced to singly- and multiply-spliced mRNA appear to be determined by the viral Rev protein, which is itself a product of multiple-spliced mRNA. The envelope gp120 and gp41 proteins are made after processing of gp160, which comes from a singly-spliced message. Gene products of other spliced viral mRNAs make up a variety of viral regulatory and accessory proteins, which affect HIV-1 replication in various cell-types. HIV-1 Tat, a transcriptional trans-activating protein, along with certain cellular proteins, interact with an the RNA loop structure termed TAR (Tat responsive element). Tat is a major moiety involved in upregulating HIV-1 transcription. The viral regulatory protein, Rev (regulator of viral protein expression), interacts with a *cis*-acting RNA loop structure called RRE, located in the viral envelope mRNA. This interaction involves cellular proteins and multimers of the Rev protein, and permits unspliced mRNA to enter the cytoplasm from the nucleus, giving rise to structural proteins required for viral progeny production.[7] Recent findings indicate that p17 and Vpr functions are involved in the translocation of viral preintegration complexes into the nucleus of infected cells.[39]

Intracellular immunization-based gene therapy strategies, which apply to HIV-1, require the delivery of HIV-1 "resistance genes" into host CD4$^+$ cells. HIV-1-specific resistance genes should satisfy three main criteria. They should be effective, nontoxic, and should not be affected by HIV-1 variation. A number of strategies have been developed for inhibiting HIV replication, using either DNA/RNA- or protein-

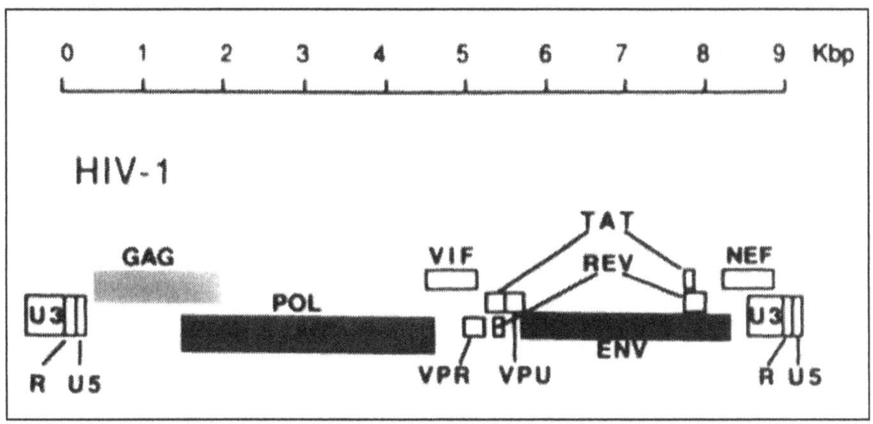

Fig. 10.1. Structure of HIV-1 genome.

based modalities. Figure 10.2 illustrates a schematic representation of the replication cycle of HIV-1 and lists various inhibition strategies, indicating the step in the viral replication cycle which might be altered.

Antibodies and Gene Therapy

One of the most promising "intracellular immunization" strategies against HIV-1 infection is specific immunoglobulin gene-based therapy. The specific and high-affinity binding properties of immunoglobulin (Ig) molecules have long been of use in the biomedical sciences as tools for identification, purification, and functional manipulation of target antigens. It is well-known that purified natural antibodies introduced into the cells by micro-injection can transiently inactivate a target protein.[10a] The concept of cloning specific immunoglobulin genes, with genetic engineering, and stably transducing the cloned Ig gene into cells, has been proposed. A recent series of articles now demonstrate the potential for using modified antibodies to interfere, in a highly specific manner, with biological processes inside cells (see below).

Antibodies possess a vast repertoire of specificity and affinity. Antibodies have a distinctive structure often depicted as a Y or T shape with two distal segments (Fab) containing the sites for antigen binding. Antibodies consist of two identical light (L) chains and two identical heavy (H) chains that fold into domains. The structure of the classic Ig folding has been established with the determination of the structure of a Fab fragment[32] and the presence of the same folding in the Fc fragment.[8] This folding and its variants have also been observed in nonantibody molecules, including T-cell receptors. The Fab contains a variable domain (V_L/V_H) and constant domain (C_L/C_H), with the two halves of each domain formed from the two H chains.

The variable domains (V_L/V_H) associate noncovalently to form a twisted antiparallel β-sheet structure. Although the framework is well-conserved between known antibody structures, variations in the packing of β-sheets do occur. The β-sheets are interspersed with six segments that are the most variable regions of the antibody, both in sequence and in structure. These regions are known as the

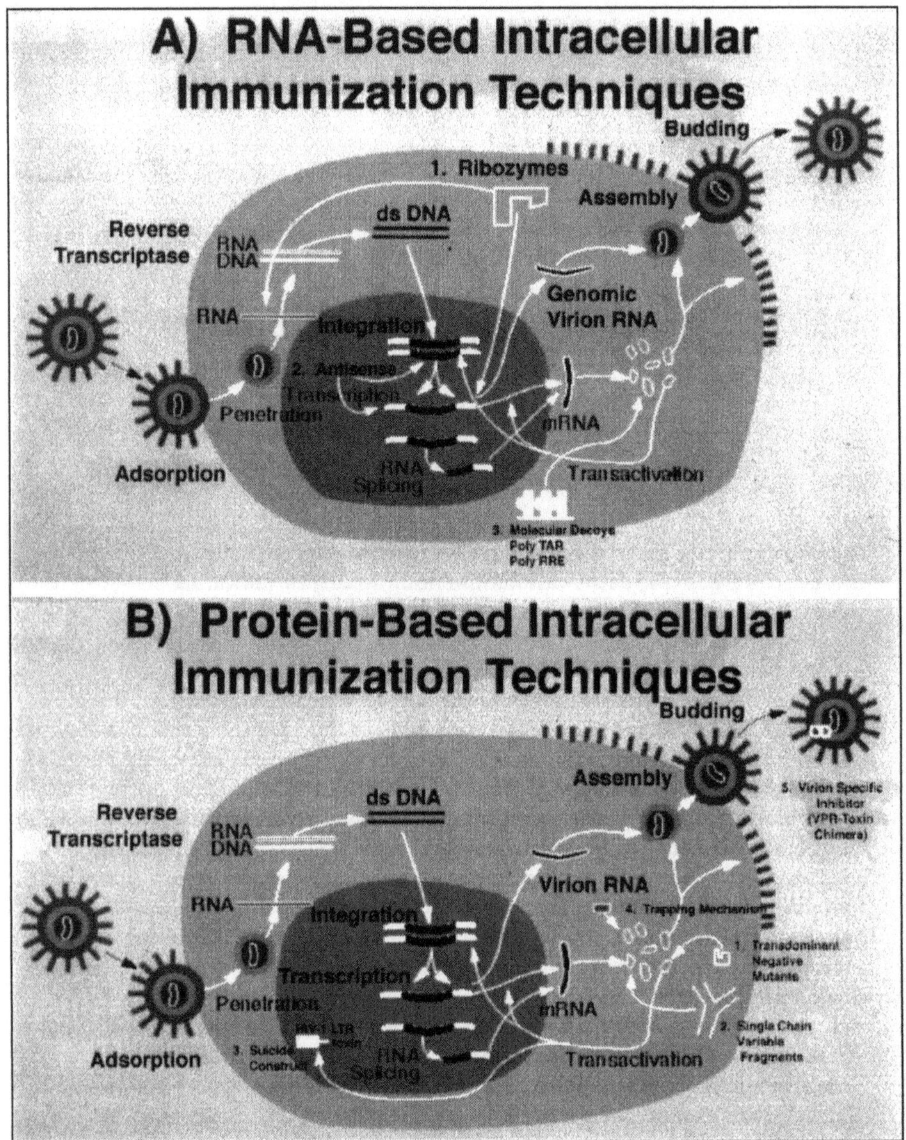

Fig. 10.2. A and B illustrate the RNA and protein based intracellular immunization strategies which have been used to interfere with the HIV-1 life cycle.

hypervariable or complementarity determining regions (CDRs). Each CDR interconnects a β-strand, with three CDRs derived from the L chain and three from the H chain. This interconnection of the anti-parallel β-strands brings the CDRs close together in space at the distal end of the antibody. Some or all of the CDR loops may be involved in antigen binding. Data indicate that the H chain CDR3 loop plays a major role in the binding of antigen.[31]

By somatic mutagenesis of the Ig gene variable region amino acid sequence, under natural selection, Ig proteins can specifically bind their corresponding antigen. Of note, though, the binding specificity and affinity of Ig proteins to an antigen is not affected by most modifications of the Ig protein.[31]

In 1988, the first report on the expression a fully functional recombinant Fab in *E. coli* was one of the initiating events in the development of the field of antibody engineering.[31] Shortly thereafter, the cloning and expression of antibody-variable domains as single-chain Fv, or scFv, was reported. This provided a method for rapid cloning of stable antibody domains from hybridoma cells into *E. coli*, in a fully functional form. More recent developments in the field, such as expression and selection of libraries of scFv or Fab on phage particles, have allowed the de novo generation of antibodies with binding properties equivalent to or, in some cases, better than those derived by traditional hybridoma methods.[17]

One of the chief benefits of antibody engineering is the ability to rapidly probe structure-function relationship as they relate to antibody-antigen binding interactions. Initially, much of the work was focused on the sequencing of heavy and light chains cloned from panels of hybridomas. Comparison of amino acid sequences of antibodies with related binding properties provided some insights into which residues in the binding pocket can participate in antigen binding. With the ability to clone and express antibody domains in *E. coli* rapidly,[29] antibody structure-function relationships could now be explored using a number of different sophisticated methods, such as site-directed mutagenesis, alanine-scanning mutagenesis, heavy- and light-chain shuffling, CDR shuffling, and CDR randomization.

Despite evidence that immunoglobulin heavy and light chains can functionally associate in the cytosol, initial attempts to manipulate cells by the intracellular expression of separate heavy and light chain genes met with limited success.[4] Recently, it was demonstrated that by construction and expression of single-chain variable fragments (scFv), in which the antigen binding region of an Ig protein heavy and light chain variable domains are synthesized as a single polypeptide in *E. coli*, those renatured and purified scFvs bound in similar manner to the antigen, as compared to their parent antibody. Based on this knowledge, it is possible to genetically engineer Ig gene variable regions with small flexible peptide linkers, such as $(Gly_4Ser)3$, to combine the heavy and light chains together, and form "single-chain antibodies" (scFv), which can be expressed intracellularly in mammalian cells.[10] Thus, one can now utilize intracellular scFvs to interdict in the complex life-cycle of human retroviruses, such as HIV-1 (Fig. 10.3).

Intracellular Immunization and the PreIntegration Phase of the HIV-1 Life Cycle

As described above, the HIV-1 life cycle can be simply divided into preintegration and postintegration stages.[43] Once proviral DNA is integrated into host chromosomes, it is much more difficult to control viral replication and it is even more difficult to completely eliminate the infected cells, in which some HIV-1 viruses are in a latent stage of infection. As such, the anti-HIV-1 strategies which block viral infection at preintegration stages are significantly more attractive for therapeutic purposes.[33] There are five major, yet simplified, steps which are required for the HIV-1 virus to finish the preintegration stages of its life cycle:

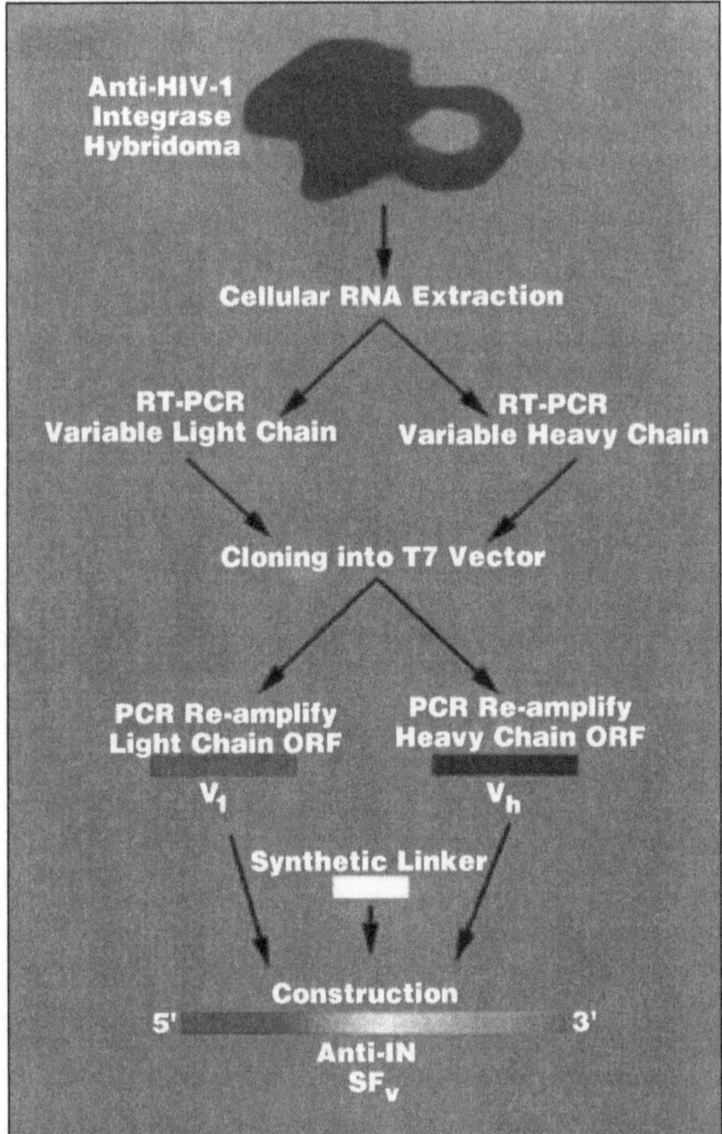

Fig. 10.3. Strategies for construction of sFv from mouse hybridomas.

1) Binding of HIV-1 viral envelope protein gp120 to the CD4 receptor. This interaction will induce conformational changes of gp120 for further interactions with either CXCR4 or CCR5 coreceptors for virus entrance into host cells.
2) Escaping intracellular endosome degradation of the viral preintegration complex.

3) Viral reverse transcription reaction.
4) Transport of the preintegration complex into nucleus, and the
5) Integration reaction.

An early event in the life cycle of all retroviruses, including HIV-1, is reverse-transcribing its RNA into double-stranded DNA, in the form of a preintegration complex.[2] After the preintegration complex actively migrates into the nucleus, then integration of a double-stranded DNA copy of the viral genome into chromosomal DNA of the host cells occurs. These steps are necessary for productive viral replication.[22] Genetic analyses have shown that the virion-encapsidated RT and integrase, as well as the host tRNA primer and the sequences at the termini of retroviral DNA, are required for the RT and/or IN reaction.[2]

The function of RT in the viral life cycle has been demonstrated by both site-specific mutagenesis of the RT protein and by using agents which specifically inhibit RT function. HIV-1 RT is capable of several enzymatic functions which include RNA-dependent DNA polymerase activity (RDDP), DNA-dependent DNA polymerization (DDDP), DNA-RNA duplex-dependent ribonuclease activity (RNase H) and RNA-RNA duplex-dependent ribonuclease activity (RNase D). In HIV-1 virions, the p66/p51 heterodimer is the only active form of RT.[2] The HIV-1 proteinase, expressed along with p66 from the pol gene is required for proteolytic cleavage of p66 to form p51.

Of importance, early after infection, retroviral DNA and proteins are found within a subviral nucleoprotein complex in the cytoplasm of infected cells.[39] In vitro, these isolated complexes can accomplish integration of viral DNA into an exogenous target molecule.[14] In a natural infection, linear viral DNA contained within preintegration complexes has been shown to be the direct precursor of proviral DNA. When viral integrase (IN) function is abrogated by site-specific mutagenesis, newly synthesized viral DNA cannot be integrated. Instead, intracellular accumulation of viral DNA with joined ends, producing a tandem arrangement of the viral long terminal repeat sequences (2-LTR), is observed.[13]

Analysis of IN activity in vitro has demonstrated that this viral product is the only protein required for HIV-1 integration.[19] The integration reaction, as defined by in vitro studies, proceeds in two critical steps. The first step (processing) is integrase-dependent cleavage of two nucleotides from the 3' end of each strand of viral DNA. The next step (joining) is a concerted cleavage and ligation reaction. In this step, a staggered cut is generated in the target DNA by nucleophilic attack involving the hydroxyl group present at the recessed 3' ends of viral DNA, which then becomes linked to the 5' ends of the target DNA at the cleavage site. The resulting gaps in the target DNA are repaired, likely by host enzymes, to create short direct repeats flanking the provirus, a hallmark of all retroviral integration reactions.

As RT and IN play a key role in the early-stages of the retroviral life cycle, these proteins have become very attractive targets for specific therapy against HIV-1. Analysis of RT and integrase reactions in vitro have demonstrated that HIV-1 RT and IN proteins can be segregated into several functional domains as illustrated in Figure 10.4. HIV-1 RT functions as a p66/p51 heterodimer and IN also works as a dimer. The amino-terminus of IN possesses zinc finger homology. The central core domain harbors catalytic activity and is characterized by the presence of three invariant acidic residues (D, D-35-E).[23] Finally, a less conserved carboxy-terminal

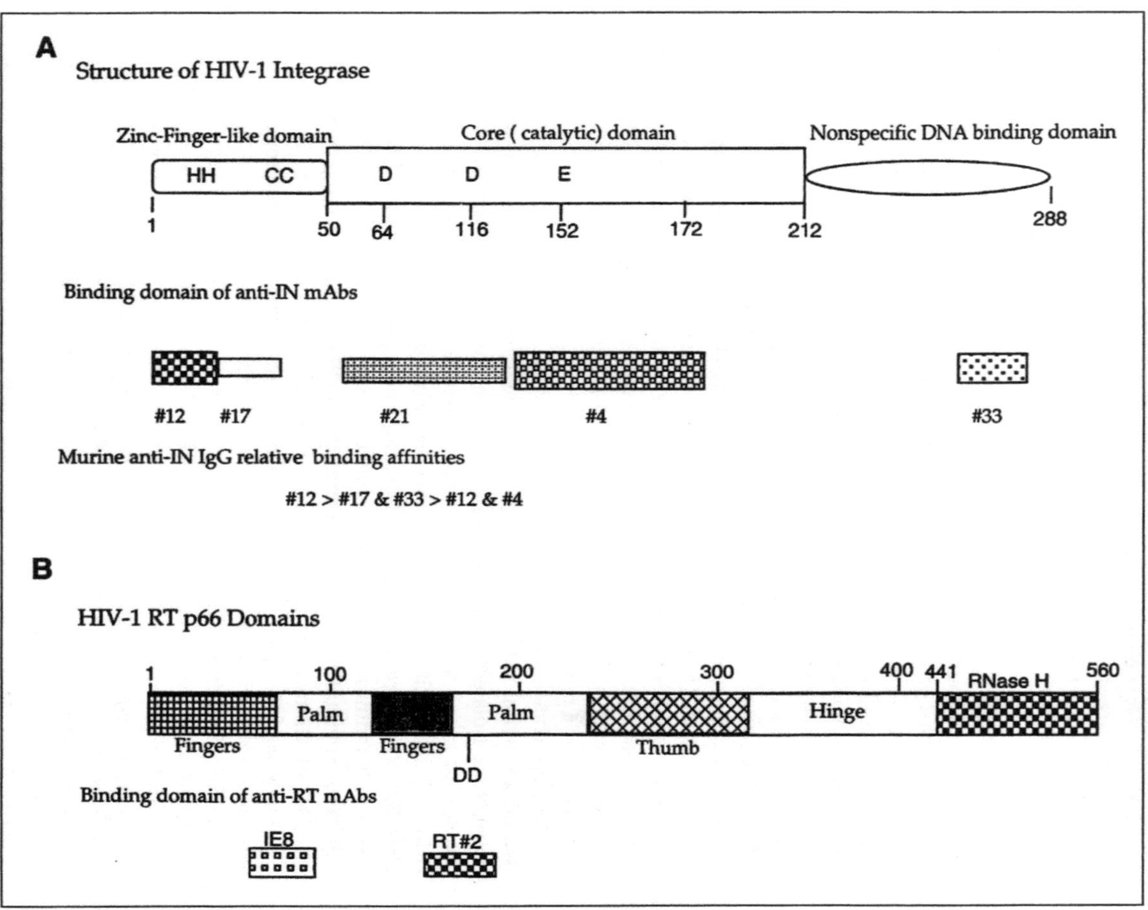

Fig. 10.4. Binding domains of anti-HIV-1 IN and RT monoclonal antibodies. Each of the anti-HIV-1 IN mAbs were purified, and in vitro binding affinities of purified antibodies to bacterially-expressed HIV-1 IN protein were analyzed via ELISA. The binding domain of each was mapped using various deleted IN proteins. Anti-HIV-1 RT mAbs binding domains were mapped via ELISA using RT peptides. Reprinted with permission from the Journal of Virology.

domain may be involved in dimerization of IN, as well as providing a nonspecific DNA binding activity.[1] Collectively, the integrity of these domains is important for precise and efficient integration of retroviral DNA into the host cell's genome.

As such, the RT and IN enzymes of HIV-1 were chosen as targets for sFv-induced inhibition, to study the early stages of HIV-1 replication, prior to the establishment of provirus. If such steps can be blocked successfully, viral replication in human cells should be halted.

Rapid PCR Cloning of Murine or Human Antibody Variable Domains as scFvs

The polymerase chain reaction (PCR) provides a rapid method for directly amplifying H and L chain antibody-variable domains from Ig RNA.[27] Using existing data bases of murine and human antibody sequences, degenerate primer sets have been designed to allow cloning of virtually any mouse or human V_H or V_L domain.[20] Additional sequences encoding a linker are placed in the primers to bind the antibody domains, prior to cloning and expression, also have been reported.[18]

To illustrate the RT-PCR cloning strategies for mouse and human antibody variable domains, the murine and human anti-HIV-1 RT and integrase antibodies have been extensively evaluated by our laboratories. The cloning of a panel of murine anti-HIV-1 RT and integrase scFv gene constructs consisting of V_L and V_H antibody domains connected in frame with a flexible $(GLY_4SER)_3$ peptide, tethering the two antibody domains together and facilitating interchain interactions to form a monovalent binding pocket, have been developed. Although the order of the antibody domains in the scFv construct does not appear to be important, and can be either V_H-linker-V_L or V_L-linker-V_H, as the H chain CDR3 loop plays a major role in the binding of antigen, and in order to maintain the flexible CDR3 structure, we usually construct scFvs using the arrangement, V_L-linker-V_H.

Identification of Aberrant κ Chain (abVκ) cDNAs from Hybridoma Sequences and Screening abVκ by Primer Selection

As reported by Carroll et al,[5] hybridoma fusion cell-lines from Sp2/o backgrounds contain one heavy chain transient and two light chain transcripts, one of which is an aberrant κ transcript with a four nucleotide delection at the VJ joining site. Utilizing standard RT-PCR with degenerate Ig primers, we were able to clone 12 out of a total 13 anti-HIV-1 antigen-specific hybridomas' heavy chain variable region cDNAs.[11] For heavy chain cloning, the PCR usually can amplify correct heavy chain cDNA, as Sp2/o cells do not express endogenous heavy chain RNA. Of note, cloning κ light chains required mini-preparation of 150 clones from X-Gal-selected cultures for our early studies of anti-HIV-1 Rev hybridoma cell-line D8, to obtain only two clones with correct anti-Rev κ chain cDNA. All of the other clones contained aberrant κ chain cDNA (Fig. 10.5). Importantly, from the 12 screened anti-HIV-1 antigen-specific hybridoma κ chain PCR-derived constructs, over 150 PCR-derived κ chain have been analyzed by DNA sequencing and more than 95% of the clones carried this endogenous κ chain cDNA.

Utilizing the DNA sequence of the Sp2/o endogenous κ chain variable region, aberrant κ chain CDR1- and CDR3-specific oligonucleotide primers (abVκ-1 and abVκ-2) (see Fig. 10.5), have been developed for quick and efficient screening of

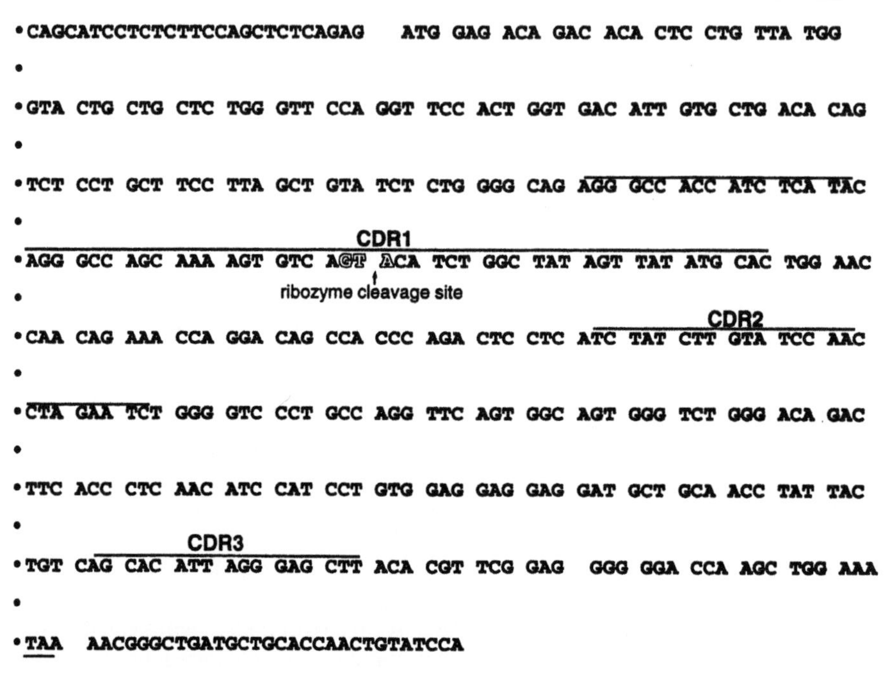

Fig. 10.5. Nucleotide sequence of the aberrant Vκ RNA in murine myeloma fusion cell partner: The complementarily defining regions CDR1, CDR2 and CDR3 are indicated in this figure. The abvκ ribozyme cleavage site (CTA) is illustrated.

the recombinant plasmids. By using these two primers, the aberrant κ chain cDNA-containing plasmids can be screened by PCR, thus dramatically reducing the DNA sequencing necessary in these studies.[11]

Specific Ribozyme Cleavage of abVκ RNA In Vitro

Importantly, even by using specific PCR screening to eliminate the aberrant κ chain cDNAs, one still had to screen more than 50 clones to obtain only a few abVκ PCR-negative clones for further DNA sequencing, to obtain correct κ chain cDNA. In order to reduce endogenous aberrant κ chain RNA levels and to enhance specific κ chain RNAs, a ribozyme DNA sequence, based on the abVκ-specific CDR1 sequence has been developed (Fig. 10.6A).

Ribozymes are small RNA molecules capable of highly specific catalytic cleavage of RNA. Ribozyme-mediated cleavage in trans was first demonstrated in vitro by Ulhenbeck[40] and, subsequently, by Haseloff and Gerlach.[16] In viroids and virusoids, the reaction is intramolecular. However, ribozymes that possess a catalytic domain and flanking sequences complementary to the target mRNA can cleave in trans, provided that a three base sequence (GUX) occurs within the target molecule.[35] The hammerhead-like ribozyme motif to specifically target the abVκ CDR1 region was selected. As such, the abVκ ribozyme only binds the abVκ CDR1 region

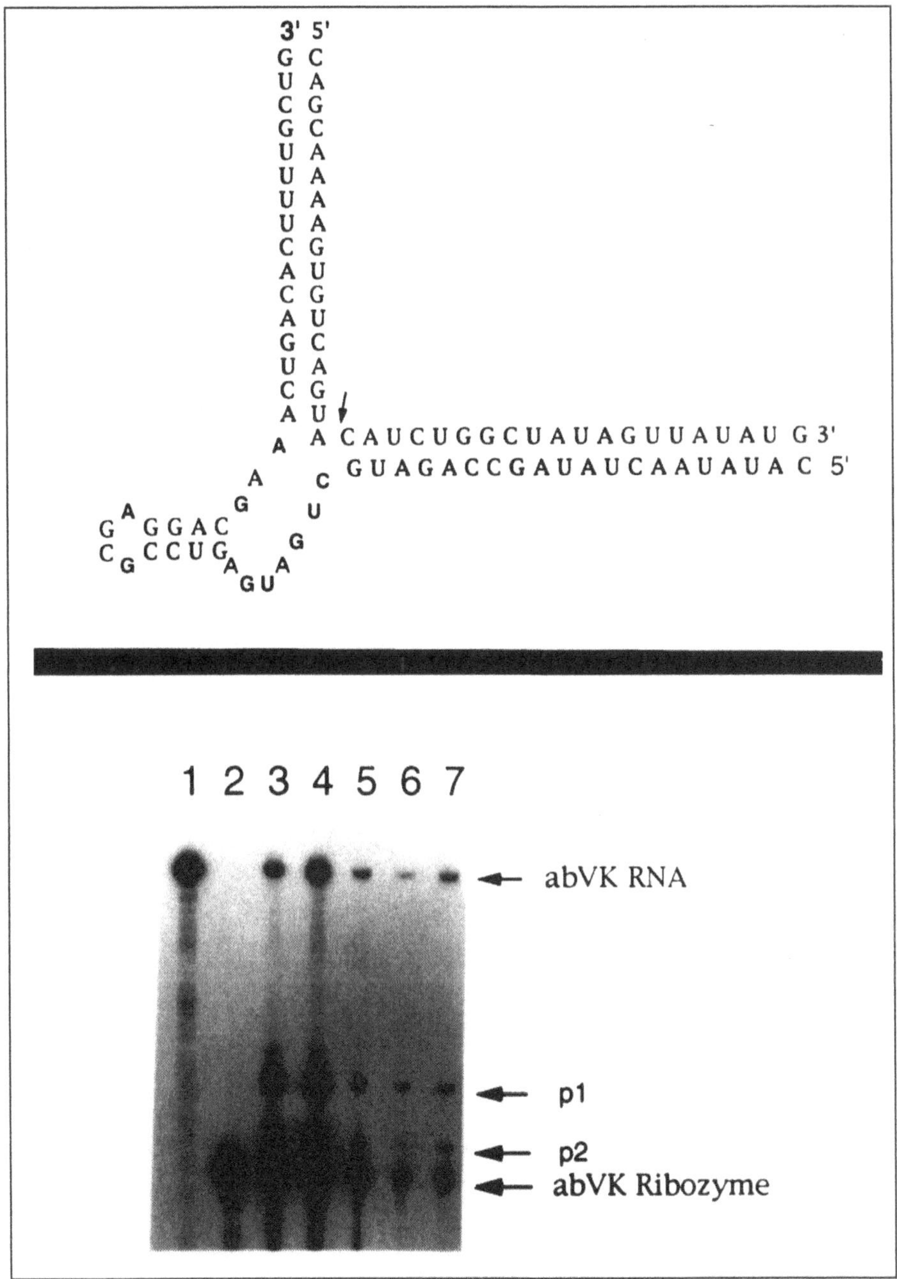

Fig. 10.6. (A, top) Cleavage structure of abVκ-specific ribozyme based on the abVκ CDR1 region GUA site. (B, bottom) In vitro cleavage reaction using an abVκ-specific ribozyme. The abVκ and ribozyme-labeled RNA were mixed and incubated at different concentrations of RT buffer for 30 minutes and resolved using 8 M urea-6% PAGE. Lane 1: abVκ RNA. Lane 2: abVκ ribozyme. Lanes 3-7 are mixtures of the two RNAs at RT buffer concentrations of 1X to 5X, p1 and p2 are cleavage products. Reprinted with permission from Nucleic Acids Research.

and specifically cleaves solely abVκ RNA. This strategy can dramatically reduce the endogenous abVκ RNA level, and increase the proper κ chain/abVκ ratio to specifically enhance the RT-PCR amplification efficiency.

As Figure 10.6b illustrates, the in vitro transcript abVκ ribozyme functions quite well in a wide range of $MgCl_2$ concentrations, within the RT buffer against abVκ RNA substrict. These conditions allowed the following RT-PCR reaction to be performed in a simple and efficient manner. Firstly, heating of 1 mg-5 mg of RNA template at 72°C for 3-5 minutes was performed. Then 1 mg of ribozyme RNA was added in 3 X RT buffer, incubated at 37°C for 30 minutes. The entire mixture was then used for the RT-reaction.

By first treating total cellular RNA with the abVκ specific-ribozyme, then performing RT-PCR amplification of murine Ig variable region cDNA, this technique can dramatically reduce abVκ RNA background, and potently enhance the specific κ chain cDNA ratio, prior to the cleavage treatment.

Retroviral Vector DNA Delivery of the abVκ Ribozyme into Hybridoma Cells

Although the preribozyme cleavage reaction and then RT-PCR amplification can reduce the abVκ background and enhance the specific κ chain RNA for RT-PCR amplification, this reaction occurs in vitro. Thus, it may sometimes cause the nonspecific degradation of total cellular RNA. In order to avoid this potential difficulty, a specific retroviral shuttle vector delivery system which utilizes either an internal cytomegalovirus (CMV) promoter or tRNA promoter have been tested for its efficiency (Fig. 10.7). By establishing a stable packaging cell-line, a viral titer of 5 x 10^5 cfu/ml, containing the murine leukemia virus vector which carries the abVκ ribozyme DNA, can be obtained. Based on this very efficient DNA delivery system, 2-3 x 10^6 hybridoma cells were centrifuged at 400 g for 5 minutes, and resuspended in 8-10 ml of packaging cell supernatant, with 8 mg/ml polybrene overnight. The cells were then cultured in 5 ml of fresh medium for another 24 hours. Total cellular RNA was then isolated and RT-PCR for Vκ cDNAs were performed. The cDNAs were then cloned into the pT7 Blue (R) vector for PCR screening and sequence analysis, after X-gal selection.

This ribozyme-transduced retrovirus vector treatment was quite efficient and reduced the abVκ RNA background, without requiring an in vitro cleavage reaction. Of note, the abVκ ribozyme-expressing retroviral vector system was even more potent than the methodology in which isolated cellular RNA was directly treated with the ribozyme.[11]

Inhibition of HIV-1 Replication by Targeting the Viral PreIntegration Complex

Several murine hybridoma cell-lines, which produce monoclonal antibodies (MAbs) against HIV-1 RT or integrase (IN), were chosen as they bind to different domains of RT and IN proteins. As indicated in Figures 10.4a and 10.4b, the hybridomas were transduced with a retroviral vector, that expresses a murine abVκ ribozyme, as described above.[3,41] Approximately 1-10 x 10^6 cells were used for preparation of total cellular RNA by lysis with guanidinium isothiocyanate buffer and a cesium chloride gradient procedure.

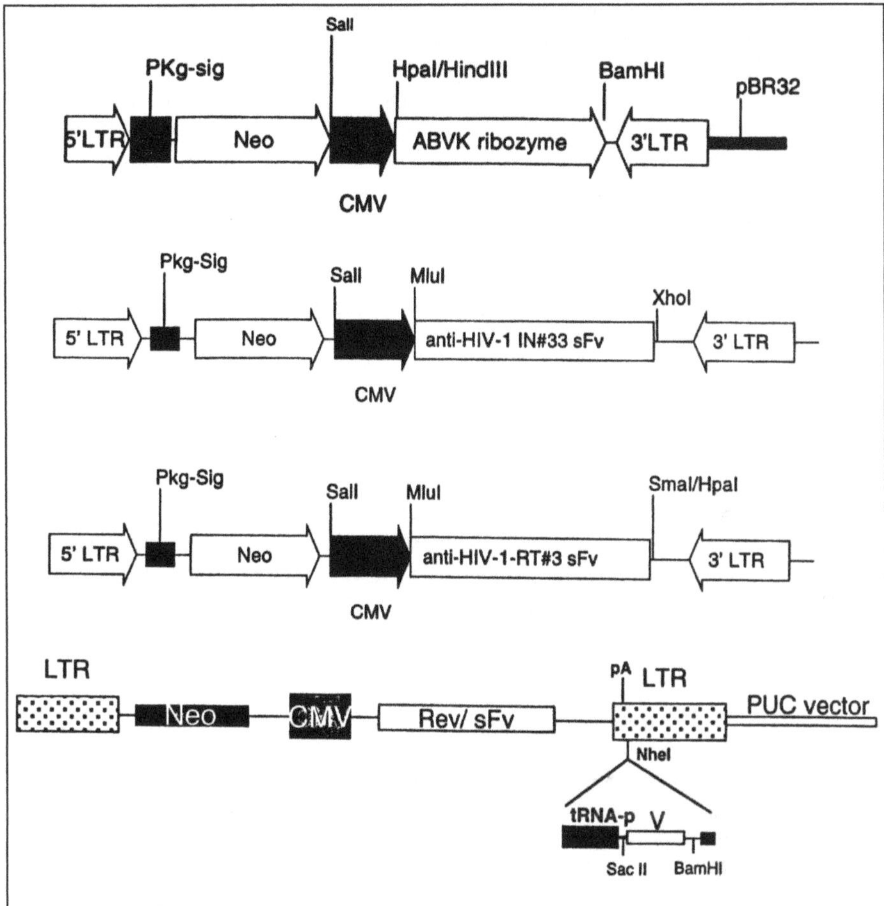

Fig. 10.7. Schematic representations of anti-HIV-1 RT sFv genes constructs or ribozymes in the retroviral expression vector. Pkg-sig: packaging signal; CMV, cytomegalovirus early promoter. Neo: neomycin resistance gene. tRNA-p: human tRNA promoter. V: anti-HIV-1 RRE ribozyme V. pA: poly-adenylation site.

The cDNAs of the variable heavy (V_H) and light chain (V_L) regions of anti-HIV-1 RT or IN monoclonal IgG transcripts were synthesized by RT-PCR. Primer sets used to amplify murine variable regions, heavy and light chains, were obtained from Novagen (Ig-Primer Methodology, Novagen, CA). For first strand cDNA synthesis, 5 mg of total RNA was mixed with the 3' primers (antisense) specific for V_L or V_H regions, heated for 5 minutes at 65°C and then incubated at 37°C for 1 hour with 200 units of AMV-RT IN in a buffer containing: 0.07 M KC_L, 0.02 M Tris (pH 8.3), 5 mM DTT, and 1.0 mM (each) deoxynucleoside triphosphates (dNTP) in 10 ml volume. After reverse transcription, 5 ml of template cDNA was subjected to PCR amplification, using appropriate primers (Novagen, CA) with 0.5 ml *Taq* polymerase (Perkin-Elmer Corp., Norwalk, CT) in PCR buffer containing: 2.5 mM MgC_{L2}

in a 50 ml total volume. Amplification was carried out in a thermal cycler (Perkin-Elmer Corp.) for 1 cycle of denaturation at 94°C for 5 minutes and then for 35 cycles with the following parameters: denaturation for 1 minute at 94°C, annealing for 90 seconds at 50°C, and extension for 60 seconds at 72°C, with a final extension of 10 minutes. Each of the PCR-amplified fragments was cloned into the pT7 Blue (R) vector (Novagen, CA).[26] Clones were screened first by PCR amplification, using primers specific to the pT7 polylinker for the correct size, and abVκ primers to remove abVκ cDNA contamination, then analyzed further by DNA sequencing.

The sequenced mouse V_L or V_H can then be analyzed against Kabat IgG gene data bases, to define the framework (FR) and CDR regions.[20] As these scFvs were used for intracellular expression, the V_L or V_H secretary signal domains were removed by designing oligonucleotides which specifically amplify V_L and V_H from FR1 to FR4 regions.

PCR primers (5' and 3' ends) containing appropriate restriction enzyme sites for further cloning were used to re-amplify each of the V_L and V_H fragments, with deletion of remaining secretory leader sequences. The DNA of light and heavy chains were joined together via a flexible linker (GLY$_4$SER)$_3$. The anti-RT or IN V_L fragments were cloned via *Nde* I and *Apa* I sites 5' to the linker and the V_H was cloned as *Bgl* II and *Eco* RI fragment 3' to the linker, to obtain the full-length scFvs as a single gene construct. For construction of scFvs that contain the HIV-1 Tat nuclear localization signal (NLS),[38] a 54 bp *Sac I-Bgl II* DNA fragment was then inserted into pT7IN-scFv #33, which was partially digested with *Sac I* and *BgL II* to generate pT7IN-scFv #33/NU.

Delivery of scFvs into Target Cells

Target gene delivery into host cells with high efficiency, is still a major challenge for the field of any gene therapy procedure. Several currently used gene delivery systems have been tested for delivery of anti-HIV-1 RT or IN scFvs into target cells. The most commonly used systems are retroviral vectors.

Retroviral vectors are modified and defective retroviruses. The genes required for viral replication have been removed and replaced by therapeutic genes and selected markers. The viral information retained, i.e. the LTR and the packaging signal, allows the packaging of recombinant RNA into viral particles and reverse transcription of the RNA upon the infection of target cells. Viral replication functions are provided by packaging cells in trans. These cells are designed to provide all proteins necessary for the assembly of viral particles.[33]

The murine leukemia virus (MLV)-based retroviral expression vector, pSLXCMV, which contains the bacterial neomycin-resistance gene (*neo*) under control of the viral LTR and scFv expression under control of an internal CMV promoter was used in these experiments, as it maintains gene expression quite adequately in human T-cell-lines.[12] Each of the anti-IN-scFv constructs was subcloned as an *Mlu* I-*Xho* I fragment in the *Mlu* I-*Xho* I sites of the polylinker region of the pSLXCMV vector. The CAT (chloramphenicol acetyl transferase) cDNA, plus D8scFv (anti-HIV-1 Rev) and anti-hepatitis B core antigen (HBcAg) scFvs have been constructed in the same vector and were used as controls in certain experiments.[25,36] Transduction of recombinant virus into human T cell-lines such as SupT1 or CEM was performed, using either packaging cell supernatant or coculturing target cells with packaging cell-lines. After rapid G418 selection, HIV-1 challenge experiments can be performed. As illustrated in Figure 10.8, cells transduced with anti-HIV-1 scFvs were resistant to HIV-1 infection.

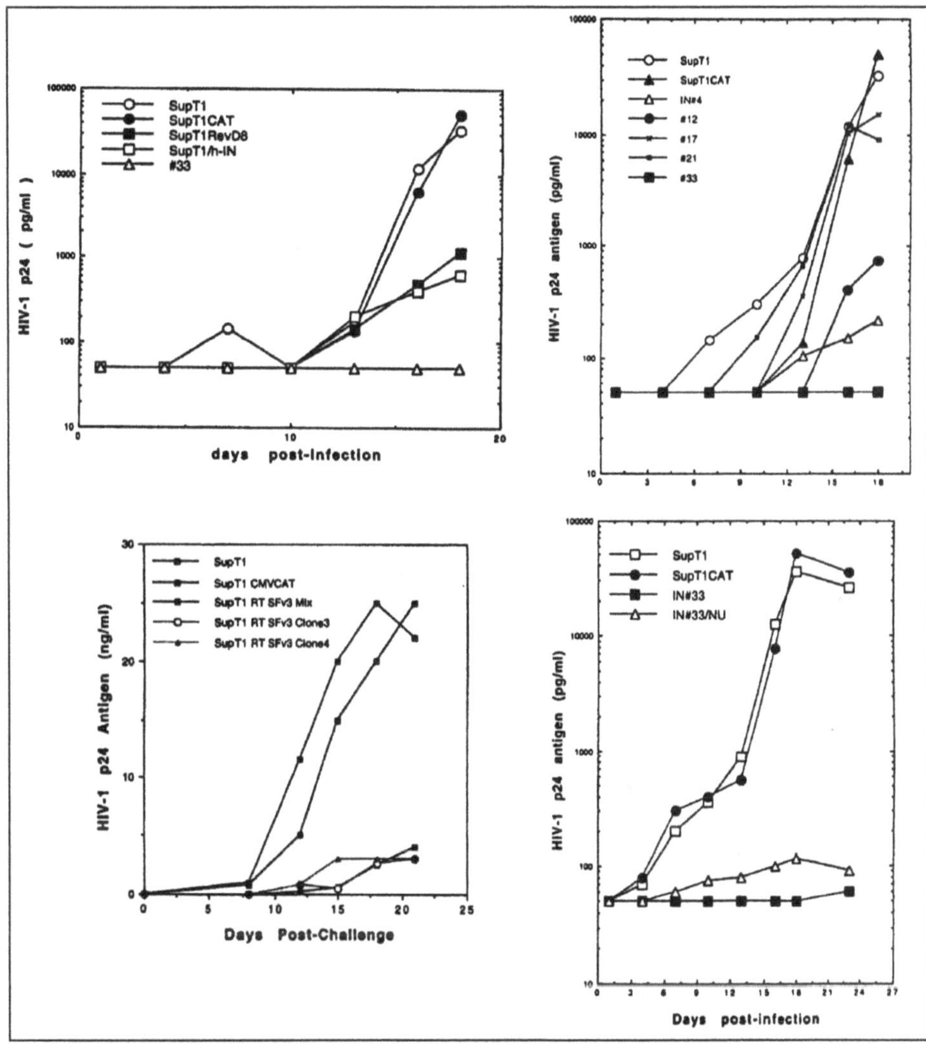

Fig. 10.8. Inhibition of HIV-1 replication in anti-RT or IN sFv transduced human T lymphoid cells. SupT1 cells were transduced with CAT, anti-Rev sFv or different anti-RT and IN sFvs (mixed population or single clones derived data are presented). Reprinted with permission from the Journal of Virology.

Human Anti-IN scFv Construction

Due to the sub-cellular compartmentalization of scFvs, complete intracellular expression of murine anti-HIV-1 RT or IN scFvs should theoretically eliminate the problem of host immune-reactions to those scFv proteins. Nevertheless, the complexities of host cell antigen presentation mechanism(s) are significant. At the present time, there are no clinical data to clearly resolve this issue. Alternative ways to overcome this potential problem are either to humanize the murine scFvs with human Ig FR regions or directly clone a human anti-HIV-1 RT or IN Fab, then to construct fully human-specific anti-HIV-1 scFvs.

In 1989, the first experiments to generate Mabs by antigen selection from combinatorial libraries expressed in phage were described.[17] This methodology was then extended to allow generation of monoclonal Fab fragments from human peripheral blood lymphocytes, using either lambda phage or a M13-based phage-display system. Foreign peptides are readily fused with a structural protein of filamentous bacterial M13, by insertion of the corresponding nucleotide sequence into the gene coding for the protein. This insertion results in the display of a protein on the phage surface, providing that the insert does not interfere with the essential functions of the protein.

For cloning human anti-HIV-1 Fabs using a phage display system, an asymptomatic HIV-1-infected patient's peripheral blood mononuclear cells (PBMC) were used for RNA extraction and RT-PCR were performed separately for both heavy and light chain Fd regions. Then, these fragments were sequentially cloned into the M13-derived phagemid vector, pABC. Infection of phagemid-bearing *E. coli* with VCSM-13 helper phage allowed production of a packaged phagemid library, with phage simultaneously expressing Fab molecules on the phage head and carrying DNA encoding the Fab molecules. Four rounds of panning phage against several HIV-1 purified antigens, and elution of phage indicated at least a 100-fold increase in the number of eluted phage compared to the first panning. DNA encoding the phage cap protein was then excised by digestion with N*he*I and *Spe*I, and the compatible ends of the vector religated to allow production of soluble Fab proteins for specific binding assays. As data illustrated in Figure 10.9 demonstrates, a panel of human anti-HIV-1 Fabs have been constructed efficiently using this system.[30] Of note, when a human anti-HIV-1 IN scFv was constructed and transduced into human T-cells, to compare with murine anti-Rev scFv, the human scFv transduced cells were found to be quite resistant to HIV-1 infection (Fig. 10.8).

We have also recently demonstrated that the binding locations (epitopes) on a target protein determine the inhibitory properties of anti-HIV-1 scFvs. Five murine anti-HIV-1-IN scFvs, which bind to different domains of IN were constructed, in the same vector, and transduced into the same target cells. As summarized in Figure 10.4b, when scFvs binds to C-terminus and catalytic domains of IN, they dramatically inhibited viral replication. This inhibition was not correlated in terms of in vitro binding affinity of each purified IgG protein.[25] The same phenomenon has also been confirmed, when an anti-Rev activation domain-specific scFv was shown to inhibit HIV-1 much more potently than an scFv which binds to the far C-terminus of Rev. Importantly, the latter scFv demonstrated a 10-fold higher in vitro binding affinity to recombinant HIV-1 Rev.[108,44]

Subcellular Compartmental Localization of scFvs

As the HIV-1 preintegration complex migrates into the nucleus to attack host DNA for proviral DNA integration, the anti-HIV-1 IN C-terminal domain of scFv #33 was modified to carry the Tat nuclear translocation signal, to direct this scFv into the nucleus. Of note, HIV-1 challenge data indicated that a cytosol location of anti-HIV-1 IN scFv #33 demonstrated better inhibition of HIV-1 replication, as compared to the same scFv localized to the cell nucleus. Functional assays of an anti-Rev scFv have shown that the complex of the scFv with Rev protein in the cytoplasm dramatically reduces Rev half-life from 16 to less than 4 hours.[21] Thus, by binding to IN in the cytosol, an scFv localized to the cytoplasm may not only directly inhibit the IN enzymatic function, but as recent data have indicated that IN

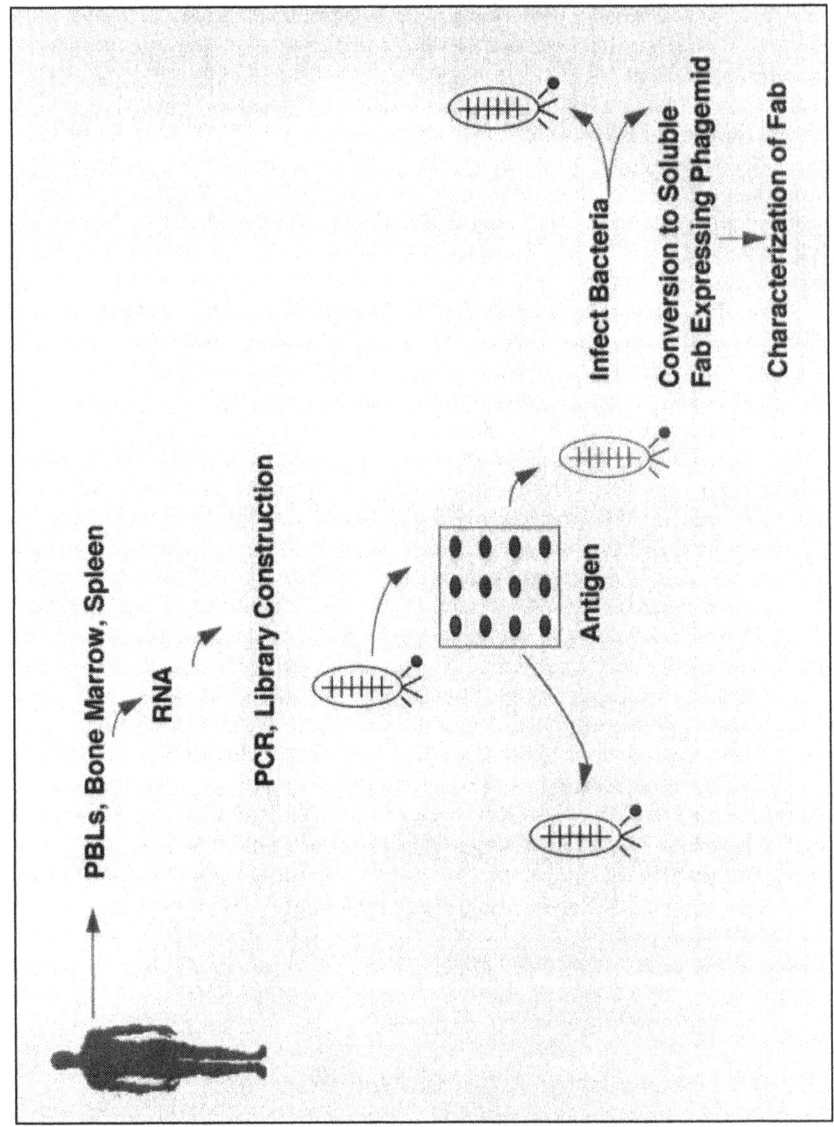

Fig. 10.9. Strategies for construction of sFvs from human bone marrow or PBMC using the phage-display screen.

may also play a role in directing the preintegration complex of HIV-1 to the nucleus,[43] such binding may also trap this complex in a cytoplasmic location and decrease its half life.

Combination of Ribozymes with scFvs

As a further development of intracellular immunization strategies against HIV-1 infection, suitable combinations of different anti-HIV-1 moieties, which specifically inhibit different key HIV-1 proteins or viral RNA elements can be constructed. These may not only be able to enhance anti-viral efficiency, but also prevent possible viral escape mutants.[33]

The HIV-1 early-gene product, Rev, which controls the expression of the structural genes of HIV-1, exerts its function through binding to an RNA element, the RRE. HIV-1 RRE, located in the envelope's RNA sequence, consists of 234 nucleotides and can assume a highly complex secondary structure consisting of a central stem and five stem-loops.[7] By binding to the RRE, specifically stem-loop II, HIV-1 Rev functions as a chaperone to transport unspliced viral RNA into the cytoplasm, for viral structural protein translation and viral genome packaging. As RRE functions in the productive HIV-1 life-cycle, disrupting the RRE structure using minimal RRE oligomers, as molecular decoys has been explored, as an intracellular immunization strategy against HIV-1 replication.[24] Of note, in vivo binding of multimers of HIV-1 Rev protein to the RRE during HIV-1 replication may form a protected RRE-Rev protein complex. As such, this complex may not be so easily targeted by ribozymes. Anti-HIV-1 Rev scFv studies have demonstrated functional Rev can be trapped in a cytoplasmic location by complexing with a cytoplasmic scFv protein, and this scFv reduces intracellular Rev half-life more than 4-fold, to less than 4 hours.[21] Thus, the HIV-1 RRE RNA region will not be efficiently bound by the Rev protein in the presence of the scFv and, the open structure (i.e. loop II of the RRE) may allow further targeting of the RRE RNA region by specific ribozymes. Fine mapping has shown that sequences within stem-loop II encode the primary determinants for Rev function and are the major binding sites for Rev. Importantly, data have also demonstrated that the interaction between Rev and the RRE is absolutely required for HIV-1 replication.[34] By targeting both the RRE RNA and Rev protein, this strategy will potentially block HIV-1 replication at both the RNA and protein levels. This forms an HIV-1-specific dual-blocking strategy to inhibit acute and chronic HIV-1 infection.

Development and Design of RRE-Specific Ribozymes

One of the parameters which determine intracellular ribozyme function is the molecular ratio of target RNA (substrate) and ribozyme RNA. Intracellular ribozyme RNA expression levels can be manipulated by either increasing RNA transcription, such as by using RNA polymerase (pol) III promoters,[42] fusing ribozyme RNA with a rRNA transcription unit,[14] or modifying the structure of the expressed ribozymes to increase stability.[37] Of note, for inhibition of HIV-1, full-length unspliced viral RNA should be an ideal target as its expression level is much lower than spliced viral RNA, and is ideal for a combination anti-Rev scFv/ribozyme dual-blocking strategy. The HIV-1 Rev regulatory protein functions not only in unspliced viral RNA translocation from the cell nucleus to the cytoplasm, but also binds to the RRE region to stabilize viral RNA. Without functional Rev, the full-length viral RNA is reduced to nearly undetectable levels.

The parameters which are involved in intracellular cleavage of target RNA by specific ribozymes are not fully understood. In order to search for the best cleavage site for an anti-RRE ribozyme, all of the GUC triplets in the HIV-1$_{IIIB}$ and NL4-3 strains' RRE region sequences were analyzed by using a computer-assisted program, which predicts the secondary structure of RNA molecules.[9] Figure 10.10 illustrates the two sequences of the trans-acting hammerhead ribozymes, designed to cleave the selected GUC triplets in RRE RNA sequences near the loop structures, which were finally selected for testing (ribo-II and V). As well, a mutant ribozyme with a two-base alteration in its functional domain is also illustrated in Figure 10.10 (ribo-Vm). Of importance, this target region in the RRE is well-conserved among different clades of HIV-1.[27]

In Vitro Ribozyme-Mediated Cleavage of HIV-1 RRE RNA

The constructed RRE-specific ribozymes have been functionally tested in an in vitro cleavage assay. When these ribozymes were incubated with the substrate RNA at a 1:1 molar ratio for 20 minutes, the substrate RRE RNA was cleaved, at the expected specific sites, into P1 and P2 fragments (see Fig. 10.10) and demonstrates that ribozymes (II and V) cleave the target RRE RNA at specific sites in a cell-free cleavage system, equally well. By combining the two transcribed ribozymes (II and V) in a single construct, additive and efficient in vitro cleavage of the targeted RRE RNA was obtained.

Inhibition of HIV-1 Replication in T lymphocytic Cells Transduced with Retroviral Vectors Expressing Ribozymes Driven by a Pol III Promoter

The anti-HIV-1 hammerhead RRE ribozymes function much more efficiently when expressed from a pol III promoter, as compared to a pol II promoter, such as a CMV promoter. Human T lymphocytic cells, SupT1, were transduced with the retroviral vector containing either ribozyme-II, -V, under the control of a pol III promoter (tRNA) in the DCt retroviral vector (Fig. 10.7). Due to the inserted DNA size limitation in the pol III transcription unit [e.g., for efficient transcription not more than 80 bases], the ribozyme-II + V fragment was not inserted into the DCt vector. As a control, an anti-mouse immunoglobulin gene κ chain (abVκ) ribozyme was cloned into the same retroviral vectors DCt. To distinguish the mechanism(s) of inhibition of HIV-1 replication, via antisense and/or ribozyme functions, the RRE ribozyme-V sequence was mutated AA to TT, in the ribozyme catalytic domain, to disrupt ribozyme function but maintain antisense arms. This mutated ribozyme DNA fragment was also inserted into the DCt vector, as a control.

After transduction and G418 selection, both pooled SupT1 mixed populations and individual resistant clones were analyzed in HIV-1 challenge experiments. Figure 10.11 illustrates, at different multiplicities of infection (MOIS) with the HIV-1 NL4-3 strain, that only the tRNA-ribozyme-V construct showed significant (40%-90%) inhibition of HIV-1 replication in retroviral-transduced, mixed SupT1 cell populations, in multiple independent experiments. This inhibition of viral replication was solely due to ribozyme function, not antisense effects, as the ribozyme-Vm construct, with two mutated nucleotides in the ribozyme catalytic domain, showed no viral inhibition. At the high level challenge (MOI—0.22), there was no difference in viral growth between the control cells and all ribozyme-expressing cells. Thus, the anti-HIV-1 effects of these ribozymes can be overwhelmed with a

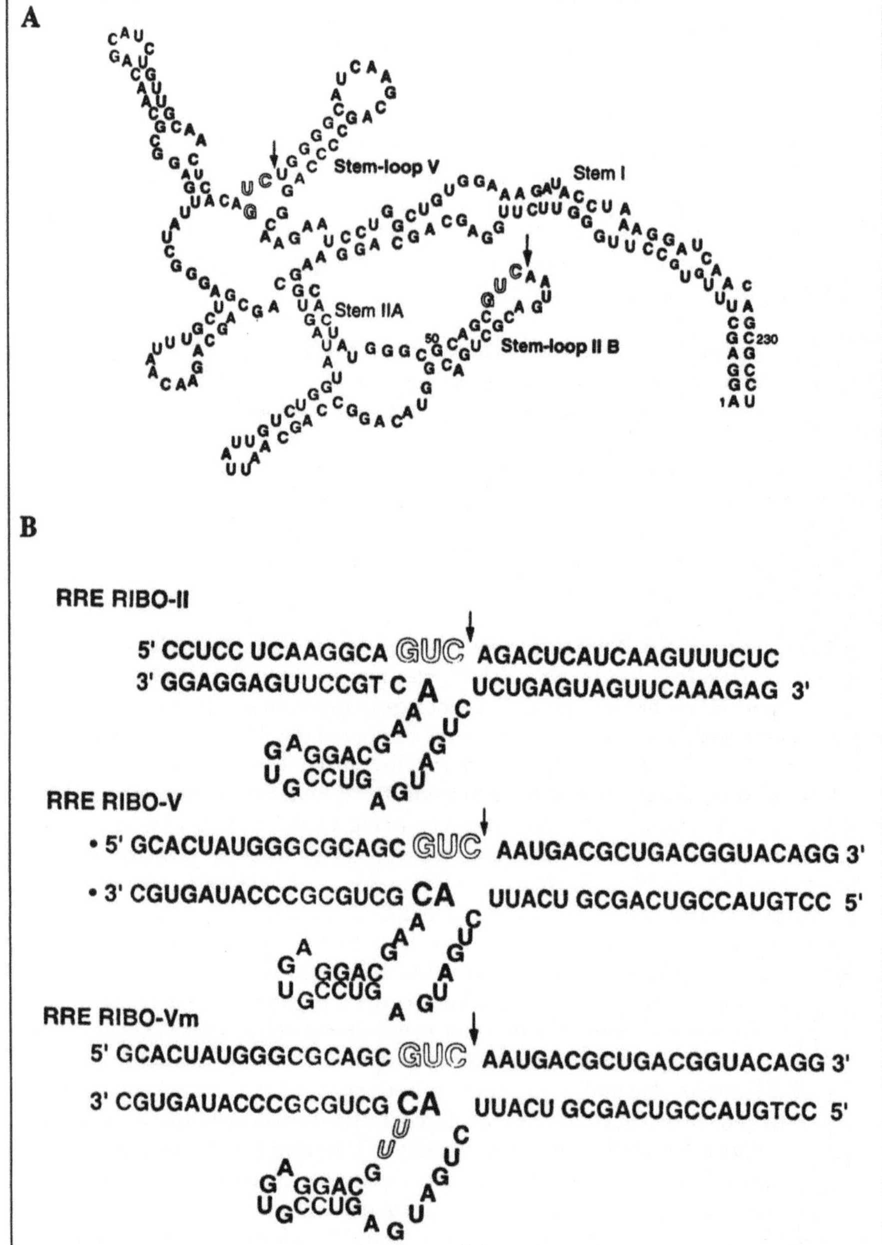

A
Stem-loop V
Stem I
Stem IIA
Stem-loop II B
50
230
1
B
RRE RIBO-II
5' CCUCC UCAAGGCA GUC AGACUCAUCAAGUUUCUC
3' GGAGGAGUUCCGT C A UCUGAGUAGUUCAAAGAG 3'
G A GGAC G AA A GUC
U G CCUG A GUAGUC
RRE RIBO-V
• 5' GCACUAUGGGCGCAGC GUC AAUGACGCUGACGGUACAGG 3'
• 3' CGUGAUACCCGCGUCG CA UUACU GCGACUGCCAUGTCC 5'
G A GGAC G AA A GUC
U G CCUG A GUAGUC
RRE RIBO-Vm
5' GCACUAUGGGCGCAGC GUC AAUGACGCUGACGGUACAGG 3'
3' CGUGAUACCCGCGUCG CA UUACU GCGACUGCCAUGTCC 5'
G A GGAC G UU U GUC
U G CCUG A GUAGUC

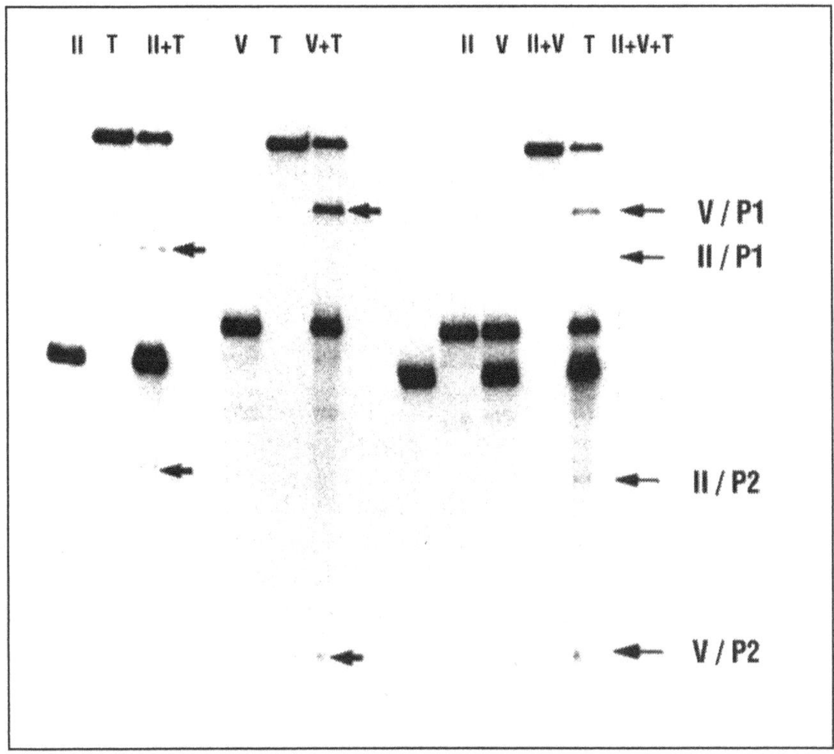

Fig. 10.10. (A, opposite) Sequences of targeted sites in the HIV-1 RRE RNA and the predicted structures of ribozymes II, V and Vm. The target GUC sequences of HIV-1 RRE RNA are located as indicated by arrows (shadowed letters) (A). Ribozymes are designed to associate with RRE RNA through their antisense sequences. Mutated ribozyme-Vm was created by specifically altering two nucleotides (AA to TT) in the ribozyme-V catalytic domain (B). (B, above) In vitro cleavage of HIV-1 RRE RNA substrate by ribozymes II, V or combination of II + V. Ribozyme-II (118 bases) and ribozyme-V (124 bases) were labeled with [^{32}P]UTP (Rz) and cleaved RRE RNA substrate (S) into two fragments; P1 (5'-cleavage product) and P2 (3'-cleavage product). Lane II: ribozyme-II; Lane T: Substrate RRE RNA; Lane II + T: incubation products of the substrate and ribozyme-II. Lane V: ribozyme-V; Lane V + T: incubation products of the substrate and ribozyme-V; Lane II + V: ribozymes II + V; Lane II + V + T: incubation products of the substrate and ribozymes II + V. Reprinted with permission from Gene Therapy.

very high viral input. In those experiments, if the ribozyme was driven by an internal CMV promoter, none of the tested constructs showed any protection in HIV-1 challenges, even with low MOIS (0.004).

Within the DCt vector there are at least two promoters. Previously, the murine leukemia virus-long terminal repeat (MLV-LTR) has been shown that it can maintain reasonable transcriptional activity in human T-cell-lines for more than six months.[36] Thus, all of tRNA promoters' transcriptional orientations in the constructs were designed in the same direction as the 5' MLV-LTR, so antisense effects of the LTR-driven transcripts would not alter tRNA-driven transcripts.

Fig. 10.11. Inhibition of HIV-1 replication in anti-RRE ribozyme- or dual functional vector-transduced T lymphoid cells. SupT1 cells were transduced with CAT-, anti-RRE ribozyme II-, ribozyme V- and II+V ribozyme-expressing retroviral vectors, using an internal CMV promoter, or ribozymes II, V, and Vm in a tRNA cassette [mixed cellular populations]. (A) Transduced SupT1 cell-lines infected with NL4-3 (MOI: 0.01). (B) Human PBMC stimulated using PHA/IL-2, were transduced with CAT-, anti-Rev D8scFv-, tRNA-derived ribozyme V-, and the dual-functioning CMV-D8scFv/tRNA-V vectors, three times with daily changes of fresh vector-containing medium. Cells were then infected with primary HIV-1 isolates for two hours. HIV-1 replication was quantified by determining HIV-1 p24 antigen levels in the culture supernatants, using an ELISA (Dupont). (C) H9/IIIB cells were transduced with pLSXN-M10, D8scFv, ribozyme-V, the dual-functioning retroviral vector, pCMV-D8scFv/tRNA-V, selected in G418 for three weeks, and then further cultured in G418-free medium for 3 weeks. The cells were standardized to 1×10^6 cells/ml and split 1:3 every third day. HIV-1 production was quantitated by assaying HIV-1 p24 antigen levels in the culture supernatants, using an ELISA (Dupont). Reprinted with permission from Gene Therapy.

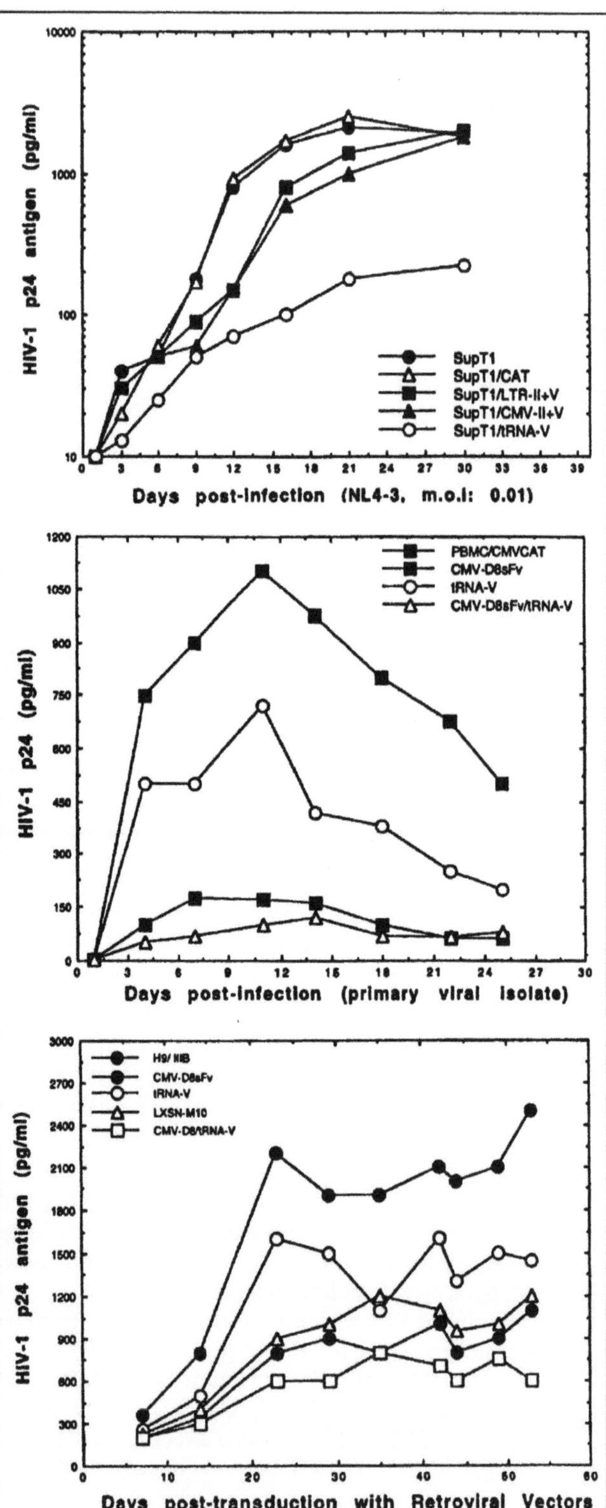

Dual-Functioning Retroviral Constructs and Inhibition of HIV-1 Replication in Cell Culture

HIV-1 challenge experiments demonstrated that ribozyme-V was significantly more potent in inhibiting viral replication, as compared to ribozyme-II with a tRNA promoter (Fig. 10.11). In the anti-HIV-1 Rev scFv studies, D8scFv has been shown to possess the ability to inhibit HIV-1 replication through specific alteration of Rev function.[12] As such, combination of the protein- and RNA-based intracellular strategies to attempt to improve viral inhibition and prevent possible escape mutations have been performed. The dual vector was constructed, in which the tRNA promoter-ribozyme-V and pol III terminal function unit, was inserted into the pSLXCMV-D8scFv 3' LTR NaeI site, with the same transcriptional orientation as the CMV promoter in this vector (Fig. 10.7). This dual-functioning retroviral vector was then transduced into SupT1 cells and PBMC for HIV-1 challenge experiments. The protein expression D8scFv from this vector was confirmed using Western blotting. In the acute HIV-1 infection experiments, this dual-functioning vector showed improved viral inhibition, as compared with D8scFv alone. Increased inhibition by this dual-functioning retroviral vector was also observed (Fig. 10.11). Of note, there was always a 0.5 to 1.0 log decrease in HIV-1 replication, comparing the D8scFv plus the ribozyme-V to D8scFv alone.[10]

The concept of anti-HIV-1 gene therapy, using intracellular immunization strategies, should not only protect the cells which have not yet been infected by HIV-1, these approaches should also be able to inhibit the replication of HIV-1 in chronically-infected cells, to reduce viral load in different stages of disease. To test the inhibitory effects of anti-Rev scFv, anti-RRE-ribozyme, and the dual-functioning expression vector on HIV-1 chronically-infected cells, each of the packaged retroviral vectors were transduced into H9/IIIB cells, which is a human T-cell chronically infected with HIV-1. After two weeks of G418 selection, the mixed populations of cells were maintained in G418-free medium for three weeks, then HIV-1 p24 antigen production from the transduced H9/IIIB cells were quantitated. Anti-Rev scFv, and the trans-dominant negative mutant Rev protein, RevM10, demonstrated 40%-50% inhibition of viral replication, but the ribozyme alone only reduced the HIV-1 p24 antigen production level by 18%-27%. When the anti-Rev and ribozyme dual-retroviral vector was transduced into the H9/IIIB cells, 75%-84% inhibition of viral replication inhibition was achieved. As such a dual protein- and RNA-based approach may be efficient for genetic therapeutics targeting chronically HIV-1-infected cells.[10]

Conclusion

Gene therapy offers perhaps one of the broadest technological platforms for medical treatment developed to date, and human somatic gene therapy has become a reality. As might be expected, this new technology has been quickly evaluated for combating AIDS. By inhibiting HIV-1 viral replication, and intracellularly, targeting specific viral proteins using single chain variable fragment moieties, promising strategies are being evaluated experimentally, and clinical trials have been initiated. As with any emerging scientific and medical technology, many technical challenges remain to be addressed and remedied before the clinical utility of scFv-derived human gene therapy against HIV-1 can be realized. Some recent advances in anti-viral vector development have shown that the gene of interest can be delivered into resting cells efficiently[28] and cell-type specific vectors can be obtained, by modified viral envelope using anti-cell membrane antigen-specific scFvs.[6] The use of cellular matrix-attachment sites and locus control elements may help maintain long-term expression of transgenes. By preventing infection of host cells with HIV-1 and the potential for gene "knock-out" of integrated HIV-1 proviral DNA from host cells[15] with combinations of powerful chemotherapeutics and scFv-based intracellular immunization strategies, these approaches may provide long term and consistent clinical benefits for patients with HIV-1 infection.

Acknowledgments

The authors wish to thank Drs. Anna Marie Skalka, Didier Trono and Mark Wainberg plus the members of the Center for Human Virology at Thomas Jefferson University for helpful discussions; and Ms. Brenda O. Gordon and Ms. Rita M. Victor for excellent secretarial assistance. This work was supported in part by U.S. PHS grant AI 36552 to R.J.P.

References

1. Andrake MD, Skalka AM. Multimerization determinants reside in both the catalytic core and C-terminus of avian sarcoma virus integrase. J Biol Chem 1995; 270: 29299-29306.
2. Arts EJ, Wainberg MA. Human immunodeficiency virus type 1 reverse transcriptase and early events in reverse transcription. Adv Virus Res 1996; 46:97-163.
3. Bender DB, Kulkosky J, Skalka AM. Monoclonal antibodies against HIV type 1 integrase: Clues to molecular structure. AIDS Res Hum Retrovir 1994; 10: 1105-1115.
4. Biocca S, Cattaneo A. Intracellular immunization: antibody targeting to subcellular compartments. Trends Cell Biol 1995; 5:249-252.
5. Carroll WL, Mendel E, Levy S. Hybridoma fusion cell lines contain an aberrant k transcript. Molec Immunol 1998; 25:991-995.
6. Cosset FL, Russell SJ. Targeting retrovirus entry. Gene Ther 1996; 3: 946-956.
7. Cullen BR, Greene WC. Regulatory pathways governing HIV-1 replication. Cell 1989; 58:423-426.
8. Deisenhofer J, Colman PM, Huber R. Crystallographic structural studies of a human Fc fragment I. An electron density map at 4 $ resolution and a partial model. Hoppe-Seyler's Z Physio Chem 1976; 357:435-445.
9. Denman RB. Using RNA-FOLD to predict the activity of small catalytic RNAs. Biotechniques 1993; 15:1090-1094.
10. Duan L-X, Zhu M, Ozaki I et al. Intracellular inhibition of HIV-1 replication using a dual protein- and RNA-based strategy. Gen Ther 1997; in press.
10a. Duan L-X, Pomerantz RJ. Intracellular immunization. Science & Medicine 1996; 3:24-33.
11. Duan L-X, Pomerantz RJ. Elimination of endogenous aberrant k chain transcription from Sp2/o-derived hybridoma cells by specific ribozyme cleavage: Utility in genetic therapy of HIV-1 infections. Nucl Acid Res 1994; 22:5433-5438.
12. Duan L-X, Zhu M, Bagasra O et al. Intracellular immunization against HIV-1 infection of human T lymphocytes: Utility of anti-Rev single-chain variable fragments. Hum Gene Ther 1995; 6: 1561-1571.
13. Engelman A, Englund G, Orenstein JM et al. Multiple effects of mutations in human immunodeficiency virus type 1 integrase on viral replication. J Virol 1995; 69:2729-2736.

14. Engelman A, Mizuuch K, Craigie R. HIV-1 DNA integration: Mechanism of viral DNA cleavage and DNA strand transfer. Cell 1991; 67:1211-1221.

15. Flowers CC, Woffendin C, Petryniak J et al. Inhibition of recombinant human immunodeficiency virus type 1 replication by a site-specific recombinase. J Virol 1997, in press.

16. Haseloff J, Gerlach WL. Simple RNA enzymes with new and highly specific endoribonuclease activities. Nature 1988; 334:585-591.

17. Huse WD, Sastry L, Iverson SA et al. Generation of a large combination library of the immunoglobulin repertoire in phage lambda. Science 1989; 246:1275-1281.

18. Huston JS, Levinson D, Mudgett-Hunter M et al. Protein engineering of antibody binding sites: Recovery of specific activity in anti-digoxin single-chain Fv analogue produced in Escherichia coli. Proc Natl Acad Sci USA 1988; 85:5879-5883.

19. Katz RA, Merkel G, Kulkosky J et al. The avian retroviral IN protein is both necessary and sufficient for integrative recombination in vitro. Cell 1990; 63:87-95.

20. Kabat EA, Wu TT, Perry HM et al. Sequences of Proteins of Immunological Interest. 5th Ed. U.S. Department of Health and Human Services. NIH Publication 1991; 91: 3242.

21. Kubota S, Duan L-X, Rika A et al. Nuclear preservation and cytoplasmic degradation of human immunodeficiency virus type I Rev protein. J Virol 1996; 70: 1282-1287.

22. Kulkosky J, Skalka AM. Molecular mechanism of retroviral DNA integration. Pharmacol Ther 1994; 61:185-203.

23. Kulkosky J, Jones Jr KS, Katz A et al. Residues critical for retroviral integrative recombination in a region that is highly conserved among retroviral/retrotransposon integrases and bacterial insertion sequence transposases. Molec Cell Biol 1992; 12:2331-2338.

24. Lee TC et al. Over-expression of RRE-derived sequence inhibits HIV-1 replication in CEM cells. The New Biologist 1992; 4:66-74.

25. Levy-Mintz P, Duan L-X, Zhang H-Z et al. Intracellular mapping of single-chain variable fragments (scFv) to inhibit early stages of the viral life-cycle by targeting HIV-1 integrase. J Virol 1996; 70:12.

26. Marchuk D, Drumm M, Saulino A et al. Construction of T-vectors, a rapid and general system for direct cloning of unmodified PCR products. Nucl Acids Res 1990; 19:1154.

27. Meyers G, Rabson AB, Berzofsky JA et al. Human retroviruses and AIDS: A compilation and analysis of nucleic acid and amino acid sequences. Los Alamos National Laboratory, Los Alamos, NM 1996.

28. Naldini L, Blomer U, Gallay P et al. In vitro gene delivery and stable transduction of nondividing cells by a lentiviral vector. Science 1996; 272:263-267.

29. Orlandi R, Gussow DH, Jones PT et al. Cloning immunoglobulin variable domain for expression by the polymerase chain reaction. Proc Natl Acad Sci USA 1989; 86:3833-3837.

30. Pilkington GR, Duan L-X, Zhu MH et al. Recombinant human Fab antibody fragments to HIV-1 Rev and Tat regulatory proteins: Direct selection from a phage library. Molec Immunol 1996; 33:439-450.

31. Pluckthun A. Strategies for the expression of antibody fragments in *Escherichia coli*. Methods: A Companion to Methods in Enzymology. Lerner R, Burton D, eds. Academic Press Inc., NY 1991; 2:88-96.

32. Poliak RJ, Amzel LM, Avery HP et al. Three-dimensional structure of the Fab fragment of a human immunoglobulin at 2.8 $. Proc Natl Acad Sci USA 1993; 70: 3305-3310

33. Pomerantz JR, Trono D. Genetic therapies for HIV infections: Promise for the future. AIDS 1995; 9:985-993.

34. Pomerantz RJ, Seshamma T, Trono D. Efficient replication of human immunodeficiency virus type 1 requires a threshold level of Rev: Potential implications for latency. J Virol 1991; 66:1809-1813.

35. Sarvar N et al. Ribozymes as potential anti-HIV-1 therapeutic agents. Science 1990; 247:1222-1225.

36. Shaheen F, Duan L-X, Zhu M et al. Targeting HIV-1 reverse transcriptase by intracellular expression of single-chain variable fragments (scFv) to inhibit early stages of HIV-1 replication. J Virol 1996; 70:3392-3400.

37. Sigurdsson ST, Eckstein F. Structure-function relationships of hammerhead ribozymes: From understanding to applications. TIBTECH 1995; 13:280-285.

38. Siomi H, Shida M, Maki M et al. Effects of a highly basic region of human immunodeficiency virus tat protein on nucleolar localization. J Virol 1990; 64:1803-1807.

39. Stevenson M. Identification of factors that govern HIV-1 replication in nondividing host cells. AIDS Res Hum Retrovir 1994; 10:S11-S15.

40. Symons RH. Small catalytic RNAs. Ann Rev Biochem 1992; 61:641-671.

41. Szilvay AM, Nornes S, Kannapiran A et al. Characterization of HIV-1 reverse transcriptase with antibodies indicates conformational differences between the RNAse H domains of p66 and p15. Arch Virol 1993; 131:393-403.

42. Thompson JD. Improved accumulation and activity of ribozymes expressed from a tRNA-based RNA polymerase III promoter. Nucl Acids Res 1995; 23:2259-2269.

43. Trono D. Molecular biology and the development of AIDS therapeutics. AIDS 1996; 10:S53-S59.

44. Wu Y, Duan L-X, Zhu M et al. Binding of intracellular anti-Rev scFvs to different epitopes of HIV-1 Rev: variations in viral inhibition. J Virol 1996; 70:3290-3297.

Index